Essential Physiological Biochemistry

Essential Physiological Biochemistry

An organ-based approach

Stephen Reed

Department of Biomedical Sciences
University of Westminster, London, UK

A JOHN WILEY & SONS, INC., PUBLICATION

This edition first published 2009
© 2009 John Wiley & Sons, Ltd

Wiley-Blackwell is an imprint of John Wiley & Sons, formed by the merger of Wiley's global
Scientific, Technical and Medical business with Blackwell Publishing.

Registered office: John Wiley & Sons Ltd, The Atrium, Southern Gate,
Chichester, West Sussex, PO19 8SQ, UK

Other Editorial Offices:
9600 Garsington Road, Oxford, OX4 2DQ, UK
111 River Street, Hoboken, NJ 07030-5774, USA

For details of our global editorial offices, for customer services and for information about
how to apply for permission to reuse the copyright material in this book please see our
website at www.wiley.com/wiley-blackwell

Library of Congress Cataloguing-in-Publication Data

Reed, Stephen, 1954-
   Essential physiological biochemistry : an organ-based approach / Stephen Reed.
      p. ; cm.
   Includes bibliographical references and index.
   ISBN 978-0-470-02635-9 (cloth) – ISBN 978-0-470-02636-6 (pbk.)
 1.   Biochemistry. 2.   Organs (Anatomy) 3.   Metabolism.   I.   Title.
   [DNLM: 1.   Biochemical Phenomena. 2.   Metabolism–physiology.   QU
34 R326e 2009]
   QP514.2.R44 2009
   612'.015–dc22

                                                             2009021620

ISBN: 978 0 470 02635 9 (HB)   978 0 470 02636 6 (PB)

A catalogue record for this book is available from the British Library.

Set in 10.5/12.5 pt Minion by Thomson Digital, Noida, India

First impression – 2009

To my parents, C and I, who made so many good things happen;
To Jessica who will be an excellent physician;
To Ele who will find success outside science.

Contents

Preface

My purpose in writing this book was not to produce a comprehensive textbook of human biochemistry; there are numerous excellent textbooks that meet that need. The impetus for writing this book was to present aspects of metabolism within an appropriate physiological context. Indeed the original working title for the project was a 'hitch-hiker's guide to metabolism', with a focus on important processes, integration and control aspects without delving too deeply into chemical mechanisms. Too often, students perceive learning biochemical pathways to be 'difficult' or 'boring', largely because they are presented with a sequence of named intermediates and enzymes which are to be learnt by heart. In fact, metabolism is neither difficult nor boring, if it is approached in the right way. This book aims to present a 'right way' and so make learning and, more importantly, understanding as easy as possible. To the reader, I say this: imagine you are about to take up a new sport or you may simply want to be a spectator. The excitement of the competition may in itself be adequate for you to derive some enjoyment but to fully appreciate the sport, you will need to learn the rules or laws of the game first. Similarly, taking time and trouble to learn the rules of metabolism and learning to recognize themes will require some intellectual application. Once mastered, appreciation of the beautiful, logical simplicity of intermediary metabolism will enhance your understanding.

Whilst it is inconceivable that any text on biochemistry could be devoid of chemistry, this book is aimed at student biologists for whom biochemistry is a component part, but not the principal part, of their studies, so word equations have been used to illustrate reactions wherever possible and chemical structures shown only where necessary. The student groups who will find *Essential Physiological Biochemistry* valuable are those following undergraduate courses in physiology, nutrition, sports science and biomedical science. This text will also act as useful primer and quick reference for medicine or for postgraduates who need a revision guide.

The chapters are arranged into two distinct sections, the first dealing with basic concepts of metabolism and its control whilst the second focuses on selected tissues, illustrating how biochemistry underpins the physiological activities of certain tissues and systems. Where pathways occur in two or more organs of the body, details are given in only one chapter with cross-referencing to other relevant sections. Decisions to describe common pathways in one chapter rather than another were pragmatic ones and doubtless some would argue, not entirely appropriate.

If, having read this text, students feel that biochemistry is accessible rather than intimidating; if they become aware that metabolism is flexible, integrated and 'logical' rather than a collection of apparently disparate reactions and pathways which must be memorized, and if they acquire even a small sense of wonderment and awe about the subtlety that the subject holds, the effort of writing the book will have been justified.

Acknowledgements

The realization of this book has been a long process and it would not have happened without the advice and support of several people. First, Nicky McGirr at Wiley-Blackwell for her unwavering encouragement and whose patience and belief in the project were crucial when my own were flagging. Secondly, Fiona Woods and Robert Hambrook also at Wiley-Blackwell for valuable help and advice in the latter stages of the project.

I am grateful to many colleagues at Westminster who offered critical and constructive comment on parts of the text during its compilation. Many of their views have been accommodated, but any errors are entirely my own.

Lastly, to Gill, MBL, for her understanding.

Stephen Reed

1

Introduction to metabolism

Overview of the chapter

In this chapter we will consider definitions of metabolism; the biochemistry–physiology continuum. The concept of metabolic pathways and their organization and control of metabolism are likened to a road map involving 'flow' of substrates but with mechanisms to accelerate or slow down pathways or to direct substrates through alternative routes.

Introductions to enzyme kinetics and bioenergetics are given with explanations of key terms such as K_m and V_{max}; coenzymes, cofactors and inhibitors; typical metabolic reactions; free energy; exergonic and endergonic reactions, catabolism and anabolism.

Guidance on how to study metabolic pathways is given using glycolysis as a model pathway.

1.1 Introduction

Movement; respiration; excretion; nutrition; sensitivity; reproduction. These are the six criteria often used by biologists to define 'life'. Whilst physiologists describe many processes of human biology at the tissue and organ level, a biochemist studies the same processes but at a 'higher magnification'. To a biochemist, the six features listed above can all be described in terms of chemical events, so a useful definition of biochemistry is 'the study of life at the molecular level'. The discipline of cell biology fits between physiology and biochemistry, but the three disciplines together form a continuum of knowledge and investigation.

Biochemical studies follow several themes. For example, investigations can be focussed on the chemical *structures* of molecules, (for example the structure of glycogen, DNA or protein conformation) or the structural inter-relationship between molecules (e.g. enzymes with their substrates, hormones with their receptors). The other branch of biochemical enquiry is into those numerous *'dynamic'* events known collectively as 'metabolism', defined here as 'all of the chemical reactions and their associated energy changes occurring within cells'. The purpose of metabolism is to provide the

Essential Physiological Biochemistry: An organ-based approach   Stephen Reed
© 2009 John Wiley & Sons, Ltd

energy and building materials required to sustain and reproduce cells and thereby the organism.

It is estimated that there are between 2000 and 3000 different types of metabolic reaction occurring, at various times, within human cells. Some of these are common to all cell types whilst others are restricted to one or two particular tissues whose specialized physiological functions reflect the specialized metabolic changes occurring within them. Metabolism is a fascinating, yet at first sight, complicated process apparently representing a daunting challenge for the learner. For most students, it is unnecessary to learn *every* reaction in *every* pathway; what is important is that there an *understanding of concepts* of metabolism so that what appears to be a complicated set of reactions and pathways can be seen in terms of relatively few chemical and thermodynamic (energetic) principles. Metabolism may be likened to a journey; there is a starting place and a destination and there will be some important intermediate stops, perhaps where a change of mode of transport will be necessary, there will be points of interest and also places *en route* which deserve little or none of our attention. This analogy will be further developed later in the text. Furthermore, it is vital to realize that metabolism is adaptable; changes in physiological situations, for example fed or fasting, resting or exercising, health or ill- health will result in changes in particular aspects of metabolism.

The purpose of this book is to present metabolism in an organ-based fashion to make clear the links between biochemistry and physiology. By presenting metabolism in an appropriate tissue-context, the significance of pathways and their inter-relationships should be more meaningful.

1.2 Metabolic pathways

The term 'intermediary metabolism' is used to emphasize the fact that metabolic processes occur via a series of individual chemical reactions. Such chemical reactions are usually under the control of enzymes which act upon a substrate molecule (or molecules) and produce a product molecule (or molecules) as shown in Figure 1.1. The substrates and products are referred to collectively as 'intermediates' or 'metabolites'. The product of one reaction becomes the substrate for another reaction and so the *concept* of a metabolic pathway is created.

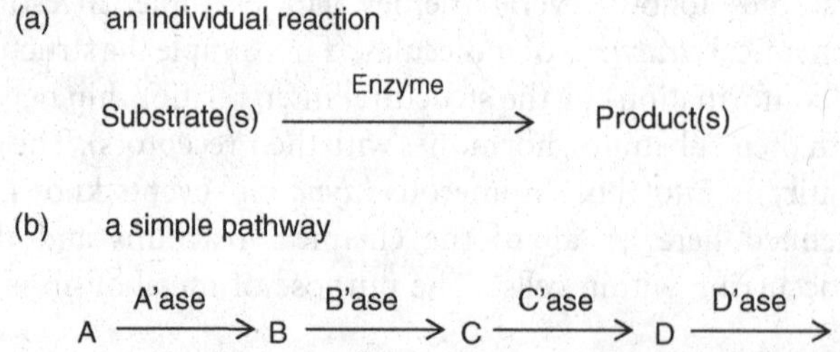

(a)   an individual reaction

$$\text{Substrate(s)} \xrightarrow{\text{Enzyme}} \text{Product(s)}$$

(b)   a simple pathway

$$A \xrightarrow{\text{A'ase}} B \xrightarrow{\text{B'ase}} C \xrightarrow{\text{C'ase}} D \xrightarrow{\text{D'ase}}$$

Figure 1.1 Simple representation of a metabolic pathway. 'Compound' B is the product of the first reaction and the substrate for the second reaction, and so on. Capital symbols represent metabolic intermediates and lower case letters with the suffix 'ase' represent enzymes

Metabolism and the individual reactions which comprise a pathway represent a dynamic process. Terms such as 'flow', 'substrate flux', 'rate' and 'turnover' are all used to communicate the idea of the dynamic nature of metabolism.

The student should be aware that a pathway is essentially a conceptual 'model' developed by biochemists in order to represent the flow of compounds and energy through metabolism. Such models are simply ways of trying to explain experimental data. A potential problem in representing metabolic pathways as in Figure 1.1 is that there is an implication that they are physically and/or topographically organized sequences. This is not necessarily true. With some exceptions (described in Section 1.3), most enzymes are likely to be found 'free' within the cytosol or a compartment of a cell where reactions occur when an enzyme and its substrate meet as a result of their own random motion. Clearly this would be very inefficient were it not for the fact that cells contain many copies of each enzyme and many molecules of each type of substrate.

Think again about making a journey; a useful analogy is a road map of a city centre where there are main and subsidiary routes, one-way systems and interchanges, where traffic flow is controlled by signals and road-side signs. Very few people would attempt to learn, that is memorize, a complete route map, but learning the 'rules of the road' coupled with basic map reading skills and knowledge of main roads will enable most people to negotiate successfully a journey from one place to another. A complete diagram of intermediary metabolism appears to be as complicated as a road map of a city or region, that is a tangle of individual reactions with numerous substrates.

Understanding biochemical pathways is somewhat similar to map reading. The flow of traffic along roads and through the city is conceptually similar to the flow of substrates within the cell. Rather than visualizing cars, vans and trucks, think about the numerous carbon, hydrogen, nitrogen, phosphorus and oxygen atoms 'flowing' as component parts of substrate molecules, through pathways within the cell. Just as the traffic flow is regulated and directed with signals and restrictions, so too is the flow of substrates. Vehicles (metabolites) join or leave a particular traffic flow at intersections (converging or diverging pathways); the rate of flow is affected by traffic signals (enzymes), by road works or accidents (defective enzymes) and by the number of vehicles using the road (concentration of substrate molecules); they may need to take short-cuts or be diverted to avoid congested areas. Similarly, substrate molecules also may be routed via alternative pathways in a manner which best serves the physiological requirements of the cell at any particular moment. At times vehicles will need to take on fuel and some molecules need to be 'activated' by attachment to coenzyme A or uridine diphosphate (UDP), for example).

A note about terminology.

Glucose-1-phosphate (Glc-1-P) means that glucose has a phosphate attached at carbon 1 in place of a hydrogen atom.

Fructose-6-phosphate (Frc-6-P) means that fructose has a phosphate attached at carbon 6 in place of a hydrogen atom.

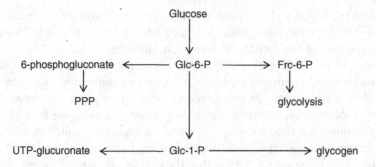

The relative activities of the enzymes which use glc-6-P as substrate determine the net flow.

Frc-6-P = fructose-6-phosphate
Glc-1-P = glucose-1-phosphate
PPP   = pentose phosphate pathway
UTP   = uridyl triphosphate

Figure 1.2 Glucose-6-phosphate is at a 'metabolic cross-roads'

1,3-bis phosphoglycerate (1,3-BPG) glyceric acid (glycerate) has 2 phosphate groups attached at carbons 1 and 3.

NB: 'bis' formerly designated as 'di'

Cells contain a large number of individual *types* of substrates; this is often referred to as the 'pool of intermediates'. One type of substrate may have a role to play in two or more pathways at different times according to the physiological demands being made on the cell. Metabolic regulation involves enzymes operating on substrates that occur at junctions of two or more pathways to act as flow-control points, rather like traffic signals. A good example of a substrate at a crossroads is glucose-6-phosphate (Glc-6-P), an intermediate that is common to glycolysis, glycogen turnover, the pentose phosphate pathway (PPP) and via UDP-glucose, the uronic acid pathway (Figure 1.2).

Clearly, substrates such as Glc-6-P do not 'belong' to a particular pathway but may occur within several routes. Channelling of the compound through a particular pathway will be determined by the relative activity of the enzymes using the substrate which in turn will be determined (regulated) by cellular requirements. Different pathways become more or less significant according to the physiological conditions (e.g. fed or fasting state, active or resting) in which the cell or organism finds itself.

1.3 Organization of pathways

Pathways can be illustrated in a metabolic map as linear, branched or cyclic processes (Figure 1.3) and are often compartmentalized within particular subcellular location: glycolysis in the cytosol and the Krebs tricarboxylic acid (TCA) cycle in

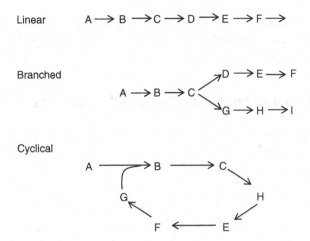

Figure 1.3 Conceptual arrangement of pathways

mitochondria are obvious examples. However, not all reactions of a particular pathway necessarily occur in the same organelle or location. Haem synthesis and urea synthesis (both described in Section 6.2) for example occur partly in the mitochondria and partly in the cytosol of liver cells.

Once an enzyme-catalysed reaction has occurred the product is released and its engagement with the next enzyme in the sequence is a somewhat random event. Only rarely is the product from one reaction passed directly onto the next enzyme in the sequence. In such cases, enzymes which catalyse consecutive reactions, *are* physically associated or aggregated with each other to form what is called a multi enzyme complex (MEC). An example of this arrangement is evident in the biosynthesis of saturated fatty acids (described in Section 6.30). Another example of an organized arrangement is one in which the individual enzyme proteins are bound to membrane, as for example with the ATP-generating mitochondrial electron transfer chain (ETC) mechanism. Intermediate substrates (or electrons in the case of the ETC) are passed directly from one immobilized protein to the next in sequence.

Biochemical reactions are interesting but they are not 'magic'. Individual chemical reactions that comprise a metabolic pathway obey, obviously, the rules of organic chemistry. All too often students make fundamental errors such as showing carbon with a valency of 3 or 5, or failing properly to balance an equation when writing reactions. Furthermore, overall chemical conversions occur in relatively small steps, that is there are usually only *small* structural changes or differences between *consecutive* compounds in a pathway.

To illustrate this point, consider the following analogy. The words we use in everyday language are composed from the same alphabet of letters. Changing even one letter within a word changes the meaning. Try converting the word WENT into COME by changing *only one* letter at a time. Each intermediate must be a meaningful English word.

This exercise is conceptually similar to biochemical conversions. One of the skills of the experimental biochemist is to identify metabolic intermediates and then to arrange them in a chemically sensible sequence to represent the pathway, that is develop the model to explain the experimental results. A model answer to the word puzzle cited above is given at the end of the chapter.

1.4 Enzymes and enzyme-mediated reactions

This section deals with the nature of enzymes and their importance in metabolic control is discussed more fully in Chapter 3. Enzymes are biocatalysts whose key characteristics are as follows;

Enzymes are:

- Proteins;

- Chemically unchanged at the end of the reaction they catalyse and so are reusable;

- Required in small amounts because they are recycled;

- Able to act upon a specific substrate or structurally very similar substrates;

- Able to act on a particular part (functional group) within the substrate;

- Able to catalyse a specific type of chemical reaction;

- Able to operate under mild conditions of pH, temperature and pressure (if gases are involved).

1.4.1 Equilibrium or steady state?

The majority of biochemical reactions are *reversible* under physiological conditions of substrate concentration. In metabolism, we are therefore dealing with chemical *equilibria* (plural). The word equilibrium (singular) signifies a balance, which in chemical terms implies that the *rate* of a forward reaction is balanced (i.e. the same as) the *rate* of the corresponding reverse reaction.

$$r \rightleftharpoons p \text{ which may also be written as } r \leftrightarrow p$$

r = reactant(s)

p = product(s)

$$r \rightarrow p \text{ is the forward reaction}$$

and

$$p \rightarrow r \text{ is the reverse reaction}$$

Many chemical reactions (especially those occurring within cells) are theoretically reversible under reasonable conditions of pressure (when gasses are involved, which is rare), temperature and concentration.

In a closed system, that is one in which there is no addition of 'r' nor any removal of 'p', the reaction will come to a perfect balance; 'the point of equilibrium'. A common misunderstanding of the concept of this point of equilibrium is that it implies an *equal concentration* of r and p. This is not true. The point of equilibrium defines the relative concentrations of r and p when the *rate* of formation of p is exactly equal to the *rate* of formation of r. The point of equilibrium value for a chemical reaction can be determined experimentally. If the starting concentration of the reactant is known, then it follows that the *relative* concentrations of r and p when equilibrium has been reached must reflect the *relative* rates of the forward and reverse reactions. For a given reaction, under defined conditions, the point of equilibrium is a constant and given the symbol K_{eq}.

Thus

$$K_{eq} = \frac{[p]}{[r]}$$

where [ ] indicates molar concentration

When the equilibrium concentration of p is *greater* than the equilibrium concentration of r, we can say that the forward reaction is favoured (faster) and $K_{eq} > 1$;

$$r \rightleftharpoons p \qquad \text{so at equilibrium, } [p] > [r]$$

NB: the weight and size of the arrows represents the relative rates of reaction

The higher the value of K_{eq}, the more difficult it is for that reaction to 'go backwards' so effectively it becomes unidirectional.

Conversely, if [r] is greater then [p], the reverse reaction is favoured and $K_{eq} < 1$ because [p] < [r] signifies that the forward reaction becomes increasing less likely and the value becomes smaller. It could be argued that a 'true' equilibrium occurs only when [r] $\approx$ [p], but K_{eq} is a measure of the relative *rates* of the forward and reverse reactions. An important consequence of the magnitude of K_{eq} is that the further away a reaction is from a true equilibrium, the greater the energy change involved in that reaction. This is explained in more detail later in this chapter and also in Chapter 2.

Most individual biochemical reactions are reversible and are therefore quite correctly considered to be chemical equilibria, but cells are not closed systems; fuel (e.g. a source of carbon and, in aerobic cells, oxygen) and other resources (e.g. a source of nitrogen and phosphorus) are continually being added and waste products removed, but their relative concentrations within the cell are fairly constant being subject to only moderate fluctuation. Moreover, no biochemical reaction exists in isolation, but each is part of the overall flow of substrate through the pathway as a whole.

Stated simply, biochemical reactions never reach a true equilibrium because the product of one reaction is the substrate for the next and so the reaction is 'pulled' towards completion achieving net formation of product. Indeed, if reactions inside a cell were *true equilibria*, there would be no net flow of substrate, no formation of end products and therefore no metabolic pathway. Biologically, this would not be very desirable! The situation which exists within cells is better described as a *steady state*. In this condition, there *is* net flow of matter but the instantaneous concentrations of intermediates fluctuate relatively little, unless a 'stress' for example the need to respond to a physiological challenge, is placed on the system.

Although there is a bewildering array of individual reactions occurring within cells they can be classified into a small number of groups. Learning the types of reactions and then identifying particular examples as and when they arise is easier than trying simply to memorize a sequence of chemical changes. Typical biochemical reactions include the following (Figures 1.4 to 1.17).

1. **Atomic and molecular rearrangements**
    Isomerization involving (a) a change in functional group or (b) the repositioning of atoms within the same molecule, for example

    a.   ribose-5-P ⎯⎯⎯⎯⎯⎯⎯→ ribulose-5-P
    (aldehyde to ketone)

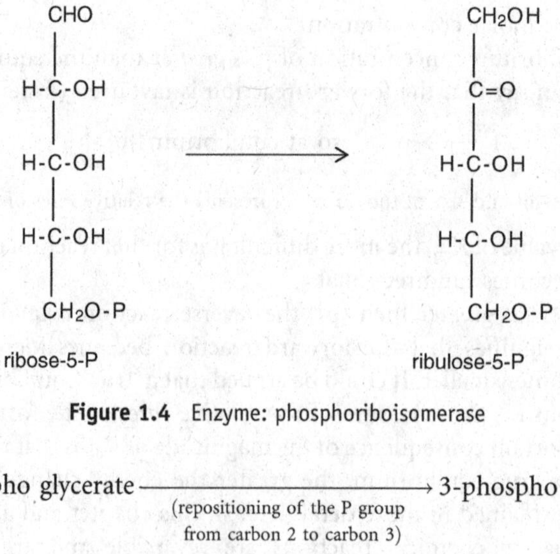

<pre>
        CHO   CH₂OH
         | |
       H-C-OH   C=O
         | |
       H-C-OH   ⎯⎯⎯⎯→   H-C-OH
         | |
       H-C-OH   H-C-OH
         | |
       CH₂O-P   CH₂O-P

     ribose-5-P   ribulose-5-P
</pre>

Figure 1.4   Enzyme: phosphoriboisomerase

    b.   2-phospho glycerate ⎯⎯⎯⎯⎯⎯⎯→ 3-phospho glycerate
    (repositioning of the P group
from carbon 2 to carbon 3)

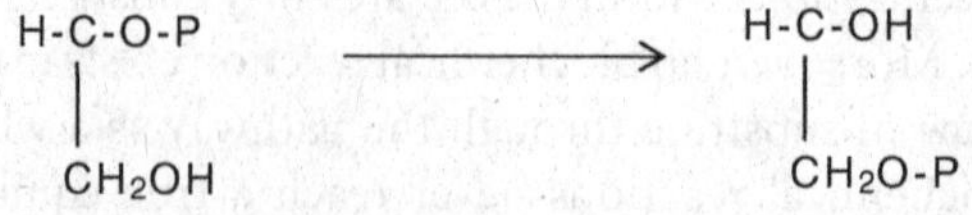

<pre>
         COO⁻   COO⁻
          | |
       H-C-O-P   ⎯⎯⎯⎯→   H-C-OH
          | |
        CH₂OH   CH₂O-P

         2-PG   3-PG
</pre>

Figure 1.5   Enzyme: phosphoglyceromutase

2. **Substitution**

   Replacement of one atom or group with another, for example, a hydrogen atom is replaced by a methyl group;

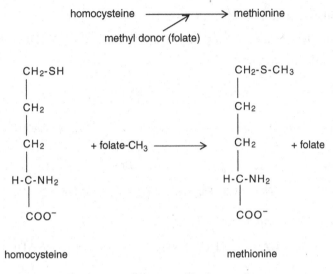

Figure 1.6 Enzyme: methionine synthase

3. **Redox reactions ***These Are Very Important ****

   Oxidation and reduction reactions always occur together and are usually easily spotted because of the involvement of a coenzyme.

   $$A\text{-}H_{(red)} + coenz(ox) \leftrightarrow B_{(ox)} + coenz\text{-}H(red)$$

   Coenz = coenzyme;

   (red) = reduced form;

   (ox) = oxidized form (i.e. fewer hydrogen atoms/electrons than the reduced form);

   for example, lactate dehydrogenase

   $$lactate + NAD^+ \leftrightarrow pyruvate + NADH + H^+$$

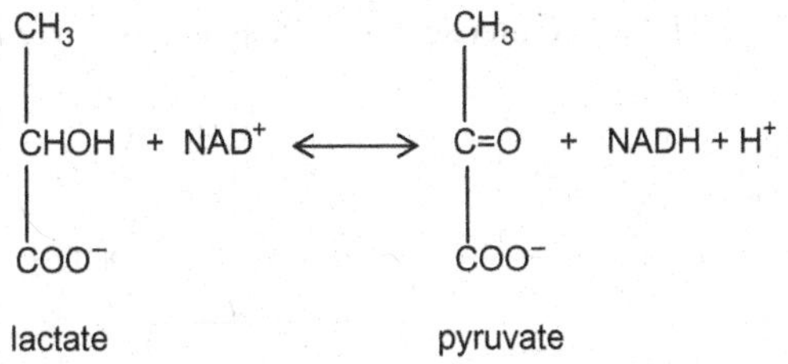

Figure 1.7 Enzyme: lactate dehydrogenase

The lactate is oxidized (two hydrogen atoms removed) and the NAD^+ is reduced to $NADH + H^+$

Oxidation sometimes occurs simultaneously with another chemical change. For example, oxidative decarboxylation or oxidative deamination.

a. Oxidative decarboxylation; CO_2 released

$$pyruvate + NAD^+ + CoASH \rightarrow acetyl\ CoA + CO_2 + NADH + H^+$$

(CoASH = co enzyme A)

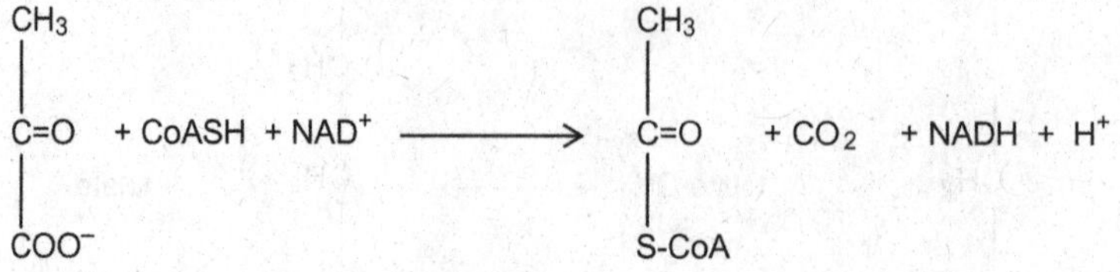

$$
\begin{array}{c}
CH_3 \\
| \\
C{=}O \\
| \\
COO^-
\end{array}
+ CoASH + NAD^+ \longrightarrow
\begin{array}{c}
CH_3 \\
| \\
C{=}O \\
| \\
S\text{-}CoA
\end{array}
+ CO_2 + NADH + H^+
$$

Figure 1.8 Enzyme: pyruvate decarboxylase

b. Oxidative deamination; NH_3 released

$$glutamate + NAD^+ + H_2O \rightarrow NH_3 + 2\text{-oxoglutarate} + NADH + H^+$$

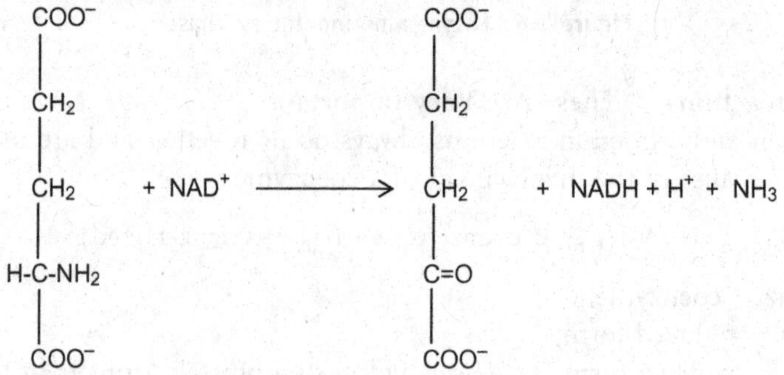

$$
\begin{array}{c}
COO^- \\
| \\
CH_2 \\
| \\
CH_2 \\
| \\
H\text{-}C\text{-}NH_2 \\
| \\
COO^-
\end{array}
+ NAD^+ \longrightarrow
\begin{array}{c}
COO^- \\
| \\
CH_2 \\
| \\
CH_2 \\
| \\
C{=}O \\
| \\
COO^-
\end{array}
+ NADH + H^+ + NH_3
$$

Figure 1.9 Enzyme: glutamate dehydrogenase

4. **Cleavage**

a. Hydrolysis, if water is used to break a bond

Glucose-6-P + H_2O → Glucose + Pi   Pi symbolises free inorganic phosphate

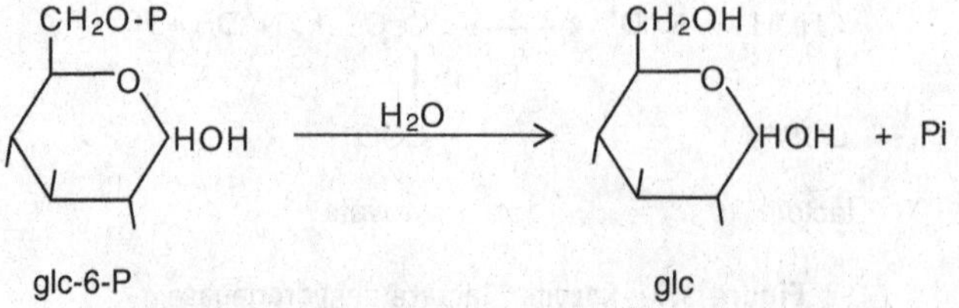

$$CH_2O\text{-}P \qquad \xrightarrow{\ H_2O\ } \qquad CH_2OH$$

glc-6-P   glc

Figure 1.10 Enzyme: glucose-6-phosphatase

b. One molecule is split into two

F-1, 6-*bis* phosphate → glyceraldehyde-3-P + dihydroxyacetone-P

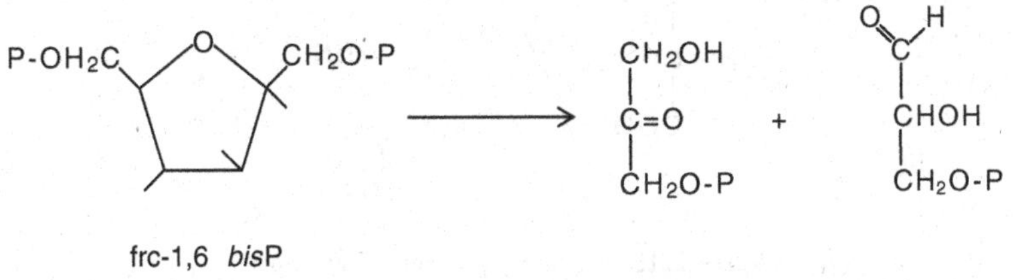

frc-1,6 *bis*P

Figure 1.11 Enzyme: aldolase

5. Condensation

Two molecules join together with the elimination of a H_2O. Condensation reactions are used when macromolecules are being formed. Amino acids are joined via peptide bonds and monosaccharides via glycosidic bonds, both of which are condensation reactions.

$$2 \text{ glucose} \rightarrow \text{maltose} + H_2O$$

$$2 \text{ amino acids} \rightarrow \text{dipeptide} + H_2O$$

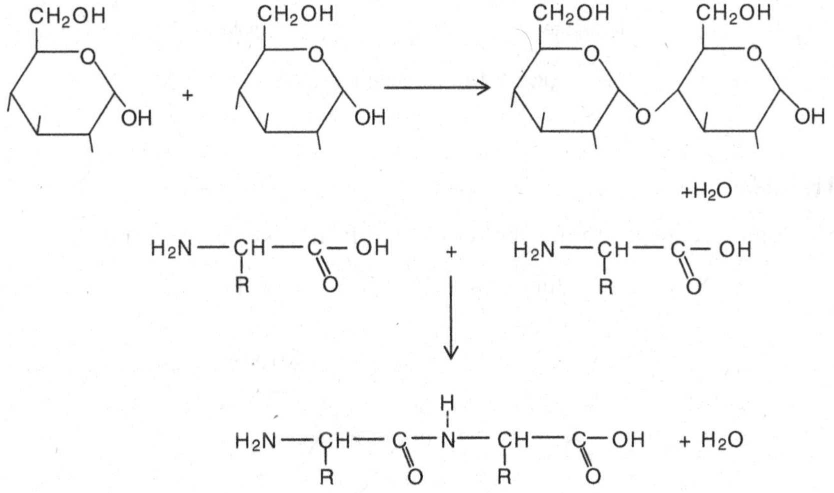

Figure 1.12 Enzymes: synthases

6. Addition

a. Two molecules are joined together but water is not eliminated. Often ATP is used to provide energy.

$$\text{glutamate} + NH_3 + ATP \rightarrow \text{glutamine} + ADP$$

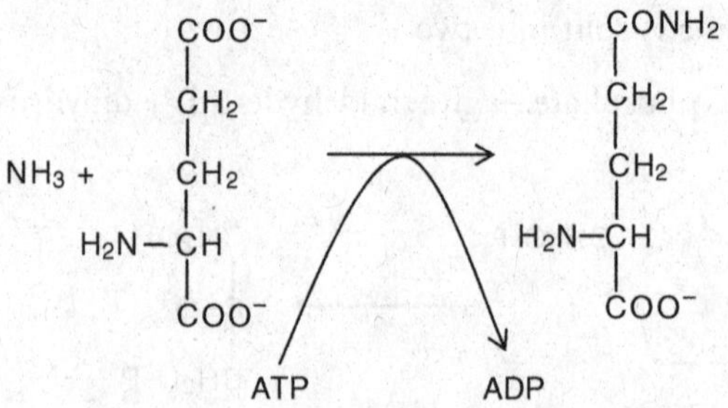

Figure 1.13 Enzyme: glutamine synthase

b. Alternatively, addition may across a double bond

$$\text{fumarate} + H_2O \rightarrow \text{malate}$$

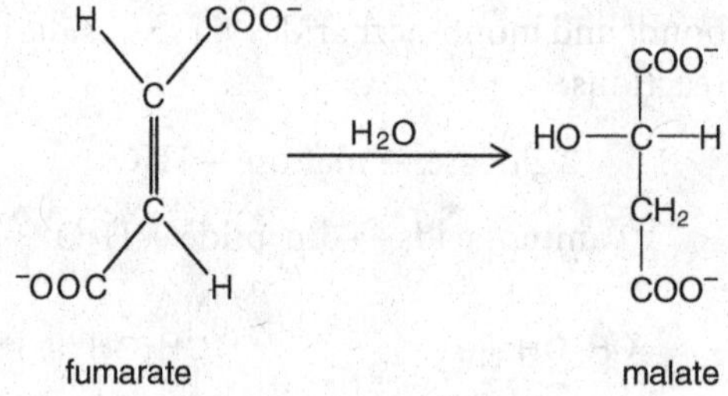

fumarate   malate

Figure 1.14 Enzyme; fumarase

7. **Transfer**

a. A phosphate group may be transferred from ATP to a substrate;

$$\text{glucose} + ATP \rightarrow \text{G-6-P} + ADP$$

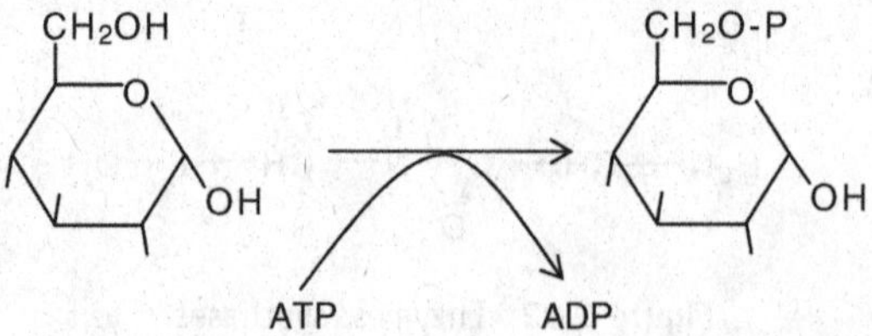

Figure 1.15 Enzyme: glucokinase or hexokinase

b. A functional group may be 'swapped' between two molecules

$$\text{glutamate} + \text{oxaloacetate} \rightarrow \text{2-oxoglutarate} + \text{aspartate}$$

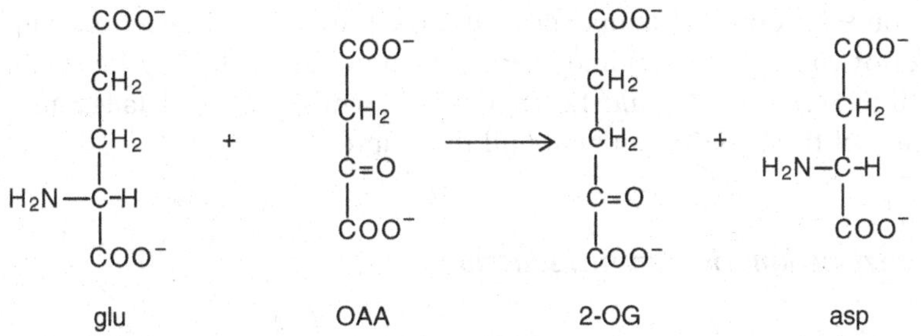

Figure 1.16   Enzyme: aspartate transaminase (= aspartate aminotransferase)

c.   Quite complex chemical groupings may be transferred

sedoheptulose-7-P + glyceraldehyde-3-P → erythrose-4-P + fructose-6-P

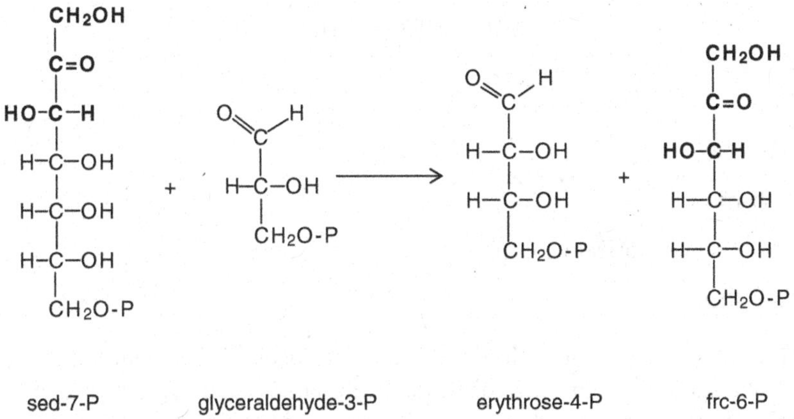

Figure 1.17   Enzyme: transaldolase

Here, a C3 unit (bold) has been transferred from sedoheptulose-7-P to glyceraldehyde-3-P.

The reactions given above illustrate the chemical changes that frequently occur in biochemistry. When you meet a reaction for the first time, it is a good idea to first of all identify the type of reaction occurring, and then look at the specific details.

Enzyme-mediated catalysis requires the breaking and making of chemical bonds between atoms; this involves changes in energy and is described by thermodynamics. Enzymes reduce the activation energy, that is make the reaction process easier to initiate but do not alter the overall energy change, which is determined by the free energy difference between the substrate(s) and the product(s). The change in free energy determines the *spontaneity* or likelihood of a reaction but the *speed* (kinetics) of an enzyme-catalysed reaction is governed by factors such substrate concentration, enzyme concentration, pH temperature and the presence of activators or inhibitors. Principles of enzyme kinetics and thermodynamics as applied to biochemistry are dealt with in Sections 1.4.2 and 1.5 respectively, whilst a more detailed analysis and explanation of these topics can be found in Chapter 2.

1.4.2   Enzyme kinetics: an introduction

Kinetics is the study of the factors which influence reaction rates. Enzyme-catalysed reactions are subject to the same principles of rate regulation as any other type of chemical reaction. For example, the pH, temperature, pressure (if gases are involved) and concentration of reactants all impact on the velocity reactions. Unlike inorganic catalysts, like platinum for example, there is a requirement for the substrate (reactant) to engage a particular region of the enzyme known as the active site. This binding is reversible and is simply represented thus:

$$E + S \rightleftarrows [ES] \rightarrow P + E$$

Where,

E = enzyme

S = substrate

[ES] = enzyme-substrate complex

P = product

The relative rates of formation and dissociation of [ES] is denoted as K_m, the Michaelis constant. Each enzyme/substrate combination has a K_m value under defined conditions. Numerically, the K_m is the substrate concentration required to achieve 50% of the maximum velocity of the enzyme; the unit for K_m is therefore the same as the unit for substrate concentration, typically µmol/l or mmol/l. The maximum velocity the enzyme-catalysed reaction can achieve is expressed by the V_{max}; typical unit µmol/min. The significance of K_m and V_{max} will be discussed in greater detail in Chapter 2.

In a cell, enzymes do not always work at their V_{max}. The precise rate of reaction is influenced by a number of physiological (cellular) factors such as:

- [S]

- [coenzyme]

- presence of activators or inhibitors.

Because enzymes are proteins, they are subject to all of the factors (e.g. pH, temperature) which affect the three-dimensional integrity of proteins in general.

The ability of some organisms to control the pH and temperature of their cells and tissues represents a major biological development. Homeothermic animals (e.g. mammals) maintain a constant temperature of about 37 °C as this corresponds to the temperature of optimum activity of most enzymes. Poikilothermic or so-called cold-blooded animals (e.g. reptiles) have to sun themselves for sometime every morning in order to raise their body temperature in order to optimize enzyme activity within their cells.

Plants and single celled organisms have no means of autoregulating their operating temperature and thus their growth and replication are influenced by external conditions. Hence, we keep food at 4 °C in a refrigerator to prevent spoilage yet we incubate bacterial cultures at 37 °C and usually in a buffered medium when we wish to cultivate the cells for further study.

Homeostatic mechanisms also allow animals to control their intracellular pH very strictly. In humans for example, blood pH (usually taken as a reliable but indirect measure of cellular pH) is 7.4 ± 0.04. At 37 °C cytosolic pH is actually slightly lower at about 7.0 but different compartments within the eukaryotic cells may have quite different pH, for example, lysosomes have an internal pH of about 5; the inside of a mitochondrion is more alkaline than the outside whilst the inside of a phagosome in a white blood cell is more acidic than its surrounding cytosol, both situations arising due to proton pumping across a membrane.

Except in a few instances, the enzyme molecule is very much larger than the substrate(s) upon which it works. The reason for this great disparity in size is not entirely obvious, but the possibility of the enzyme binding with more than one small molecule (e.g. regulator molecules, see Section 1.4.3) arises when we are dealing with large structures.

1.4.3 Enzyme ligands: substrates, coenzymes and inhibitors

As we saw earlier in this chapter, substrates are the molecules which undergo chemical change as a result of enzyme activity. Many enzymes will only operate when in the presence of essential co-factors or coenzymes. The term 'coenzyme' is not entirely appropriate as it implies that, like enzymes themselves, these compounds do not undergo chemical change. This is not true and more accurate terminology would be co-substrate. Coenzymes are always much smaller than the enzymes with which they operate and are not heat sensitive as are the proteins.

Examples of coenzymes: vitamin-derived nucleotides; for example adenosine phosphates; ATP, ADP, AMP; nicotinamide derivatives; NAD^+, NADH, $NADP^+$, NADPH; flavin derivatives; FAD, $FADH_2$; coenzyme A (abbreviated to CoA, CoASH or CoA-SH).

Not all vitamin coenzymes need to be in the form of a nucleotide (base, sugar, phosphate). For example; thiamine; biotin; pyridoxine; vitamin B_{12}.

Some enzymes also require inorganic factors to achieve full activity. Such co-factors include metal ions, Mg^{2+}, Mn^{2+}, Zn^{2+} and non-metals, Cl^-.

Inhibitors are compounds which reduce the efficiency of an enzyme and are important in directing and regulating the flow (or flux) of substrates through a pathway. Inhibitors which bind strongly to the enzyme for example, poisons such as cyanide, cause irreversible effects, but inhibition is rarely 'all or nothing' in a cell. Most inhibitors bind reversibly (as does the substrate of course) to the enzyme. Inhibitors which are structurally very similar to the true substrate effectively 'block' the active site and are called competitive inhibitors, because they compete with the true substrate for binding to the enzyme. Here, the ratio of substrate [S] to inhibitor [I] is critical in determining the quantitative effect of the inhibitor. Non-competitive inhibitors are also act reversibly by preventing the release of the product or by distorting the shape of the enzyme so preventing the substrate accessing the active site.

1.5 Bioenergetics: an introduction to biological thermodynamics

Thus far, our discussion has considered the chemical changes which constitute metabolism. We must now introduce some fundamental ideas of bioenergetics. Further details can be found in Chapter 2.

All molecules have an amount of energy determined mainly by their chemical structure. Metabolism involves chemical change. Inevitably therefore, energy changes always accompany the chemical changes which occur in metabolism. Our understanding of bioenergetics arises from physics and the laws of thermodynamics.

The First Law of Thermodynamics states that energy can be neither created nor destroyed but different forms of energy can be interconverted. The three forms of energy which are important to us are enthalpy (heat or 'total energy', represented by the symbol H), free energy ('useful energy' symbol G, in recognition of Josiah Gibbs) and entropy ('wasted energy', symbol S). Free energy is termed 'useful' energy because it can bring about useful work such as biosynthesis, transmembrane secretion or muscle contraction. Entropy however is not available for work but is the energy associated with chaos, disorder, loss of organization or an increase in randomness. Imagine a building, a castle, a tenement, or an office block which has not been maintained and thus shows the ravages of time and neglect. The building has lost its initial organization and structure because insufficient energy has been expended on its upkeep. You are now imagining entropy.

These three energy terms we have met are related by the following equation:

$$\Delta H = \Delta G + T\Delta S$$

where Δ indicates 'change in' and T is absolute temperature (Kelvin; $°C + 273$).

Rearranging this equation gives $\Delta G = \Delta H - T\Delta S$, which shows that as entropy increases as a function of temperature, free energy decreases.

The Second Law of Thermodynamics states that the entropy of the universe is constantly increasing. Cells are of course highly organized, a state which like the building referred to above, can only be maintained if free energy is expended. In other words, metabolism provides via catabolic (energy liberating, degradative) reactions free energy to prevent cells falling into disrepair by ensuring that biosynthesis and other cellular work can occur via anabolic (energy consuming, synthetic) reactions. *Homo sapiens*, like all animals is a heterotroph, meaning that the energy and raw materials required to maintain cellular structure and integrity are derived from the diet.

Figure 1.18a illustrates a phenomenon known as 'coupling'. Energy liberated from one process is used to drive forward an energy-requiring process. Individual biochemical reactions may be viewed similarly. Reactions which occur with a net loss of free energy ($-\Delta G$, termed exergonic) are spontaneous and favourable whereas those in which the products have more free energy than the reactants ($+\Delta G$, termed endergonic) will *not* occur spontaneously. An endergonic reaction can be driven forward by utilizing some of the energy liberated by the previous reaction in the pathway. Alternatively, ATP, the 'universal energy currency' of the cell can be called upon to provide the energy needed to overcome an endergonic reaction. The hydrolysis of ATP to ADP and inorganic phosphate (Pi) lies very far from equilibrium, has a very large Keq, and so is associated with a large energy change. It is not strictly true to state that energy is liberated from the *bond* (often incorrectly referred to as a 'high energy' bond) between the second and third phosphate groups of ATP. Figure 1.18b illustrates the coupling of energy liberating (catabolic) and energy consuming (anabolic) processes.

1.6 Enzyme-mediated control of metabolic pathways

Previously, the analogy was drawn between substrate flow in metabolism and traffic flow in towns, with enzymes acting as the 'traffic signals'; let us return to that image and consider where and how enzymes fulfil their control function. In the absence of traffic control measures, it is not difficult to imagine a situation of complete 'grid-lock' arising with no vehicles moving anywhere. Enzymes fulfil a regulatory function and prevent metabolic grid-lock by directing substrates along one pathway or another, by accelerating or slowing a particular pathway. Further details of metabolic control are given in Chapter 3.

Theoretically, all enzyme reactions are reversible but the overall flux (flow) of substrate in a pathway is unidirectional. To extend our road map analogy, this type of reaction acts as a control point in a pathway, rather like a one-way street, allowing substrates to flow in only one direction.

Such a 'metabolic one-way street' comes about in large part due the fact that certain chemical reactions are associated with a large energy change, which in chemical terms mean that the reaction is operating far away from its true equilibrium. Reactions of this nature are difficult to reverse under the conditions of pH, temperature and substrate concentration which exist inside cells and so become 'physiologically irreversible'.

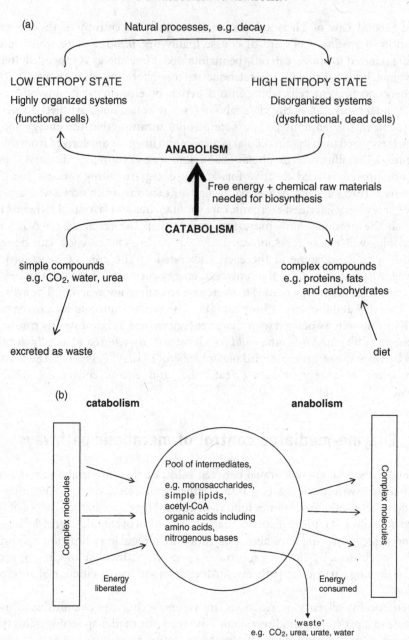

Figure 1.18 Energy flow in biological systems

Reactions operating far from their equilibrium position are not easy to identify merely by looking at the metabolic map, although the involvement of ATP is often a significant clue to a reaction being irreversible. Stated simply, hydrolysis of ATP 'energizes' a reaction which is normally irreversible. However, we can often predict the existence of flow-control enzymes at or near to the beginning of a pathway or at branch points (junctions) in pathways.

Enzyme activity is affected by changes in pH, temperature, substrate concentration, enzyme concentration and the presence of activators or inhibitors. Inside cells, both pH and temperature are normally tightly regulated so neither is able to influence greatly the physiological action of enzymes (a notable exception being the marked pH change seen in vigorously exercising muscle, see Chapter 7). Substrate concentration certainly does vary considerably within cells due, for example, to recent food intake or physical activity. Finally, enzyme concentration and the presence of activators or inhibitors also affect the rate of a reaction.

Enzymes are proteins (gene products) synthesized by DNA transcription and messenger RNA (mRNA) translation. Many enzymes are described as being 'constitutive', meaning they are present at all times. Others are 'inducible', meaning that their synthesis can be increased on-demand when circumstances require. By increasing the concentration of certain enzymes, induction allows more substrate to undergo chemical reaction and the pathway accelerates.

Activators and inhibitors regulate not the amount of enzyme protein but the activity ('efficiency') of that which is present. Two principal mechanisms of control are (i) competitive and (ii) allosteric. Competitive control (inhibition) occurs when a compound which is structurally similar to the true substrate binds to the active site of the enzyme. This is how a number of drugs and poisons bring about their effect. For example, a group of therapeutic drugs called statins are used to treat heart disease because by inhibiting a key enzyme called HMGCoA reductase, they reduce the hepatic synthesis of cholesterol and therefore the plasma concentration of that lipid.

Allosterism (Greek 'other place') is the name given to the mechanism whereby endogenous regulators, compounds found within or associated with the pathway in which the target enzyme occurs or from a related pathway, control a particular reaction. These regulators, allosteric activators and allosteric inhibitors, bind to the enzyme at identifiable allosteric sites, not the active site. The activity of the target enzyme changes as the cellular concentration of the allosteric regulators rise or fall. Details of allosteric control are given in Chapter 3.

Fluctuation in regulator concentration reflects the metabolic status of the cell and so the regulators themselves are acting as intracellular 'messengers'. For example, ATP, ADP and AMP act as allosteric regulators in glycolysis. When the cytosolic concentration of ATP in the liver or muscle is high, the cell has enough 'energy currency' so to process more glucose through to pyruvate would be wasteful. It is more useful to divert the glucose in to glycogen synthesis, an effect which is achieved by the allosteric inhibition of phosphofructokinase (PFK). Conversely, if the cytosolic concentration of ADP is high, PFK activity is accelerated, allowing more pyruvate to be synthesized leading to increased production of acetyl-CoA to be used in the Krebs TCA cycle and ultimately the synthesis of ATP. Here we have a good example of biochemical feedback.

Allosteric regulators bind to the target enzyme in a non-covalent manner. An entirely different enzyme control mechanism is covalent modification. Here, the conformation of the enzyme protein, and thereby its activity, is changed by the attachment of, usually, phosphate donated by ATP. Reversible phosphorylation is itself mediated via protein kinases (which transfer inorganic phosphate from ATP to a substrate) and protein phosphatases (which remove, by hydrolysis, inorganic phosphate from a substrate).

Not all allosteric proteins are enzymes. In fact, probably the best-known and characterized allosteric protein is haemoglobin, which like an enzyme binds ligands (small molecules) to itself, for example, oxygen rather than a substrate.

1.7　Strategy for learning the details of a pathway: 'active learning' is essential

When asked to learn a pathway, the temptation is to sit down and memorize each step in turn from top to bottom. A common failing in students who are new to metabolic biochemistry is in trying to memorize the whole of a pathway at the outset: *Rule 1 Resist the temptation to memorize*! This approach leads to 'knowing' but not really 'understanding'. Moreover, memorizing individual reactions/pathways is not always helpful. To use a microscopical analogy, begin with a low power view, try to see the pathway in relation to others and be clear about the physiological purpose of the pathway. Metabolism is a mosaic of component parts; pathways do not exist in isolation and taking the time see the broad picture at the start of the learning process will make the learning process more meaningful and therefore easier.

Rule 2 Be positive: *Don't* think about pathways simply as information gathered from experiments carried out in test tubes. *Don't* think about biochemistry as a body of knowledge that has to be mastered to pass an exam. *Do* think about what is going on inside your own cells and tissues at various times; having read a chapter in this book or after attending a lecture, think about how reactions and pathways in your own cells and tissues respond to your changing physiological circumstances, such as sleeping, sitting, walking, running, fasting, after food. Biochemistry is dynamic and its about you.

The following strategy should help put the pathway into its proper context.

The overview

1.　**WHERE** does the pathway occur?
　　that is in which cell types (prokaryotic or eukaryotic or both, in which tissue(s) of a multicellular organism, in which compartment of the eukaryotic cell, (cytosol, mitochondria, lysosomes etc.).

2.　**WHAT** is the biochemical purpose of the pathway?
　　for example, to release energy; to produce reducing power, to produce a key functional molecule, to synthesize a macromolecule.

3.　**WHAT** links are there between the pathway and any others?

4. **WHERE** are the control points within the pathway?

5. **WHEN** does the pathway operate? Is it always active or is it an 'adaptive' pathway?

The answers to all of these questions may not be evident immediately, but are usually to be found by diligent study active learning.

Once the overview is clear, begin to look in more detail at the chemistry and mechanisms of process. Here are some more suggestions of points to look for when studying an unfamiliar pathway in more detail.

The details

Skeletal view

What are the first and last substrates?
Is the pathway linear, branched or cyclical?
How many intermediate substrates are present?
Learn the names of the intermediates.
Which coenzymes are involved and where?

Chemistry of the intermediates and the reactions:

Look at them as organic chemicals;
How many carbon atoms are present and what types of functional groups are present?
What structural similarities and differences are there between the intermediates?
What sort of chemical reactions are occurring, for example, oxidation, condensation, hydrolysis.
Don't worry about getting the right sequence at this stage

Learn the names of the enzymes in sequence.

Use of the EC naming system will help you deduce the name of the substrate and the chemical change occurring

Learn the structures of the intermediates.

Use cue cards with the structure of an intermediate on one side and its name on the other;
Test yourself by selecting at random the name of an intermediate substrate and then draw from memory its structure.

Redraw the pathway in a different way

Include structures and all names in a different way;
Design a different image thus avoiding merely reproducing diagram from a book or the one given during a lecture.
Be creative; make the diagram as vivid and memorable as possible.

1.7.1   An Example: glycolysis as a model pathway

You will probably be familiar with glycolysis (the Embden–Meyerhof pathway, Figure 1.20) from previous studies at school perhaps, so let's use this important pathway to illustrate some points in the recommended strategy.

1.   **The overview**

Where?	Universal, occurs in all cell types. Cytosolic
What purpose?	
	To begin the oxidative catabolism of glucose. The production of ATP is small so this is not a prime role in most tissues. The end products pyruvate (or lactate) are important compounds for other pathways.
What are the links to other pathways?	
	Pentose phosphate pathway and glycogen metabolism (both are linked via glucose-6-P); glycerol from lipids may enter at the level of triose phosphate
Where are the control points? Reactions catalysed by	
	Hexokinase/Glucokinase Phosphofructokinase Pyruvate kinase
When does the pathway operate?	
	All of the time (constitutive).

---

2.   **The details**

Skeletal view

glucose → 2 × pyruvate($C_3H_3O_3$) if operating aerobically
(or 2 × lactate, $C_3H_5O_3$, if anaerobic)

Carbon balance:

$C_6H_{12}O_6 \rightarrow 2 \times C_3H_3O_3$ compounds
(or 2 × $C_3H_5O_3$ if anaerobic)

number of intermediates = 11 including glucose and pyruvate
10 enzyme-catalysed reactions

Coenzymes   2 molecules of NADH + H^+ are generated per molecule
of glucose oxidized;
net gain of 2 molecules of ATP per molecule of glucose oxidized,
that is, 2 molecules ATP consumed and 4 molecules produced
per molecule of glucose.

Chemistry of the intermediates	4 hexoses 3 of which are phosphorylated, one of which is bis-P i.e. two phosphates on different carbons within the same molecule)
	one aldehyde/ketone combination, both phosphorylated
	5 organic acids (all have 3 carbon atoms) 4 of these are phosphorylated
Reactions:	2 phosphorylations directly from ATP + 1 oxidative phosphorylation when Pi is added
	3 isomerizations
	1 cleavage
	2 dephosphorylations
	1 rearrangement

Names of the enzymes

Hexokinase/glucokinase	(HK/GK)
Phosphohexoisomerase	(PHI)
Phosphofructokinase	(PFK)
Aldolase	(ALDO)
Triose phosphate isomerase	(TPI)
Glyceraldehyde-3-P dehydrogenase	(Gly'ald-3-P D'ase)
Phosphoglycerokinase	(PGK)
Phosphoglyceromutase	(PGM)
Enolase	(ENO)
Pyruvate kinase	(PK)

Let's try applying the active learning model approach. The chemical structure of each glycolytic intermediate substrate is shown in Figure 1.19. Remembering that each individual reaction in any pathway brings about a small chemical change, arrange the structures in a logical sequence. The names of the intermediates are given in Figure 1.20.

Hint: think back to the word puzzle in which you changed the word 'went' into 'come'. The same process of small discrete changes of chemical structure can be seen to apply here.

First, name the intermediates using knowledge of simple organic chemistry and chemical nomenclature.

Start with the easy ones! Glucose [compound (iv)] should be familiar to you *and* it is one of only two substrates in glycolysis which is not phosphorylated; the other one being pyruvate [compound (i)].

From glucose, we can easily identify glucose-6-P (Glc-6-P) [compound (v)].

Similarly, fructose-6-P, one of the five-sided furan ring sugars we meet in metabolism. [Compound (x)] and fructose,-1,6-bis P [compound (xi)] should be obvious from their structures.

There is only one compound which carries an aldehyde group, so glyceraldehyde-3-P must be compound (viii) and acetone you may already know as a ketone, so compound (ii) is dihydroxyacetone phosphate, DHAP.

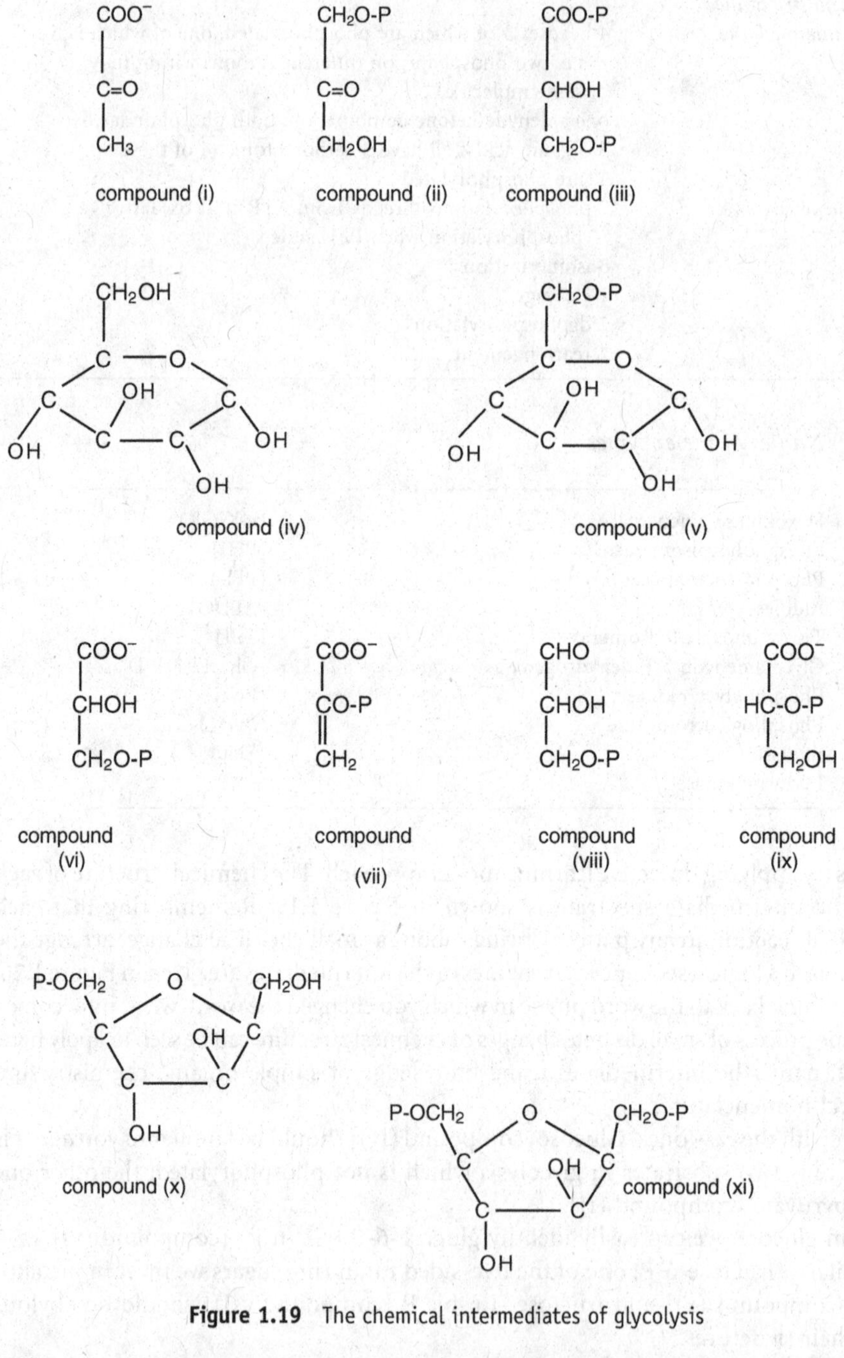

$$\begin{array}{c} COO^- \\ | \\ C=O \\ | \\ CH_3 \end{array}$$

compound (i)

$$\begin{array}{c} CH_2O\text{-}P \\ | \\ C=O \\ | \\ CH_2OH \end{array}$$

compound (ii)

$$\begin{array}{c} COO\text{-}P \\ | \\ CHOH \\ | \\ CH_2O\text{-}P \end{array}$$

compound (iii)

compound (iv)

compound (v)

$$\begin{array}{c} COO^- \\ | \\ CHOH \\ | \\ CH_2O\text{-}P \end{array}$$

compound (vi)

$$\begin{array}{c} COO^- \\ | \\ CO\text{-}P \\ || \\ CH_2 \end{array}$$

compound (vii)

$$\begin{array}{c} CHO \\ | \\ CHOH \\ | \\ CH_2O\text{-}P \end{array}$$

compound (viii)

$$\begin{array}{c} COO^- \\ | \\ HC\text{-}O\text{-}P \\ | \\ CH_2OH \end{array}$$

compound (ix)

compound (x)

compound (xi)

Figure 1.19 The chemical intermediates of glycolysis

Now for the glycerates. 1,3 bis-phosphoglycerate [compound (iii)] is the only molecule with two attached P groups. When we number the carbon atoms in an aliphatic organic compound we invariably start at the most oxidized carbon (drawn at the top of the chain), so carbon 2 of the glyceric acid derivatives must be the middle

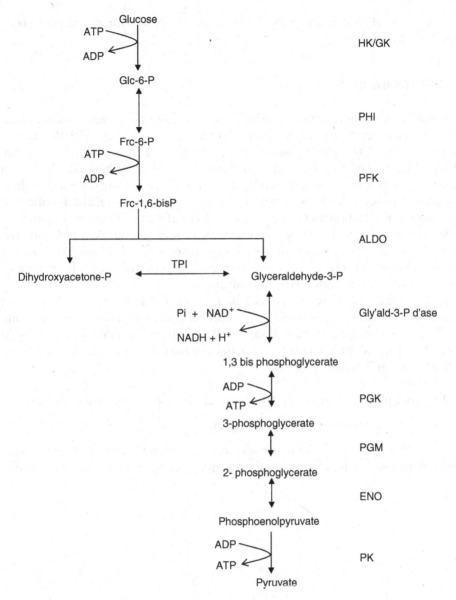

Figure 1.20 Glycolysis

one, so **2**-phosphoglycerate is compound (ix), and so **3**-phosphoglycerate must be compound (vi).

This leaves only one compound which must be phospho enol pyruvate (PEP) as compound (vii).

Metabolic pathways are better learnt as an exercise in logic than pure memory work!! Working from first principles with a firm underpinning knowledge will seldom let you down, whereas rote learning is superficial. We all suffer from 'memory blank' at various times!

Chapter summary

Metabolism describes the processes which allow energy to be utilized to maintain the integrity of an organism. Catabolic reactions usually liberate energy which the cell uses to drive forward anabolic reactions. Energy changes are associated with chemical changes which would normally occur far too slowly to be of biological use to an organism, so enzymes are used to accelerate reactions. Enzymes are catalysts but share few characteristics with inorganic catalysts such as platinum. The relative specificity of each enzyme for its substrate(s) means that each cell of the body requires hundreds of different types of enzyme and each type must be present in multiple copies. Enzyme-catalysed reactions are arranged into pathways; sequences of individual reactions in which each enzyme brings about a small chemical change. Keep in mind the road traffic analogy. Pathways are controllable and adaptable.

Learning metabolism requires a step back to focus, initially at least, not on the minute details but on the biological purpose(s) of a pathway. Look for patterns and similarities between pathways and always ask the questions 'what does this pathway do for *me*?' and 'how does this pathway adapt to changing physiological situations?' Be an active learner and make it personal!

The word puzzle. There are probably several ways to do this, here is one way:

went → want → wane → cane → came → come

Notice that apart from the number of letters, the first and last words are structurally very different and indeed have opposite meanings yet there is a logical progression.

Problems and challenges

1. Distinguish between ... free energy, entropy and enthalpy

2. Define the terms endergonic and exergonic

3. What information is given by the sign ($+$ or $-$) of the free energy value?

4. Why does metabolism *not* grind to a resounding halt when an endergonic reaction occurs within a pathway?

5. *Without* performing any calculation, state with reasons if the following reactions are likely to be strongly exergonic, weakly exergonic, strongly endergonic or weakly endergonic:

    i. $R \rightarrow P\ K_{eq} = 0.005$

    ii. $R \rightarrow P\ K_{eq} = 127$

    iii. $R \rightarrow P\ K_{eq} = 2.5 \times 10^{-4}$

    iv. $R \rightarrow P\ K_{eq} = 0.79$

    v. $R \rightarrow P\ K_{eq} = 1.27$

6. Like glucose-6-P, pyruvate and acetyl-CoA are at metabolic cross-roads. Consult a metabolic map and identify these important compounds and note the ways in which they may be formed and metabolized.

7. Refer to Section 1.4. What type of enzyme-catalysed reaction is occurring in each of the following examples?

    a. Glucose-6-phosphate $\rightarrow$ Fructose-6-phosphate

    b. Fructose-6-P $+$ ATP $\rightarrow$ Fructose-1,6 *bis*phosphate $+$ ADP

    c. pyruvate $+$ CO_2 $\rightarrow$ oxaloacetate

    d. Fructose-1,6-*bis*phosphate $+$ H_2O $\rightarrow$ Fructose-6-phosphate $+$ Pi

        (NB: Pi is an abbreviation for inorganic phosphate)

2

Dynamic and quantitative aspects of metabolism: bioenergetics and enzyme kinetics

---

Overview of the chapter

An understanding of the mathematical basis of enzyme activity and of the energy changes which occur during biochemical reactions is important to appreciate fully the control of metabolism. This chapter provides definitions and explanations of key concepts such as free energy, entropy, $K_m K_i$, and V_{max}. Worked examples of calculations and graphical derivations are provided and the results interpreted. The chapter ends with an overview of energy producing processes.

Bioenergetics: free energy (symbol G); entropy (symbol S); standard and physiological conditions; equilibrium constant for a reaction under physiological conditions, symbol K'_{eq}; calculation of free energy from equilibrium and redox data; endergonic and exergonic reactions. High energy compounds, substrate level phosphorylation and oxidative phosphorylation.

Enzyme kinetics: Michaelis constant, symbol K_m; maximum velocity of an enzyme catalysed reaction, V_{max}; inhibitor constant, symbol K_i; Michaelis–Menten equation and graph in the absence and the presence of inhibitors. Lineweaver–Burke and Eadie–Hofstee plots.

2.1 Introduction

To the non-mathematically minded, the essentially qualitative nature of biology as compared with pure chemistry or physics is an attraction. It is a common fallacy to believe that biology is a nothing more than a descriptive subject. As outlined in Chapter 1, there are facets of metabolism which can only really be appreciated when

---

analysed quantitatively. Fortunately, the mathematical knowledge required to understand metabolic processes is fairly straightforward and the skills we will use in this
chapter are little more than those of basic arithmetic, the occasional use of logarithms
and the confidence to rearrange a formula. The commonest failing is not with the
computation, but a failure to take appropriate care with use of units. The most
important understanding to be gained from this chapter is how to interpret the data
rather than how to generate them.

2.2 Bioenergetics: the application of thermodynamic principles to biological systems

The study of energy changes occurring in cells is fundamental to a sound understanding
of metabolism, but it is also one which students often find the most challenging. The
difficulties arise due to the conceptual nature of the topic and of the terms used to
describe it. Whilst it is easy to picture in one's mind eye the basic structure of a
metabolic intermediate such as glucose or cholesterol and one can easily imagine a
small amount of, say, the amino acid alanine in the palm of the hand, to conjure up an
image of energy is not so easy.

By virtue of their very existence, all substances are considered to possess energy.
The amount of energy will however vary from one compound to another due to the
nature and number and type of atoms within a molecule and the chemical bonds
which hold those atoms together. During any chemical reaction, the total energies of
the individual reactants will become redistributed: some part of the total is used, for
instance, to make and break chemical bonds; some of the overall energy may be 'lost'
(transferred) to the environment. Occasionally we encounter reactions in which the
total energy of the reactants is insufficient to initiate the reaction. To overcome this
situation, energy usually from the hydrolysis of ATP may be used to drive the
reaction forward or one of the reactants will need to be 'activated', often with
coenzyme A, often referred to as 'active acetate'. To continue our road traffic analogy
from Chapter 1, both situations are somewhat like a vehicle taking on fuel at a filling
station.

A measure of the overall energy change which occurs during a reaction is given by the
enthalpy, symbol H which is a function of the entropy (S) and free energy (G) of that
reaction. Entropy is 'wasted' energy, associated with disorder and randomness; free
energy is that energy which can be utilized to perform useful biological work, such as
driving metabolism in the right direction, transporting molecules across membranes or
causing muscles to contract. Knowledge of the change in free energy of a reaction allows
biochemists to make predictions about that reaction and its significance in a metabolic
pathway.

In practice, the *actual* values for the free energy of a given reaction are difficult to
measure experimentally. However, during a chemical reaction when one compound

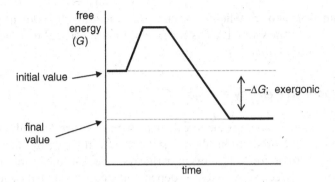

Figure 2.1 Reaction progress graph: exergonic reaction

(reactant, r) is converted to another (product, p), the *difference* in free energy (ΔG) between the reactant and the product can be measured. Thus, the *change* in free energy (ΔG) for the reaction r $\rightarrow$ p is simply;

$$\Delta G = G_p - G_r \qquad (2.1)$$

G_P is the free energy of the product(s) of the reaction

G_r is the free energy of the reactant(s) of the reaction.

Energy can be neither created not destroyed, but the total energy of the compounds at the end of the reaction (G_p) will be less than that at the start (G_r) as energy is 'lost' (transferred) to the environment ΔG is negative). Such reactions are termed exergonic and occur relatively easily ('spontaneous'). See Figure 2.1.

Alternatively, sometimes the products have more free energy than the reactants so the ΔG value is positive and the reaction is said to be endergonic and the reaction does not occur spontaneously (Figure 2.2).

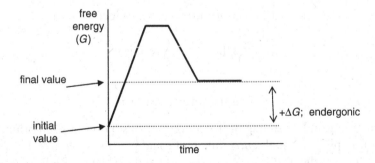

Figure 2.2 Reaction progress graph: endergonic reaction

The account given above applies to any chemical reaction, but as our attention is on reactions inside cells, the symbol $\Delta G'$ is adopted to indicate 'actual free energy change under *physiological conditions*'.

2.2.1   Standard free energy change

The main significance of being able to calculate or to measure $\Delta G'$ values is that we are then able to make predictions about reactions and in particular identify which reactions are likely to be control points in pathways. The absolute numerical value of the *actual* change in free energy ($\Delta G'$) is dependant upon the actual concentrations of the reactant(s) and product(s) involved in the reaction. Comparisons of values for different reactions are meaningless unless they have been determined under identical and standardized experimental conditions. The term *standard free energy* (symbol $\Delta G°$) is used to specify just such conditions.

The standard free energy change is the value obtained when the reactants and products (including H^+) are at molar concentration and gasses (if present) are at 1 atmosphere of pressure. Such conditions are quite unphysiological, especially the proton concentration, as 1 molar H^+ concentration gives a pH 0; biochemical reactions occur at a pH of between 5 and 8, mostly around pH 7. So a third term, $\Delta G°'$, is introduced to indicate that the reaction is occurring at pH 7.

If the reaction conditions are fixed (and standard), the value for $\Delta G°'$ must be a constant for any given biochemical reaction. The value for $\Delta G°'$ may be seen as a 'benchmark'; the further away $\Delta G'$ is from $\Delta G°'$ the further away the real reaction is from standard conditions.

2.2.2   Equilibrium reactions

So far the discussion has assumed that the reactions under study are unidirectional $r \rightarrow p$. The majority of biochemical reactions are however reversible under typical cellular conditions; in metabolism, we are therefore dealing with chemical equilibria.

$$r \leftrightarrow p \quad r = \text{reactant(s)}, \ p = \text{product(s)}$$

$$r \rightarrow p \text{ is the forward reaction}$$

and

$$p \rightarrow r \text{ is the reverse reaction}$$

A common misunderstanding of the concept of this 'point of equilibrium' is the belief that it implies an *equal concentration* of r and p. This is *not true*. The point of equilibrium defines the *relative concentrations* of r and p when the RATE of formation of p is exactly equal to the RATE of formation of r. For a given reaction, under defined conditions, the point of equilibrium is a constant and given the symbol K_{eq}.

Thus,

$$K_{eq} = \frac{[p]}{[r]} \qquad \text{right-hand side of the chemical equation}$$

right-hand side of the chemical equation
left-hand side of the chemical equation

When a reaction contains two or more reactants or products, for example

$$a + b + c \leftrightarrow u + v$$

the equation is modified thus

$$K_{eq} = \frac{[u] \times [v]}{[a] \times [b] \times [c]}$$

As with free energy, a 'dash' (prime) on K'_{eq} signifies 'physiological conditions'. Again the greater the numerical difference between K_{eq} and K'_{eq} the further away the reaction is from its true equilibrium position and the greater the actual change in free energy.

The *actual* change if free energy ($\Delta G'$) can then be calculated if the *actual* concentrations of r and p are known:

$$\Delta G' = \Delta G^{\circ\prime} + RT\, 2.303 \log_{10} K'_{eq} \qquad (2.2)$$

or,

$$\Delta G' = \Delta G^{\circ\prime} + RT \log_e K'_{eq} \qquad (2.2a)$$

where,

$\Delta G^{\circ\prime}$ = a constant, specific for each reaction;
R = the gas constant (8.314 J/mol);
T = absolute temperature (Kelvin; K = °C + 273);
$\log_e$ = natural logarithms; 2.303 converts $\log_e$ to $\log_{10}$;
K'_{eq} = concentration of product(s)/concentration of reactant(s).
NOTE; it is the log of the K_{eq} (i.e. the RATIO of p and r) that must be used.

Thus a useful parameter ($\Delta G'$) can be calculated for any reaction (not only those operating under standard conditions) providing the relative concentrations of r and p, the temperature and the standard free energy change $\Delta G^{\circ\prime}$ are known.

For explanatory purposes, the two reactions which make up an equilibrium can be viewed separately. The forward reaction, r $\rightarrow$ p, will have a certain value for $\Delta G'$; the reverse reaction p $\rightarrow$ r will have a $\Delta G'$ which is *numerically* the same as that for the forward reaction but of *opposite sign*. So, at equilibrium $\Delta G'$ = zero. Energy 'lost' in one direction is 'gained' in the other direction

Therefore, if at equilibrium,

$$\Delta G' = \Delta G^{\circ\prime} + RT\, 2.303 \log_{10} K'_{eq} \qquad (2.3)$$

and substituting, if $\Delta G' = 0$

$$0 = \Delta G^{\circ\prime} + RT\, 2.303\, \log_{10} K'_{eq} \tag{2.4}$$

then transposing the formula;

$$\Delta G^{\circ\prime} = -RT\, 2.303\, \log_{10} K'_{eq}$$

or

$$\Delta G^{\circ\prime} = -RT\, \log_e K'_{eq} \tag{2.5}$$

Equation 2.5 is important and one you should remember.
Note the minus on the right-hand side of the equation.

Summarizing, the numerical value and the sign of $\Delta G^{\circ\prime}$ allow us to make comparisons of different reactions under identical conditions but $\Delta G'$ values reveal the real energy change occurring in a metabolic pathway. $\Delta G^{\circ\prime}$ and $\Delta G'$ values are expressed in units of kJ/mol.

It is apparent from Equation 2.2 that a very large or a very small K_{eq} is associated with a large energy change. As a general rule, if K'_{eq} is greater than $\sim 10^3$ the reaction is strongly exergonic and where K'_{eq} is less than approximately 10^{-3}, the reaction is endergonic. Reactions operating far from their true K'_{eq} are essentially irreversible under physiological conditions (see Section 2.2.4).

It is important to dissociate rate from free energy change. A large value for $\Delta G^{\circ\prime}$ (or $\Delta G'$) does not imply a fast reaction, nor does a small value suggest a slow reaction. Furthermore, the presence of an enzyme does not alter the point of equilibrium (K'_{eq}) or therefore $\Delta G^{\circ\prime}$, but merely the time taken to attain equilibrium. Put another way, $\Delta G^{\circ\prime}$ is a measure of the thermodynamic probability that the reaction will occur, *not an indicator of its speed.* Therefore the presence of an enzyme does not alter values for $\Delta G^{\circ\prime}$ or $\Delta G'$ as the overall energy change is not affected. An analogy would be to consider yourself standing on the tenth floor of a building. By virtue of the height, you will possess energy which is greater than the energy you would have at ground level. If you now move from say, the tenth floor to the ground the difference in energy change is due to the height difference and not dependant upon whether you, walk down the stairs (slow reaction), use the lift (faster reaction) or indeed jumping from the window (fast reaction)!

Remember, an enzyme is just a 'tool' for making and breaking chemical bonds to bring about structural changes. The chemical mechanism of those changes does not affect the *overall* energy change involved in the reaction.

Worked example 1

Calculation of $\Delta G^{\circ\prime}$ from K'_{eq} value.
 The following reaction is part of glycolysis:

$$\text{2-phosphoglycerate} \leftrightarrow \text{3-phosphoglycerate}$$
$$\text{2-PG} \qquad\qquad\qquad \text{3-PG}$$

If at 37 °C [2-PG] = 4.4 µmol/l and [3-PG] = 1.46 µmol/l, calculate the free energy change. NB: 1 µmol/l = 1 × 10^{-6} mol/l

First, we need to calculate the K'_{eq}. As written above, 2-PG is the reactant and 3-PG is the product, so in this simple example, with only one molecule of reactant and one of product

$$K'_{eq} = \frac{[3\text{-PG}]}{[2\text{-PG}]} = \frac{1.46 \times 10^{-6}}{4.4 \times 10^{-6}}$$

$$K'_{eq} = 0.33 \quad (\text{no units as this is a ratio})$$

Second, calculate the absolute temperature: 273 + 37 °C = 310 K

Finally, substituting into Equation 2.5, we get,

$$\Delta G^{\circ\prime} = -RT\,2.303\log_{10}K'_{eq}$$
$$\Delta G^{\circ\prime} = -8.314 \times 310 \times 2.303\log_{10}0.33\ (\text{NB: R is in joules per mol, not kilojoules per mol})$$
$$\log_{10}0.33 = -0.479$$
$$\Delta G^{\circ\prime} = -8.314 \times 310 \times 2.303 \times -0.479\ (\text{NB: there are two minus signs}$$

so expect a positive answer) $+ 2856.5\,\text{J/mol} = +2.9\,\text{kJ/mol}$

When performing calculations such as this it is advisable *always* to show the sign to ensure clarity but there is no advantage in expressing answers to more than one decimal place.

The calculated value is a small positive number; the reaction is endergonic would occur easily without the input of energy.

Worked example 2

Aldolase catalyses the following reaction;

Fructose-1,6-bisphosphate ↔ glyceraldehyde-3-phosphate + dihydroxyacetonephosphate

(DHAP)

If, at equilibrium, [Frc-1,6bisP] = 0.15 × 10^{-6} mol/l and the products are both at 4 × 10^{-6} mol/l, what is the free energy change at 25 °C?

First, calculate K'_{eq}; there are two products in this reaction, so

$$K'_{eq} = \frac{[\text{glyceraldehyde-3-P}] \times [\text{DHAP}]}{[\text{fructose-1,6 bis P}]}$$
$$= \frac{4 \times 10^{-6} \times 4 \times 10^{-6}}{0.15 \times 10^{-6}}$$
$$= 1.07 \times 10^{-4}$$

Second, T = 273 + 25 = 298

Finally, $\Delta G^{\circ\prime} = -8.314 \times 298 \times 2.303\log_{10}1.07 \times 10^{-4}$
$$\Delta G^{\circ\prime} = -8.314 \times 298 \times 2.303 \times -3.97$$
$$= +22652\,\text{J/mol} = +22.7\,\text{k J/mol}$$

2.2.3 Redox reactions

An important and particular type of biochemical equilibrium is that which involves oxidation and reduction that is a redox reaction. Oxidation is the loss of electrons, the loss of hydrogen or gain of oxygen and that reduction is the gain of electrons, gain of hydrogen or the loss of oxygen. Redox reactions involving electron transfer are very important in many biochemical reactions: (subscripts *red* and *ox* indicate the reduced and oxidized forms respectively):

$$A_{red} \leftrightarrow A_{ox} + e^-$$

More realistically, when two substrates (A and B) are involved, an electron is transferred from A to B. Substrate A is oxidized by the loss of the electron; B is reduced by gaining that electron.

The ease with which compounds donate one or more electrons is given by the redox potential, E'_0, (sometimes shown as E'^0) expressed in volts. The likelihood of electron transfer between two redox compounds is determined by the difference in their E'_0 values, that is $\Delta E'_0$

$$\Delta G^{\circ\prime} = -nF\Delta E'_0 \qquad (2.6)$$

Where $F =$ faraday (a constant, 96 500 joules/volt/mole)

$n =$ the number of electrons involved in the reaction

This is also an important formula to remember.
NOTE the minus sign.

Thus, $\Delta E'_0$ represents a constant and so is analogous to K'_{eq} for non-redox equilibria in that it gives us some quantitative information about the reaction.

As an example calculation, we can consider the passage of a pair of electrons along the mitochondrial respiratory chain from NADH to oxygen during oxidative phosphorylation. The process can be viewed as two half reactions, each with a redox potential;

$$NAD+ + 2H^+ + 2e^- \rightarrow NADH \quad \ldots\ldots \quad (1) \; E'_o = -0.32 \, V$$

$$\tfrac{1}{2}O_2 + 2H^+ + 2e^- \rightarrow H_2O \quad \ldots\ldots \quad (2) \; E'_o = +0.816 \, V$$

Half reactions are by convention, always written as reductions, that is showing addition of electrons going from left to right. Take careful note of the signs. When *balancing* the two half reactions however, one of them must be written as an oxidation.

So, combining the two half reactions;

$$NADH + \tfrac{1}{2}O_2 \longrightarrow H_2O + NAD^+$$

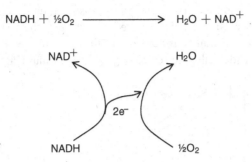

Clearly in this example, $n = 2$; $F =$ the Faraday constant (96 500 J/volt/mol).

The overall energy change is given by the redox difference between the NADH and oxygen

$$So, \Delta E_o' = (\text{the more positive } E_o') - (\text{the more negative } E_o')$$
$$\Delta E_o' = (+0.816 \text{ V}) - (-0.32 \text{ V})$$
$$\Delta E_o' = +0.816 + 0.32 \text{ V}$$
$$= +1.136 \text{ V}$$

Substituting into Equation 2.6 $\Delta G^{\circ\prime} = -nF\Delta E_o'$

$$\Delta G^{\circ\prime} = -2 \times 96\,500 \times 1.136$$
$$= -219\,248 \text{ J/mol} = -219 \text{ kJ/mol}$$

This is a very large negative $\Delta G^{\circ\prime}$ so the reaction is very strongly exergonic under standard conditions. Theoretically, enough free energy is liberated to phosphorylate seven molecules of ADP to ATP (ΔG for ADP phosphorylation $\sim +30.5$ kJ/mol). In practice, oxidative phosphorylation is less than 50% efficient so only three ATP molecules are formed, the remainder of the energy is 'lost' as heat.

2.2.4 Reaction probability and spontaneity

Individual reactions which have a small $\Delta G'$ are operating close to a true equilibrium and are thus easy to reverse. An analogy would be to consider a weight on a shallow inclined plane (Figure 2.3).

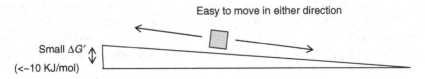

Figure 2.3 Reaction probability is high if $\Delta G'$ is small

Here, it is fairly easy to push or to pull the weight in either direction. Some reactions however operate far from their true equilibrium position that is have a large K'_{eq}, $\sim 10^3$ or $\sim 10^{-3}$). There is a large change of free energy and the reaction is said to be *physiologically irreversible*, that is under the conditions of temperature and concentration which prevail inside cells, the reaction is unidirectional (Figure 2.4).

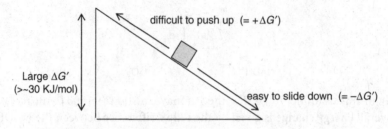

Figure 2.4   Reaction probability is high if $\Delta G'$ is large

Thus the numerical value of $\Delta G'$ (which is itself directly derived from K'_{eq} or $\Delta E'_o$) indicates both the probability (or feasibility) of a particular reaction occurring and the direction (forward or reverse) in which the reaction will proceed.

If in simple terms, A to G represent intermediates, a pathway can be illustrated thus:

$$A \longrightarrow B \leftrightarrow C \leftrightarrow D \leftrightarrow E \longrightarrow F \leftrightarrow G$$

The two reactions $A \rightarrow B$ and $E \rightarrow F$ shown with heavy single headed arrows are physiologically irreversible (large *positive* $\Delta G'$ for the reverse reaction), whilst all of the others are assumed to be relatively easily reversible. Any reaction such as $A \rightarrow B$ operating far from its equilibrium will be a 'one-way street' and so act a set of 'traffic signals' within the pathway allowing substrate to flow unidirectionally. What happens if a reaction has a large positive $\Delta G'$, indicating 'unfavourability', for the *forward* reaction? The answer is that energy must be supplied to drive the reaction forward. This may be achieved by coupling or by the use of ATP, the 'universal energy currency of the cell'.

2.2.5   Reaction coupling

A physiologically irreversible reaction as described above with a positive ΔG would effectively stall the pathway because the reactants would not have enough energy to form product. This undesirable situation can be overcome by putting energy into the reaction to drive it forward. The energy is provided by another reaction occurring within the cell. If the highly exergonic reaction (large negative $\Delta G'$) immediately precedes the highly endergonic (large positive $\Delta G'$), the two reactions become coupled and the overall energy change is the sum of the two individual $\Delta G'$ values, thus:

$$P \xrightarrow[\Delta G' = +20 \text{ kJ/mol}]{} Q \xrightarrow[\Delta G' = -25 \text{ kJ/mol}]{} R$$

overall,

$$P \xrightarrow[\Delta G' = -5 \text{ KJ/mol}]{} R \tag{2.7}$$

Not all endergonic reactions will be so conveniently placed adjacent to an exergonic reaction. In such a situation, the required energy is usually supplied by ATP:

$$G + H + ATP \rightarrow L + M + ADP + Pi \tag{2.8}$$

In effect, this is also an example of reaction coupling:

i. $G + H \rightarrow L + M$ endergonic

ii. $ATP \rightarrow ADP + Pi$ exergonic.

2.3 Enzyme kinetics

The various chemical mechanisms of enzyme action will not be discussed here but an overview of enzyme kinetics is essential to allow a full understanding of metabolic control. Enzymes accelerate biochemical reactions. The precise rate of reaction is influenced by a number of physiological (cellular) factors:

- [S]

- [E]

- presence of activators or inhibitors

- [coenzyme]

NB: pH and temperature are excluded from this list as they are taken to be reasonably constant in *most* physiological circumstances.

When the rate of reaction is measured at fixed [E], but varying [S] and the results plotted, the Michaelis–Menten graph is obtained (below). This rectangular hyperbola indicates saturation of the enzyme with substrate.

In zone 'a' of Figure 2.5, the kinetics are first order with respect to [S], that is to say that the rate is limited by the availability (concentration) of substrate so if [S] doubles the rate of reaction doubles. In zone c however, we see zero order kinetics with respect to [S], that is the increasing substrate concentration no longer has an effect as the enzyme is saturated; zone 'b' is a 'transition' zone. In practice it is difficult to demonstrate the plateau in zone 'c' unless very high concentrations of substrate are used in the experiment. Figure 2.5 is the basis of the Michaelis–Menten graph (Figure 2.6) from which two important kinetic parameters can be approximated:

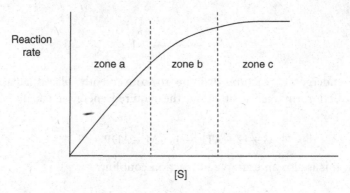

Figure 2.5 Saturation kinetics

V_{max}	the maximum possible rate of reaction achieved when [S] is saturating. Expressed in units of activity (rate of substrate conversion)
K_m	[S] at $^1/_2 V_{max}$, that is 50% maximum rate indicating that 50% saturation of enzyme with substrate has occurred. Expressed in units of [S]
	Strictly speaking, K_m is the dissociation constant for {ES} the enzyme-substrate complex.

Clearly, the measured V_{max} must be related to the concentration of enzyme present (more enzyme = faster reaction rate). A third parameter, K_{cat}, can be derived from the V_{max} to take into account the enzyme concentration thus:

$$K_{cat} = V_{max} \div [E] \quad \text{where [E] is the concentration of the enzyme}$$

Commonly, K_m (the Michaelis constant) is used to assess the affinity between the enzyme and the substrate. Another way to describe K_m would be as a 'saturatability' factor. When K_m is high there is a low affinity because a high concentration of substrate

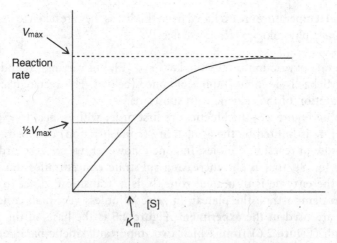

Figure 2.6 Michaelis–Menten graph

is required to achieve 50% saturation of the enzyme. Conversely, when K_m is low there is high affinity. The shape of the Michaelis–Menten graph is defined mathematically by

$$v_0 = \frac{V_{max} \times [S]}{K_m + [S]} \tag{2.9}$$

Where v_0 (initial velocity) is the instantaneous rate of reaction at the stated [S].

More accurate means of determining K_m and V_{max} are offered by either the Lineweaver–Burke (double reciprocal) plot or the Eadie–Hofstee plot (Figure 2.7).

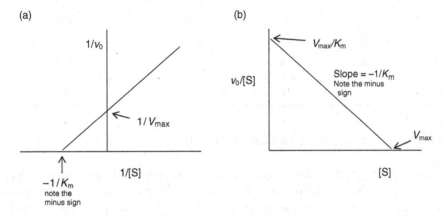

Figure 2.7 (a) Lineweaver–Burke plot (b) Eadie–Hofstee plot

The numerical values for K_m and V_{max} take the units of substrate concentration ([S], usually mmol/l or µmol/l) and velocity, respectively. Typical units of enzyme activity are shown in Table 2.1.

Table 2.1 Units of enzyme activity

Name of unit	Definition of unit
International Unit (IU)	that amount of enzyme protein which brings about the conversion of 1 µmol of substrate to product per minute under stated conditions
Katal (kat), the true SI unit of enzyme activity	that amount of enzyme protein which brings about the conversion of 1 mole of substrate to product per second under stated conditions.
Specific activity	enzyme activity (as IU or kat) per mg of total protein. This is a useful measure of the purity of an enzyme preparation

Semi-worked example: linear graphs

Use the data in Table 2.2 to plot Michaelis–Menten, Lineweaver–Burke and Eadie–Hofstee graphs to determine K_m and V_{max} values. Answers are given at the end of the chapter.

Table 2.2 Experimental data: no inhibitor present

Raw experimental data		Derived data for L–B graph		Derived data for E–H graph
[S] (mM)	v_0 µmol/min	1/[S] (mM^{-1})	$1/v_0$ (µmol/min)$^{-1}$	v_0/[S] (min/µmol)
0.1	0.48	10	2.08	4.8
0.2	0.85	5	1.18	4.25
0.4	1.40	2.5	0.71	3.5
0.5	1.60	2.0	0.63	3.2
0.8	2.05	1.25	0.49	2.56
1.0	2.31	1.0	0.43	2.31
1.2	2.45	0.83	0.41	2.04
1.5	2.67	0.67	0.37	1.78
2.0	2.92	0.5	0.34	1.46
2.5	3.10	0.4	0.32	1.24
5.0	3.5	0.2	0.29	0.7
10.0	3.72	0.1	0.27	0.372

2.3.1 Enzyme inhibitors (see also chapter 3)

Any compound which reduces the activity of an enzyme is an inhibitor. Such compounds may be endogenous or exogenous (i.e. drugs or toxins) and the impact on the enzyme may be a temporary or permanent effect. Temporary or reversible inhibitors are important in metabolic control whereas permanent or irreversible inhibitors are in effect enzyme poisons.

$$E + I \rightleftarrows \{EI\} \text{ reversible (competitive or non-competitive)}$$

$$E + I \rightarrow EI \text{ irreversible}$$

In just the way that K_m defines the dissociation of an enzyme from its substrate, K_i (the inhibitor constant) defines the 'strength' of binding of a reversible inhibitor to the enzyme.

The mechanisms which underlie enzyme inhibition are described more fully in Chapter 3. Suffice to say here that reversible inhibitors which block the active site are called competitive whilst those which prevent release of the product of the reaction are non-competitive. By preventing the true substrate accessing the active site, competitive inhibitors *increase* K_m (designated by K'_m or $K_m^{apparent}$). A non-competitive inhibitor *decreases* V_{max}, represented as V'_{max} or $V_{max}^{apparent}$. Do not confuse the use of the prime symbol (') here to imply 'physiological' as it does for energy change.

The extent or degree of inhibition brought about by inhibitor binding can be calculated if K_i is known or can be determined experimentally. The Michaelis–Menten equation can be rewritten to include a correction factor $\{1 + ([I] \div K_i)\}$. For the sake of clarity, let the correction factor be represented by ϕ which will always have a value greater than 1. For K_m to *increase* in the presence of a competitive inhibitor, the correction must be ϕK_m, and it follows that $\phi K_m = K'_m$. To *reduce* the V_{max} the correction must be $V_{max} \div \phi$, thus $V'_{max} = V_{max} \div \phi$

Therefore, the apparent initial velocity (symbolized variously as v_i, v'_0 or $v_0^{apparent}$) for a competitive inhibitor:

$$v'_0 = \frac{V_{max} \times [S]}{\phi K_m + [S]} \tag{2.10}$$

and for a non-competitive inhibitor

$$v'_0 = \frac{\{V_{max} \div \phi\} \times [S]}{K_m + [S]} \tag{2.11}$$

Figure 2.8a and b illustrate the effects of a competitive and a non-competitive inhibitor on the Lineweaver–Burke graph. Note the changes in the slope of the lines in the presence of the inhibitor and how this affects values for K_m and V_{max}.

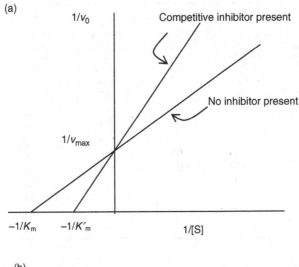

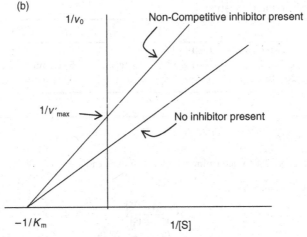

Figure 2.8 (a) Effect of a competitive inhibitor on the. Lineweaver–Burke graph. (b) Effect of a non-competitive inhibitor on the Lineweaver–Burke graph

Semi-worked example

Table 2.3 is similar to Table 2.2 but includes data for the reaction in the presence of an inhibitor.
Use the data in the table above to plot Michaelis–Menten, Lineweaver–Burke and Eadie–
Hofstee graphs to determine K_m and V_{max} values. State the type of inhibitor which is present.
Calculate the K_i based on Equations 2.10 (for competitive inhibition) or 2.11 (non-competitive
inhibition) asappropriate assuming the $[I] = 10$ mmol/l.

Table 2.3 Experimental data: inhibitor present

raw experimental data			derived data for L–B graph			derived data for E–H graph
$[S]$ (mM)see Table 2.2	v_0 (µmol/min) see Table 2.2	v_0' (µmol/min)	$1/[S]$ (mM^{-1}) see Table 2.2	$1/v_0$ (µmol/min)$^{-1}$ see Table 2.2	$1/v_0'$ (µmol/min)$^{-1}$	$v_0'/[S]$ (min/µmol)
0.1	0.48	0.25	10	2.08	4.0	2.5
0.2	0.85	0.47	5	1.18	2.12	2.35
0.4	1.40	0.81	2.5	0.71	1.23	2.03
0.5	1.60	1.0	2.0	0.63	1.0	2.0
0.8	2.05	1.4	1.25	0.49	0.71	1.75
1.0	2.31	1.6	1.0	0.43	0.63	1.60
1.2	2.45	1.78	0.83	0.41	0.56	1.48
1.5	2.67	2.0	0.67	0.37	0.50	1.33
2.0	2.92	2.3	0.5	0.34	0.43	1.15
2.5	3.10	2.5	0.4	0.32	0.40	1.0
5.0	3.5	3.08	0.2	0.29	0.32	0.62
10.0	3.72	3.48	0.1	0.27	0.29	0.35

Table 2.4 Michaelis–menten equation

	K_m	V_{max}	$[S]$	$[I]$	K_i	
a	12 µmol/l	2.4 µmol/min	3.6 µmol/l	–	–	no inhibitor
b	125 ng/ml	175 ng/h	100 ng/ml	35 µg/ml	52 µg/ml	competitive
c	9.3 mmol/l	26.5 µmol/h	5.0 mmol/l	300 µmol/l	220 µmol/l	non-competitive

Worked examples

Use the data in Table 2.4 to determine v_0 (or v_0') in each case.

a. No inhibitor present, so:

$$v_0 = \frac{V_{max} \times [S]}{K_m + [S]}$$

$$v_0 = \frac{2.4 \times 3.6}{12 + 3.6} = \frac{8.64}{15.6}$$

$$= 0.55 \text{ µmol/min}$$

b. In the presence of a *competitive* inhibitor, first calculate ϕ thus

$$\phi = (1 + \{[I] \div K_i\}).$$
$$\therefore \phi = 1 + \{35/52\} = 1.67$$
$$\text{and } \phi K_m (= K'_m) = 1.67 \times 125$$

$$v'_0 = \frac{175 \times 100}{(1.67 \times 125) + 100} = \frac{17500}{208.75 + 100}$$
$$= 56.7 \text{ ng/h}$$

this compares with 77.8 ng/h in the uninhibited reaction, derived using the same figures for V_{max} and K_m but without the correction factor ϕ.

c. In the presence of a *non-competitive* inhibitor, first calculate ϕ thus

$$\phi = 1 + \{300/220\} = 2.36$$
$$\text{and, } V_{max}/\phi = 26.5/2.36 = 11.22$$

$$v'_0 = \frac{11.22 \times 5}{9.3 + 5} = \frac{56.1}{14.3}$$
$$= 3.9 \, \mu\text{mol/h}$$

this value compares with 9.3 μmol/h uninhibited reaction, derived using the same figures for V_{max} and K_m but without the correction factor ϕ.

2.4 Energy generating metabolic processes

We will end this chapter by taking our first systematic look at some real metabolism, by considering how energy released from chemicals within the cell is trapped as ATP and the roles played by of some of the molecules involved with energy changes inside cells. Figure 2.9 shows an overview of key pathways.

The first law of thermodynamics tells us that 'energy can be neither created nor destroyed'; it can however be liberated (by exergonic reactions), transferred and utilized (by endergonic reactions). Two types of important molecules in the process are (i) coenzymes and (ii) high-energy phosphates. The coenzyme group includes the nicotinamide dinucleotides, NAD^+/NADH, $NADP^+$/NADPH, the flavin coenzymes flavin adenine dinucleotide (FAD/FADH$_2$) and flavin mononucleotide (FMN), co-enzyme A (represented as CoA or CoASH). The nicotinamide and flavin nucleotides are involved with oxidation and reduction reactions, hydrogen atom and/or electron transfers within the cell, which are invariably highly energetic. As a general rule, NAD^+

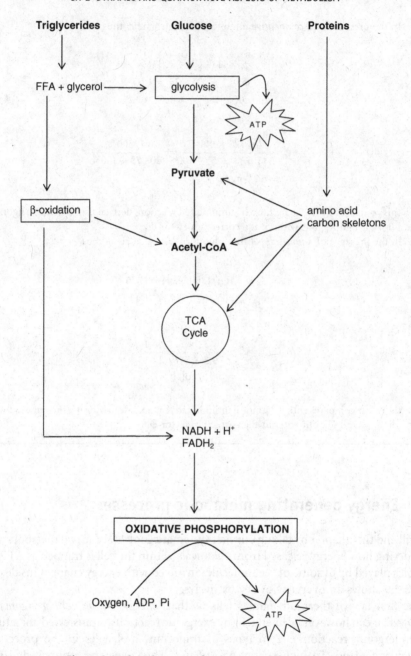

Figure 2.9 Overview map of energy generating pathways

and FAD are usually associated with catabolic reactions, for example oxidation of carbohydrate and fat, whilst $NADP^+$ is more commonly associated with anabolic reactions, one notable exception being the pentose phosphate pathway described in Chapter 5. Coenzyme A is used to 'activate' compounds which would not otherwise have sufficient energy to take part in reactions. The two best known examples being acetyl-CoA ('activated acetate') and succinyl-CoA ('activated succinate').

Creatine phosphate (also called phosphocreatine), phosphoenolpyruvate (PEP) and 1,3-bis phosphoglycerate (1,3BPG, formerly called 1,3 diphosphoglycerate, 1,3 DPG) are examples of high energy phosphates, defined, rather arbitrarily, as compounds whose hydrolysis is highly exergonic ($\Delta G^{\circ\prime}$ approximately -31 kJ/mol). The most well-known high energy-associated molecule is of course ATP which, along with similar molecules such as GTP and UTP for example, is *both* a nucleotide coenzyme *and* a high energy phosphate compound. It is actually a common misconception to believe that energy is released due to the breaking of a P–P bond within ATP. It is more accurate to state that the large energy change is because the equilibrium between ATP and ADP and Pi is very far from a true equilibrium, that is it has a large K'_{eq} and thus a large $\Delta G^{\circ\prime}$.

$$ATP \rightleftharpoons ADP + Pi$$

Catabolism provides the energy to re-phosphorylate ADP used in cellular activities. Given that dephosphorylation of ATP to ADP releases is associated with an energy change of approximately 31 kJ/mol, it follows that rephosphorylation of ADP must require the same amount of energy, forcing the equilibrium backwards. In the course of a day, an adult will turn-over several kilograms of ATP so we can conclude that re-phosphorylation is a very efficient process.

A skeletal overview of energy producing pathways is shown in Figure 2.9. Notice that catabolic pathways of carbohydrates, triglycerides and proteins meet at 'focal points', pyruvate and acetyl-CoA. Most cell types have at their disposal two mechanisms for the generation of ATP; these are oxidative phosphorylation and substrate level phosphory-lation. Red blood cells (erythrocytes), which lack organelles including mitochondria, are entirely dependent on ATP generation by the substrate level phosphorylation in glycolysis.

2.4.1.1 Substrate level phosphorylation

Substrate level phosphorylation involves the transfer of phosphate directly from a high energy compound (shown below as X-P) to ADP. This illustrates the concept of reaction coupling (Section 2.2.5) thus:

i. $X - P \rightarrow Y$ highly exergonic

ii. $ADP + Pi \rightarrow ATP$ highly endergonic.

This process may also be shown like this:

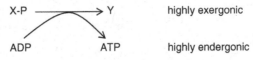

X-P ————————→ Y   highly exergonic

ADP   ATP   highly endergonic

Examples of substrate level phosphorylation are to be found in glycolysis. Phosphoglycerate kinase (PGK) and pyruvate kinase (PK) catalyse the following reactions:

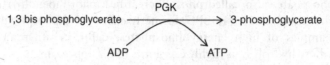

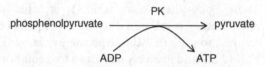

Because glucose is cleaved into two triose phosphates on its journey through glycolysis, these two reactions generate four ATP per mole of glucose. However the net gain in the pathway is only two ATP as hexokinase/glucokinase and phosphofructokinase each use one ATP to phosphorylate their substrates. Glycolysis is not therefore a very efficient means of producing ATP from glucose.

If we study the PGK reaction in isolation, we would find that it is extremely exergonic (−50 kJ/mol), and this sufficient to phosphorylate ADP (+30.5 kJ/mol) by coupling, and also pull forward the preceding reaction catalysed by glyceraldehyde-3-phosphate dehydrogenase which is overall weakly endergonic (Figure 2.10).

Because the $\Delta G'$ for step 2 (see Figure 2.10) is numerically greater than the $\Delta G'$ for step 1, the overall reaction is pulled in the direction of 1,3 BPG formation (Figure 2.11).

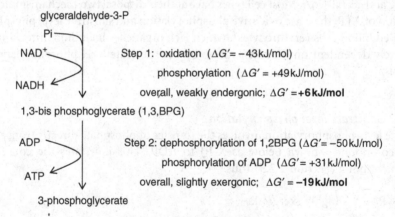

Figure 2.10 Reaction coupling

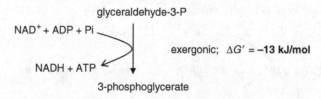

Figure 2.11 Coupled reactions shown in fig 2.10

(a) *Inter-membrane space of mitochondrion*

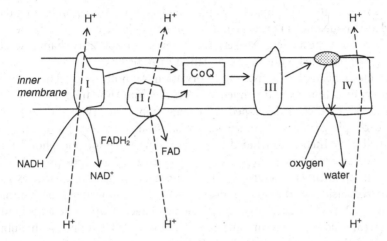

Key: I   complex I   contains FMN and Fe-S; accepts electrons and H^+ from NADH
       II   complex II   succinate dehydrogenase; has FAD as an attached coenzyme
       CoQ   cytochrome Q
       III   complex III   cytochromes b and c_1
              cytochrome c
       IV   complex IV   cytochromes a + a_3
       Solid arrows indicate electron flow; dashed arrows show H^+ pumping.

(b) High [H^+] established by proton pumping

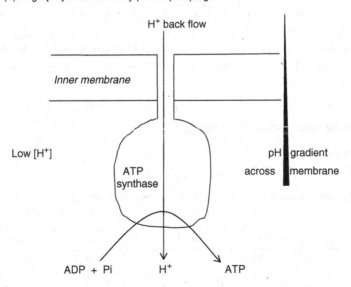

Figure 2.12   (a) Respiratory chain and electron flow (b) ATP synthesis

2.4.1.2 Oxidative phosphorylation

In contrast to substrate level phosphorylation in glycolysis, mitochondrial oxidative phosphorylation is an efficient process in that it generates in excess of 30 ATP per mole of glucose. In essence, the movement of electrons along the respiratory chain or electron transport chain is coupled with phosphorylation of ADP.

The exact mechanism of oxidative phosphorylation was hotly debated from the time of the discovery by Lehninger and Kennedy in 1948 that mitochondria are the site of the process, until Peter Mitchell developed the chemiosmotic theory in the early 1960s. Several direct coupling processes, somewhat similar to substrate level phosphorylation, were proposed but Mitchell concluded that the driving force for phosphorylation is dependent upon an electrochemical gradient established across the inner mitochondrial membrane. This gradient is generated by H^+ pumping, in which protons are extruded from the matrix (inside) of the mitochondrion as electrons flow along the respiratory chain (Figure 2.12a). The electrochemical gradient is equated with an energy gradient and as protons are allowed to re-enter the matrix by controlled leak through membrane pores, sufficient energy is released to drive ATP synthase (Figure 2.12b).

Chapter summary

The laws of thermodynamics govern the spontaneity of biochemical reactions. The magnitude and sign (+ or −) of the change in the free energy (ΔG) allows predictions to be made about the feasibility of individual reactions and their importance in pathways, for example control points. For comparative purposes, $\Delta G^{\circ\prime}$ is used to define the dynamics of a reaction but $\Delta G'$ is more informative of real reaction conditions inside cells. Energy released from strongly exergonic reactions may be conserved as ATP or used to drive forward endergonic reactions. There are two processes, substrate level phosphorylation and oxidative phosphorylation, which rephosphorylate ADP generated in metabolism.

Kinetic parameters V_{max} and K_m give information about the relative speed of biochemical reactions and the ease of interaction between the enzyme and its substrate respectively. Inhibitors may increase K_m or decrease V_{max} and metabolic control often relies on these effects.

Answers to semi-worked examples

Table 2.2

Raw experimental data		Derived data for L–B graph		Derived data for E–H graph
[S] (mM)	v_0 (μmol/min)	$1/[S]$ (mM^{-1})	$1/v_0$ (μmol/min)$^{-1}$	$v_0/[S]$ (min/μmol)
0.1	0.48	10	2.08	4.8
0.2	0.85	5	1.18	4.25

0.4	1.40	2.5	0.71	3.5
0.5	1.60	2.0	0.63	3.2
0.8	2.05	1.25	0.49	2.56
1.0	2.31	1.0	0.43	2.31
1.2	2.45	0.83	0.41	2.04
1.5	2.67	0.67	0.37	1.78
2.0	2.92	0.5	0.34	1.46
2.5	3.10	0.4	0.32	1.24
5.0	3.5	0.2	0.29	0.7
10.0	3.72	0.1	0.27	0.372

	M–M plot	L–B plot	E–H plot
K_m for this enzyme-substrate combination (mmol/l)	0.65	0.8	0.72
V_{max} for this enzyme-substrate combination (μmol/min)	3.72	4.0	3.9

Table 2.3

	Raw experimental data			Derived data for L–B graph			Derived data for E–H graph
[S] mM	v_0 (μmol/ min)	v_0' (μmol/min)	1/[S] (mM^{-1})	$1/v_0$ (μmol/ min)	$1/v_0'$ (μmol/min)$^{-1}$	$v_0'/$[S] (min/μmol)
0.1	0.48	0.25	10	2.08	4.0	2.5
0.2	0.85	0.47	5	1.18	2.12	2.35
0.4	1.40	0.81	2.5	0.71	1.23	2.03
0.5	1.60	1.0	2.0	0.63	1.0	2.0
0.8	2.05	1.4	1.25	0.49	0.71	1.75
1.0	2.31	1.6	1.0	0.43	0.63	1.60
1.2	2.45	1.78	0.83	0.41	0.56	1.48
1.5	2.67	2.0	0.67	0.37	0.50	1.33
2.0	2.92	2.3	0.5	0.34	0.43	1.15
2.5	3.10	2.5	0.4	0.32	0.40	1.0
5.0	3.5	3.08	0.2	0.29	0.32	0.62
10.0	3.72	3.48	0.1	0.27	0.29	0.35

$K_m' = 1.65$ mmol/l which is higher than the true K_m (taken as the mean of the three values above $= 0.72$ mmol/l), so this is a *competitive inhibitor*.

$$K_m' = \{1 + ([I]/K_i)\} \times K_m \quad (\text{i.e. } K_m' = \phi K_m)$$

Substituting, $1.65 = (1 + 10/K_i) \times 0.72$

Rearrange and solve for K_i, thus $K_i = 7.8$ mmol/l

Problems and challenges

1. '$\Delta G'$ not $\Delta G^{\circ\prime}$ is the appropriate criterion on which to judge the spontaneity of a reaction'. Explain.

2. Calculate $\Delta G^{\circ\prime}$ for each of the following reactions where [r] and [p] represent equilibrium concentrations of reactant and product; r1, r2, p1 and p2 indicate different reactants and products respectively in the reaction

   a.   [r] = 3.4 mmol/l
         [p] = 2.5 mmol/l
            at 37 °C

   b.   [r] = 0.18 mmol/l
         [p] = 5.4 mmol/l
            at 37 °C

   c.   [r] = 0.95 μ mol/l
         [p] = 1.05 μ mol/l
            at 30 °C

   d.   [r₁] = 1.25 μ mol/l   [r₂] = 0.85 μ mol/l
         [p] = 4.55 μ mol/l
            at 25 °C

   e.   [r] = 65.8 μ mol/l
         [p₁] = 21.5 μ mol/l   [p₂] = 3.5 μ mol/l
            at 37 °C

   f.   [r₁] = 4.75 mmol/l   [r₂] = 2.5 mmol/l
         [p₁] = 8.6 mmol/l   [p₂] = 1.1 mmol/l
            at 37 °C

3. Using your values for $\Delta G^{\circ\prime}$ for (a) to (f) above, calculate $\Delta G'$ for each reaction given the following data

   a. $K'_{eq} = 3.65$; temperature = 37 °C

   b. $K'_{eq} = 0.3$; temperature = 37 °C

   c. $K'_{eq} = 28.5$; temperature = 37 °C

   d. $K'_{eq} = 15.7$; temperature = 37 °C

   e. $K'_{eq} = 0.015$; temperature = 37 °C

   f. $K'_{eq} = 32.5$; temperature = 37 °C

4a. Given that the K_{eq} for a reaction is 1.45, which of the following statements is/are true?

   •   the forward reaction is faster than the reverse reaction;

   •   the concentration of p is higher than the concentration of r;

- the forward reaction is slower than the reverse reaction;
- the concentration of r is lower than the concentration of p.

4b. What biochemical change might cause K'_{eq} to shift to the left (net formation of r), that is the reverse reaction to accelerate?

5. The standard free energy change for the hydrolysis of ATP is estimated to be -30.5 kJ/mol.

$$ATP + H_2O \rightarrow ADP + Pi \quad \text{(Pi stands inorganic phosphate)}$$

a. estimate the relative concentrations (i.e. ratio, K'_{eq}) of ATP to ADP and Pi at $37\,^\circ C$. Assume [ADP] = [Pi]

b. calculate the actual free energy change if the actual cellular concentrations (in liver) are:

$$ATP = 4.0\,mmol/l$$

$$ADP = 1.35\,mmol/l$$

$$Pi = 4.65\,mmol/l$$

(NB: for simplicity, the concentration of water is usually ignored in hydrolysis reactions)

6. Fill in the missing values indicated by ?? in the table below.

v_0 (μmol/min)	Km (mmol/l)	V_{max} (μmol/min)	[S] (mM)	v_i (μmol/min)	[I] (mM)	K_i (mmol/l)	type of inhibitor
a ??	0.825	25	1.25				None
b ??	2.5	1.0	2.5				None
c 1.82	1.5	??	1.25				None
d 6.0	??	18.0	5.0				None
e ??	2.0	2.0	3.0	??	0.2	0.6	competitive
f ??	15	25	10	??	3.5	3.5	non-competitive
g 0.86	2.0	2.0	1.5	0.57	??	0.6	non-competitive

7. A 75 kg rower competing in the annual University Boat Race uses 52.5 kJ/min. Assuming that the race lasts 18 min,

i. *calculate* the total energy consumption during the race.

Given that the standard free energy change for the hydrolysis of ATP is -30.5 kJ/mol,

ii. *calculate* the total amount (in grams) of ATP required to furnish the energy to sustain the competitor during the race.

(HINT: The molecular weight of ATP = 507 to convert grams to moles and back use:

Actual weight of substance = molecular wgt. of the substance multiplied by number of moles)

Given that the maximum energy liberated from the complete oxidation of glucose to CO_2 and H_2O) is 2866 kJ/mol and that the molecular weight of glucose is 180

    iii.   *calculate* the mass of glucose (in grams) which would have been oxidized during the race assuming that the actual energy yield from glucose is 40% of theoretical the maximum (i.e. 2866 kJ/mol).

8.   Test your IT skills. Try creating Excel spreadsheets based on the Michaelis–Menten equation (Equation 2.9) and its variants (Equations 2.10 and 2.11). Insert into your spreadsheets your own values for K_m, V_{max}, [S], [I] and K_i and use Excel to plot Michaelis–Menten, Lineweaver–Burke and Eadie–Hofstee graphs.

3

Principles of metabolic control: enzymes, substrates, inhibitors and genes

Overview of the chapter

Metabolism is a complex yet highly ordered process able to adapt to changing physiological circumstances such as feeding and fasting, exercising and resting. This chapter describes organization of enzymes within cells, compartmentalization of pathways and how enzyme catalysed reactions may be regulated. The importance of enzyme induction, allosterism; covalent modification and the roles of inhibitors in feedback are considered. Control is also mediated through substrate availability and coenzyme ratios.

Key pathways

Glycolysis and the tricarboxylic acid (TCA) cycle are used as model pathways to illustrate the processes.

3.1 Introduction

In order to function efficiently and to survive, a cell must adapt quickly to changing circumstances and to channel intermediates along pathways which are the most appropriate for the conditions at the time. The facility to increase or reduce the rate of an enzyme catalysed reaction is a crucial part of metabolic control and therefore the adaptability of metabolism as this allows optimal utilization of possibly scarce resources. In short, a cell must be able to control its metabolic activities in order to meet a challenge from the environment. Loss of biological or metabolic control is likely to be detrimental to the cell as is illustrated by certain abnormal conditions such as cancer, genetically determined inborn errors of metabolism or following the

Essential Physiological Biochemistry: An organ-based approach   Stephen Reed
© 2009 John Wiley & Sons, Ltd

introduction of toxins. To return to the simple analogy used in Chapter 1, imagine the traffic chaos which would result if a town or city was devoid of road signs, traffic signals or if the drivers simply chose to ignore totally the conventions of the road.

3.2 General principles

We think about metabolic pathways as linear or cyclical sequences of reactions as described in Chapter 1. Individual reactions within a pathway are often dependant upon at least one other reaction. For example, we know from our studies of enzyme kinetics in Chapter 2 that the rate of an enzyme catalysed reaction is determined in part by the concentration of substrate. Remember, the substrate for one reaction is usually the product of a previous reaction, so the activity of an enzyme is affected by the activity of the preceding enzyme in the sequence.

Our knowledge of enzymes also tells us that under usual physiological conditions (i.e. at typical cellular concentrations of substrate) most metabolic reactions are reversible. Energetically irreversible reactions, i.e. those with a large positive free energy change, effectively act as 'one-way' valves allowing substrate flow in the forward direction only.

The overall flux of a pathway can be controlled by regulation of just one or two key reactions. As you study metabolic control, look out for the following principles:

i. Key control points are often

- Near the beginning of a pathway,

- At branch points where two pathways share the same substrate,

- Physiologically irreversible reactions.

ii. Irreversible reactions can only be reversed by the use of an enzyme which is different from the one which catalyses the forward reaction; for example, kinases, which phosphorylate a substrate and phosphatases which remove, by hydrolysis, phosphate from a substrate.

Metabolic control is exerted at three levels:

- The substrate, including coenzyme availability (Section 3.2.1);

- The enzyme (Section 3.2.2);

- The gene (Section 3.2.3).

3.2.1 Substrate and coenzyme availability

Substrate availability is clearly a crucial factor in determining the flux through the pathway. To understand metabolic control fully we will need to apply our knowledge of enzyme kinetics from Chapter 2. Recall that the rate of a enzyme catalysed reaction is affected by substrate concentration ([S]) and that K_m is defined in terms of [S]. Consequently where, say, at a branch point, *two enzymes are competing* for the same substrate, the enzyme with the lower K_m for that substrate will be expected to bind preferentially with the substrate. However, remember that it is also V_{max} or K_{cat} and not simply binding affinity that determines how much substrate is converted per unit time.

If the concentration of a substrate which is common to two pathways should fall below a certain threshold, one of the enzymes (and therefore the pathway in which it occurs) will slow down, assuming the enzymes have different kinetic characteristics. Conversely, when [S] is high, both pathways will operate with ease because both enzymes are essentially 'saturated' with substrate. Thus the abundance of substrate or the rate of its import into the cell necessarily affects the flux (i.e. substrate flow) through the pathway.

Substrate availability to the cell is affected by the supply of raw materials from the environment. The plasma membranes of cells incorporate special and often specific transport proteins (translocases) or pores that permit the entry of substrates into the cell interior. Furthermore pathways in eukaryotic cells are often compartmentalized within cytoplasmic organelles by intracellular membranes. Thus we find particular pathways associated with the mitochondria, the lysosomes, the peroxisomes, the endoplasmic reticulum for example. Substrate utilization is limited therefore by its localization at the site of need within the cell and a particular substrate will be effectively concentrated within a particular organelle. The existence of membrane transport mechanisms is crucial in substrate delivery to, and availability at, the site of use.

Examples of such intra cellular membrane transport mechanisms include the transfer of pyruvate, the symport (exchange) mechanism of ADP and ATP and the malate–oxaloacetate shuttle, all of which operate across the mitochondrial membranes. Compartmentalization also allows the physical separation of metabolically opposed pathways. For example, in eukaryotes, the synthesis of fatty acids (anabolic) occurs in the cytosol whilst β oxidation (catabolic) occurs within the mitochondria.

Substrate availability for certain reactions can be optimized by anaplerotic ('topping-up') reactions. For example, citrate synthase is a key control point of the TCA cycle. The co-substrates of citrate synthase are acetyl-CoA and oxaloacetate (OAA) and clearly, restriction in the availability of *either* substrate will decrease the rate of the citrate synthase reaction. Suppose, for example, a situation arises when acetyl-CoA concentration is significantly higher than that of OAA, the concentration of the latter can be 'topped-up' and the concentration of acetyl-CoA simultaneously reduced by diverting some of the pyruvate away from acetyl-CoA synthesis (via pyruvate dehydrogenase) to OAA synthesis (via pyruvate carboxylase) as shown in Figure 3.1. The net effect is to balance the relative concentrations of the two co-substrates and thus to promote citrate synthase activity.

3.2.1.1 *Coenzyme ratios*
Most intermediate substrates are converted into another compound in the process of metabolism, but some, notably the coenzymes, are recycled. The total *amounts* of

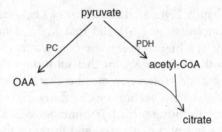

Figure 3.1 Pyruvate carboxylase (PC) diverts some pyruvate to OAA to ensure its concentration is maintained (PDH = pyruvate dehydrogenase)

adenine coenzymes (ATP, ADP and AMP) and of nicotinamide coenzymes (nicotinamide adenine dinucleotide $(NAD)^+$, NADH) are constant. During metabolic processes, these compounds are subject to reversible change. For example, ADP is reversibly changed into ATP and vice versa; NAD^+ is reduced into NADH which is then re-oxidised to NAD^+. Thus, the ratio of NAD/NADH or ATP/ADP will have a profound effect upon reactions that utilize one or other of the two forms. For example, glycraldehyde-3-phosphate, an enzyme within glycolysis, requires NAD^+ as coenzyme. If the availability of the coenzyme falls, the activity of the dehydrogenase will diminish and the glycolysis as a whole will be compromised. The mitochondrial metabolism of pyruvate offers a good example of the effect of $[NAD^+]$:$[NADH]$ ratio. Pyruvate may be converted by pyruvate carboxylase (PC, see Figure 3.1) into oxaloacetate, as described above. Alternatively, pyruvate may be decarboxylated by pyruvate dehydrogenase (PDH) yielding acetyl CoA. When the $[NAD^+]/[NADH]$ ratio is low, the PC reaction is favoured and OAA is formed. When the $[NAD^+]/[NADH]$ is high, the PDH reaction is favoured. To explain this, lets consider metabolism in a skeletal muscle cell. At rest or under mild exercise, skeletal muscle derives its energy from the metabolism of fatty acids. This process generates sufficient NADH and acetyl CoA to keep the TCA cycle and oxidative phosphorylation operating adequately. The high NADH concentration simultaneously, inhibits PDH and activates PC and glucose catabolism is reduced, allowing its use in, for example, glycogen synthesis. If however, a sudden burst of energy is needed, the NADH concentration will diminish as it is oxidised by complex I of the electron transport chain; PC activity will be reduced and PDH activity will increase, i.e. glucose is being used to supplement energy provision.

The term *coenzyme* is a poor one. Compounds such as ATP, ADP, NADH are better seen as co-*substrates* because clearly they do not conform to two of the criteria which define an enzyme catalyst, that is:

- Coenzymes are NOT proteins;

- Coenzymes ARE chemically changed at the end of the reaction.

If necessary, refer back to Chapter 1 for a formal definition of an enzyme.

3.2.2 Control at the level of the enzyme

In many sequential processes, the overall rate of a pathway is determined by the rate of the slowest individual reaction. This is called the rate determining step (RDS) or rate limiting step (RLS). Thus if the rate of an 'early' enzyme catalysed reaction is regulated (increased or decreased) in response to physiological conditions, then the overall rate of the pathway and substrate utilization will be subject to control.

It is too simplistic to consider that particular enzymes and pathways are either highly active or totally inactive. Many enzymes, especially those at control points in pathways, may retain some (perhaps very low) activity even under conditions which appear to be physiologically inappropriate, for example fuel substrate during fasting. This low activity keeps the enzyme ticking over and so able to respond more quickly and sensitively to sudden changes in conditions. It is easier to accelerate an enzyme which is 'ticking over' than to start one that is totally inactive.

The activity of an enzyme can be enhanced or retarded by the presence of other, usually small, compounds present in the cell. Generically, such compounds are called activators or inhibitors.

3.2.2.1 Enzyme inhibition

Inhibition can be defined as 'the reduction in the efficiency of an enzyme'. Note that the term does not imply that the enzyme is completely non-functional. In fact, the retention of some residual activity is quite important to achieve metabolic control. By analogy, consider how much easier it is for a runner to reach full speed from a jog than from a standing start.

Inhibitors may act reversibly or irreversibly to limit the activity of the enzyme. Irreversible inhibitors are enzyme 'poisons' and indeed many of them are poisonous in the common sense of the word; cyanide for example, is an irreversible inhibitor of one of the cytochromes in oxidative phosphorylation.

Reversible inhibitors are more subtle and act usefully to control the rate of particular enzymes. Often, reversible inhibitors are substrates found at or near the end of a pathway. These compounds act in a 'negative feedback' manner to slow down the activity of an enzyme at or near the beginning of the same pathway. Occasionally, feedback inhibitors may be substrates found within a pathway which is functionally related to the one in which the 'target' enzyme can be found. Furthermore, the products of an enzyme-catalysed reaction are often inhibitory to the enzyme that generated them (Figure 3.2). This is should not be surprising from a structural point of view because the product must 'fit into' the active site of the enzyme and so block the binding of substrate.

Many examples of product inhibition are to found. Some dehydrogenases are inhibited by NADH (a co-product of the reaction), e.g. PDH and isocitrate dehydrogenase (ICD), which are involved with the glycolysis and the TCA cycle are two such examples. Hexokinase isoenzymes in muscle (but not liver) and citrate synthase are inhibited by their products, glucose-6-phosphate and citrate respectively offering a very immediate 'fine tuning' of reaction rate to match cellular requirements and possibly allowing their substrates to be used in alternative pathways.

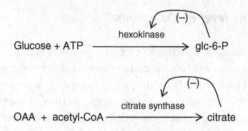

Figure 3.2 Product feedback inhibition

In terms of kinetic parameters, reversible inhibitors act to alter the rate of certain enzyme-catalysed reactions by changing the K_m or V_{max} values (designated $K_m^{apparent}$ or K'_m and $V_{max}^{apparent}$ or V'_{max} respectively to distinguish these changed values from the 'real' values. (See Chapter 2 for a more detailed account.)

Various mechanisms of reversible inhibition have been proposed. Competitive inhibition is conceptually the easiest to understand. Recall that the active site of an enzyme is complementary in shape to the shape of the substrate (crudely, the lock and key hypothesis). Suppose a compound which is not the true substrate, but structurally similar to it 'blocks' the active site by binding to it. The true substrate cannot bind and so no reaction will occur. Hence, there is competition between the true substrate and the inhibitor for binding at the active site.

Shown diagrammatically, where E = enzyme; S = substrate and I = inhibitor

$$E + S \leftrightarrow (ES) \rightarrow E + P \quad \{active\}$$
$$E + I \leftrightarrow (EI) \quad \{inactive; no\ product\ formed\}$$

In this situation, it is the *ratio* of S to I and the relative affinity coefficients (K_m and K_i) that determines the extent of inhibition.

An alternative model to competitive inhibition is called non-competitive. In this model, the inhibitor binds to the enzyme but not at the active site. Substrate binding is not affected but product release is slowed down.

$$
\begin{array}{ccccc}
E + S & \longleftrightarrow & (ES) & \longrightarrow & E + P \quad \{active\ ES\} \\
\updownarrow +I & & \updownarrow +I & & \\
EI & \xleftrightarrow{\ +S\ } & (ESI) & & \{inactive; no\ product\ formed \\
& & & & directly\ from\ ESI\}
\end{array}
$$

A third means of control of enzyme activity is achieved by altering the conformation of the protein. The actual mechanisms involved are:

- Allosteric effects

- Covalent modification.

In either case, the effect may be to enhance or diminish the activity of the enzyme.

3.2.2.2 Allosteric effects (allos = other, steros = place)

In addition to the binding of substrate (or in some cases co-substrates) at the active site, many enzymes have the capacity to bind regulatory molecules at sites which are usually spatially far removed from the catalytic site. In fact, allosteric enzymes are invariably multimeric (i.e. have a quaternary structure) and the allosteric (regulatory) sites are on different subunits of the protein to the active site. In all cases, the binding of the regulatory molecules is non covalent and is described in kinetic terms as non-competitive inhibition.

Exactly how an allosteric effector molecule (activator or inhibitor) works is explained by one of two molecular models:

- The concerted model developed by Monod, Changeux and Wyman (1965) (Figure 3.3a);

- The sequential model developed by Koshland and Nemethy (1966) (Figure 3.3b).

Both models suppose that binding of a ligand molecule, i.e. a substrate or regulator molecule, to one of the subunits of the enzyme causes that subunit to change its shape (conformation). The terms concerted and sequential refer to the manner in which the biochemists who developed the models imagine the separate subunits of the multimeric proteins to change shape. *Either*, all subunits change simultaneously (concerted model) *or* they change one at a time (sequential model). The MWC concerted model, assumes that all of the enzyme subunits have the same conformation (shown as squares or circles in Figure 3.3) and that substrate binding is easier when the protein is in the relaxed (R) state. Initially, with the protein in the tense (T) state substrate binding is 'difficult', but once one ligand molecule has bound, the conformational switch to the R state allows subsequent binding much easier. Thus the whole protein is in either the R or the T state.

Koshland's model also assumes two states, R and T, but of each subunit. The binding of one ligand causes only one subunit to switch completely but induces a slight change in a neighbouring subunit which can now bind another ligand molecule easily. Because each subunit changes in turn, hybrid forms of R and T conformations are possible.

A Michaelis–Menten type graph for an allosteric enzyme shows not the usual hyperbolic shape as shown in Section 1.4, but a sigmoidal relationship between [S] and activity.

Allosteric regulatory molecules are small molecular weight compounds which may be coenzymes (NAD^+, ATP, etc.) or intermediate substrates, possibly generated by enzymes found within the same pathway as the regulated enzyme. Alternatively, the allosteric modulator may be generated within another, perhaps complementary, pathway. For example, a regulator may be stimulatory for a catabolic route and at the same time inhibitory for the opposing anabolic pathway.

Where the pathway generates an allosteric regulator, the opportunity for feedback or indeed feed-forward control arises. Feedback control is exerted by a product of the pathway and acts to 'switch off' a key control enzyme when the concentration of the product reaches a threshold. This indicates the cell contains enough of that particular product and so no more needs to be synthesized, thus allowing valuable substrates to be used elsewhere. Using our road-way analogy from Chapter 1, a street has been closed and traffic is being diverted.

A good example of allosteric inhibition is given by hexokinase (HK) isoenzymes of muscle. The product of the HK reaction, glucose-6-P allosterically inhibits the enzyme, so matching the phosphorylation of glucose to its overall metabolism, helps to regulate

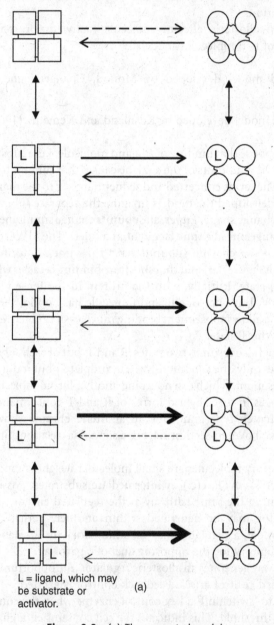

L = ligand, which may
be substrate or
activator.

(a)

Figure 3.3 (a) The concerted model

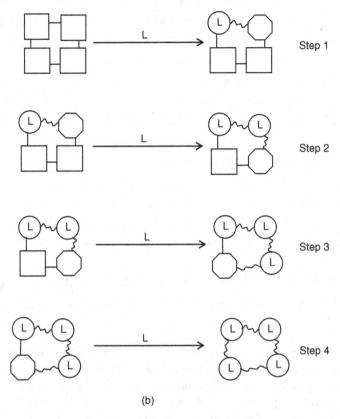

(b)

Figure 3.3 (b) The sequential model

the uptake of glucose by myocytes ensuring that only that amount of glucose which can be utilized is imported. The situation in liver is somewhat different as the main hepatic HK isoenzyme (also known as glucokinase (GK) or hexokinase IV) is not inhibited by glucose-6-P. Furthermore, the K_m of GK for glucose is much higher, indeed above typical cellular concentrations, than that for HK indicating that hepatic GK is rarely, if ever, saturated with substrate and so glucose-6-P production is allowed to continue as glucose concentrations rise and when HK is fully saturated with substrate. See also Section 3.2.1.5.

Feed-forward control is more likely to be focused on a reaction occurring at or near the end of a pathway. Compounds produced early in the pathway act to enhance the activity of the control enzyme and so prevent a back log of accumulated intermediates just before the control point. An example of feed-forward control is the action of glucose-6-phosphate, fructose-1,6-bisphosphate (F-1,6bisP) and phosphoenol pyruvate (PEP), all of which activate the enzyme pyruvate kinase in glycolysis in the liver.

One of the best known and well described allosteric enzymes is phosphofructokinase-1 (PFK-1). The interconversion of fructose-6-phosphate (F-6-P) and F-1,6-bisP is a pivotal step in glycolysis and gluconeogenesis (Figure 3.4).

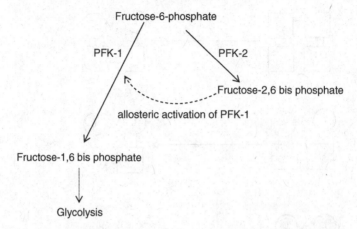

Figure 3.4   PFK-2 activity enhances glycolysis by activating PFK-1

PFK-1 is allosterically *inhibited* by (among others) ATP and citrate;

PFK-1 is allosterically *activated* by AMP, ADP and fructose-2,6 bis phosphate (F2,6-bisP).

Fructose-2,6 bis phosphate is generated by the action of PFK-2 on F-6-P. Thus there are two enzymes competing for the same substrate. When F-6-P concentrations rise, some of this substrate is used by PFK-2 and the product F2,6-bisP enhances the activity of PFK-1 so channelling more F-6-P through glycolysis. Furthermore, high cytosolic concentrations of AMP and ADP indicate that the energy status of the cell is low, so accelerating glycolysis is clearly beneficial. Conversely, when ATP and citrate (part of the TCA cycle) concentrations are high, the cell has sufficient energy, so glycolysis can be slowed. See Figure 3.4 and Section 3.3.1.2. Unlike the hexokinase example above, only ATP is actually involved with the kinase reaction and all other allosteric modifiers arise from other pathways and processes indicating how metabolism can be integrated. The reactions of glycolysis are described in Sections 1.7 and 3.3.

Competitive, non-competitive and allosteric effects all rely on weak binding between the enzyme and the various ligands (small molecules which attach to a bigger molecule). Such weak binding attractions are not covalent bonds, unlike the situation described below.

3.2.2.3   Reversible covalent control

We are very familiar with the idea of a large enzyme using a small molecule as its substrate. Here now we see an example of an enzyme that uses a protein as its substrate. Covalent modification, most frequently by phosphorylation, occurs when a controlling *enzyme is itself a substrate for another enzyme*, the 'modifying enzyme'.

For example, in the case of a phosphorylation reaction, a kinase modifier transfers the terminal phosphate group from ATP to the target enzyme, forming a covalent phosphoester with an –OH containing amino acid residue within the target enzyme.

The added phosphate group may then be removed by a phosphatase to return the target enzyme to its original level of activity (Figure 3.5).

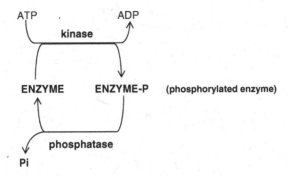

Figure 3.5 Reversible covalent phosphorylation mediated by kinase and phosphatase

Only three amino acids have a hydroxyl functional group in their side chain; tyrosine, serine and threonine. Some kinases target only tyrosine residues (tyrosine kinases) whereas others may phosphorylate serine or threonine (Ser/Thr kinases). An enzyme protein (the substrate for the kinase) may have several tyrosine, serine or threonine residues within its primary sequence, but only some of these are subject to phosphorylation by a particular kinase (see Figure 3.6)

The change in the conformation of the control enzyme brought about by covalent modification alters the activity of the control enzyme and so regulates substrate flux through that step. This fact underlines the importance of the three-dimensional structure of an enzyme. The inclusion of phosphates may bring about quite a small architectural change to the protein structure but it is sufficient to affect substrate binding and therefore enzyme activity.

Reversible phosphorylation of proteins and enzymes is an important and widely used means of enzyme control. Phosphorylations may activate or deactivate the target enzyme, leading to the opportunity for reciprocal control. Imagine two enzymes regulating opposing pathways, for example, one anabolic and the other catabolic, both enzymes are substrates for the same protein kinase and same phosphatase. Simultaneous phosphorylation of both enzymes results in one being activated and the other inactivated but when the phosphatase removes inorganic phosphate from both enzymes, the first becomes inactive and the second active. One 'switch' has opposing effects on two pathways; this is reciprocal control.

Phosphorylation/dephosphorylation cycles of regulation are usually initiated by hormonal action at the cell surface. A classic example of control mediated by reversible de/phosphorylation is glycogen metabolism. This process illustrates beautifully the symmetry that we often find in intermediary metabolism. Reciprocal, simultaneous, control of the two opposing pathways of glycogen synthesis (anabolic, involving the enzyme glycogen synthase) and degradation occurs (catabolic, involving glycogen phosphorylase, see Figure 4.17). For glycogen synthesis, activation of the control enzymes requires phosphorylation by appropriate kinases, but simultaneous phosphorylation

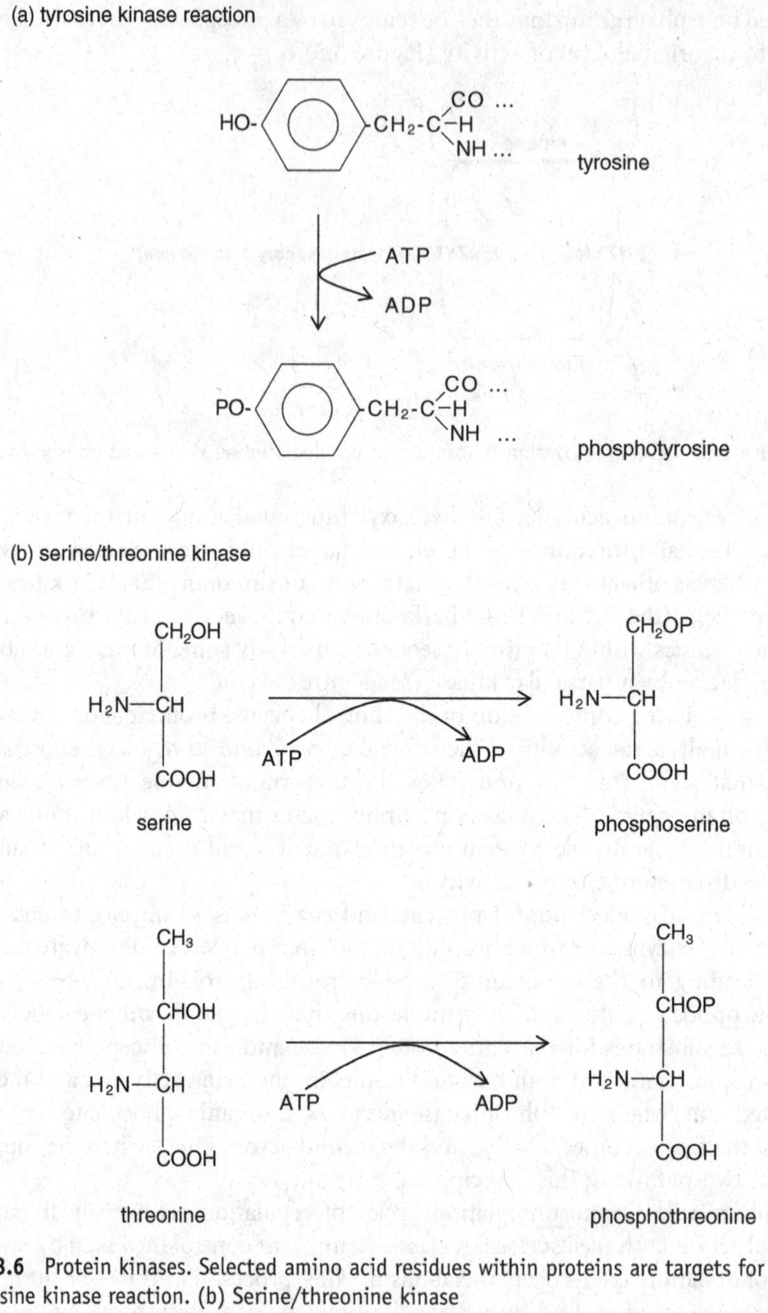

(a) tyrosine kinase reaction

tyrosine

phosphotyrosine

(b) serine/threonine kinase

serine

phosphoserine

threonine

phosphothreonine

Figure 3.6 Protein kinases. Selected amino acid residues within proteins are targets for kinases. (a) Tyrosine kinase reaction. (b) Serine/threonine kinase

of the degradative enzymes by the same kinases causes inactivation. Other important examples of enzyme targets for de/phosphorylation cycles are pyruvate kinase (PK) and PDH. Both enzymes are optimally active when dephosphorylated; the action of specific kinases inactivates these enzymes.

Some enzymes are controlled by both allosterism *and* covalent modification often brought about by hormone stimulation of the cell. Allosteric effects will take effect immediately because the enzyme is responding to local intracellular conditions of substrate or coenzyme concentrations, but covalent effects because they are driven by hormonal stimulation may take a little longer to have an impact but will be part of a coordinated response in several tissues of the body sensitive to the hormone.

3.2.2.4 Isoenzymes

As indicated in the previous discussion, the control of enzyme activity is understood in terms of kinetic parameters. Differences in K_m and or V_{max} can also arise when the same chemical (metabolic) reaction is catalysed by two structurally different enzymes. Such is the case with isoenzymes or isoforms of enzymes.

Isoenzymes arise when there exist different genes coding for the different proteins. Isoforms arise when the enzyme protein undergoes post translational modification, for example amino acid residue excision or addition of a non protein moiety such as sialic acid (a carbohydrate-like molecule).

The isoenzymes within a particular 'family' will operate under slightly different circumstances or may respond differently to metabolite feedback regulation. In this case there is some degree of structural similarity between the different isoenzymes. The usual example used to illustrate this point is lactate dehydrogenase (LD), which has five isoenzymes, each composed of four sub-units. The subunits are of two types, H or M, so the five forms arise as follows:

HHHH   HHHM   HHMM   HMMM   MMMM

The H-rich forms found in heart and red blood cells tend to operate at low lactate concentration, whilst the M-rich forms work better when lactate concentration is high. Further details about LD isoenzymes are given in Chapters 6 and 7.

Sometimes, it is convenient for a cell to use two totally different, that is structurally unrelated, enzymes to catalyse the same chemical reaction, but under different physiological conditions. Using the same example we had a little earlier, the conversion of glucose into glucose-6-phosphate in most mammalian tissues, is via one of the HK isoenzymes. HK is continually active because it has a low K_m ($\sim$1 mmol/l) for glucose, far below the expected cytosolic glucose concentration assuming a blood glucose concentration of approximately 5 mmol/l. Following a heavy carbohydrate meal, the blood glucose concentration will rise steeply and so too will the concentration within the liver cells (hepatocytes). As HK is already working almost to its maximum capacity, another enzyme, GK, (K_m for glucose $\sim$10 mmol/l), an isoenzyme of HK is used as a 'back-up' to ensure that as much of the glucose which enters the hepatocyte is converted to glucose-6-phosphate as quickly as possible, probably for storage as glycogen.

3.2.2.5 Substrate cycling

Substrate cycling (also called futile cycles) illustrates the importance of keeping key enzymes in a low activity state. Here again, PFK in the liver provides a good example.

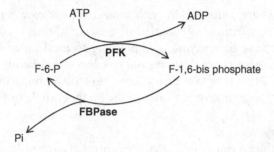

Key: PFK   phosphofructokinase
        FBPase   fructose-1,6-bis phosphatase

Compare this diagram with figure 3.5

Figure 3.7   Futile cycle involving PFK and FBP

The interconversion of fructose-6-phosphate and fructose-1,6 bis phosphate is a control point in glycolysis and gluconeogenesis. Gluconeogenesis is a pathway which allows carbon atoms from substrates such as lactate, glycerol and some amino acids to be used for the synthesis of glucose, so it is in effect physiologically the opposite of glycolysis.

As shown in Figure 3.7, PFK catalyses the forward reaction (F-6-P to F-1,6-bis phosphate) and fructose-1,6-bis phosphatase (FBPase) catalyses the reverse reaction. To have both PFK and FBP operating simultaneously appears to be illogical (in fact, futile) because ATP, a valuable commodity is consumed with no apparent net flow through the pathway. The answer to the dilemma is given when we consider the relative rates of the two reactions under various conditions, for example fed and fasting. If we consider that under resting conditions, PFK is operating at a velocity of, say, 50 (arbitrary units) and the reverse reaction (FBPase) is operating at a velocity of 40 then the net velocity is 10 in favour of phosphorylation.

If now there is a demand for glycolysis to accelerate (to provide more pyruvate for conversion into acetyl CoA) and we assume a twofold change in *each* enzyme then PFK operates at an arbitrary velocity of 100 whilst the activity of FBPase decreases simultaneously to 20 then the net velocity is 80, an increase of eightfold over the resting state of 10.

Conversely, let us suppose for the sake of argument that in the fasting state gluconeogenesis accelerates to supplement blood glucose concentration and that the rate of PFK drops to 25 arbitrary units and that of FBPase increases to 80 (again a twofold relative change in each enzyme). The net velocity is now 55 but in the reverse direction favouring formation of F-6-P. Because the relative activities of the two enzymes are controlled simultaneously, partly by allosteric feedback, careful balancing of the opposing reactions determines the overall substrate flux and brings about an immediate response to changes in physiological circumstances.

3.2.2.6 Enzyme cascades

A different and equally important way of achieving a rapid and significant change of flux is the amplification inherent in enzyme cascades. Here, each enzyme uses as its substrate the next enzyme in the sequence. In even a simple three-step cascade, the amplification achieved can be significant. For example assuming each step has an amplification factor of 5, then

$$
\begin{aligned}
\text{Step 1} \quad & 1 \rightarrow 5 \quad && \text{active enzymes} \\
\text{Step 2} \quad & 5 \rightarrow 25 \quad && \text{active enzymes} \\
\text{Step 3} \quad & 25 \rightarrow 125 \quad && \text{active enzymes}
\end{aligned}
$$

This sort of control is usually achieved by either covalent modification (phosphorylation or de phosphorylation as in glycogen metabolism) or by proteolytic cleavage (e.g. activation of digestive enzymes in the gut, or blood clotting mechanism.

In the case of glycogen metabolism, the initiating factor to glycogenolysis is hormone activation of the liver or muscle cell. This step too contributes to the amplification cascade because each hormone-receptor interaction may activate, via appropriate second messenger, several primary cascade enzymes (step 1). Simultaneously, the same hormone, adrenaline in muscle and glucagon in liver, will activate key enzymes in glycolysis and so ensure that once glucose is available in the cytosol of the cell, there is efficient and co-ordinated metabolic processing of substrate.

3.2.3 Control at the gene level

The example of HK and GK quoted earlier illustrates how a cell can adapt to relatively sudden (minutes to hours time frame) fluctuations in its environment or physiological circumstances. Suppose, however, that the stimulus to change was more prolonged (hours to days), or that a particular substrate previously absent, becomes available in abundance. Perhaps a preferred substrate becomes unavailable and the cell has to adapt to use an alternative, or a toxin needs to be metabolized quickly. In such circumstances it makes sense to change the concentration of the enzyme itself in order to cope with the new substrate load.

More enzyme molecules can be made available by modifying the expression of the gene which codes for that enzyme protein. This may be by increased transcription (additional mRNA for translation), or by increasing the stability of basal levels of the messenger RNA. The consequent increase in enzyme protein synthesis will necessarily utilize amino acids which might otherwise have been used for some other purpose. So here then is an example of the cell actively channelling its essential and perhaps limited resources through an alternative, normally quiescent, route in order to adapt to a biochemical challenge.

The *Lac* operon (Figure 3.8, described in *Esherichia coli* bacteria by Jacob and Monod), illustrates how gene expression can be switched on or off according to sudden changes in environmental conditions. Glucose (a monosaccharide) is the preferred

(a)

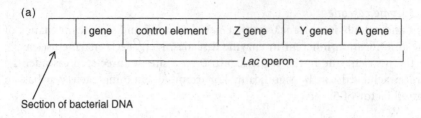

Section of bacterial DNA

(b)

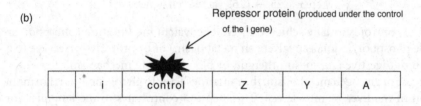

(a) glucose present so the *Lac* control element is blocked; genes A,Y and Z
     are **not** expressed

(c)

Repressor protein disengaged

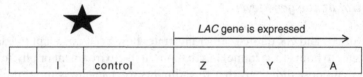

(b) lactose present, the repressor's conformation is changed so that it can no longer
bind to the control element. The A Y and Z genes **can now** be expressed; mRNA is
produced which leads to the synthesis of the enzymes which will metabolise lactose

Figure 3.8   *Lac* operon

carbon source for the bacterium *E. coli*, but if glucose supply is limited, the organism
will switch to metabolise lactose (a disaccharide). Hence, *Lac* stands for lactose.

The *Lac* operon is made up of three genes (designated A, Y and Z), which code for
enzymes which metabolize lactose and the control element whose function is to
'activate' transcription of the A,Y and Z genes. Normally (i.e. when there is sufficient
glucose available), a protein called a repressor blocks the control element and so the A, Y
and Z genes are 'off'.

The repressor is itself produced under the control of another gene (i for inhibitory)
located adjacent to the A,Y and Z control element.

When lactose but not glucose is available, a chemical derivative of the disaccharide binds to the repressor, changing its shape. This in turn prevents the repressor binding to the control element. Once the control element is 'open', it directs the expression of the A, Y and Z genes which leads to the production of the enzymes which allow the efficient utilization of the lactose.

The *Lac* operon is but one example of the genetic adaptations which allow bacteria to respond to their environment. Other examples are to be found in amino acid metabolism, for example the TRP operon which regulates tryptophan metabolism.

This process where enzymes are produced in increased quantities on demand is called enzyme induction; although eukaryotic cells do not possess operons as such, several examples of enzyme induction are to be found in humans. Eukaryotic genes have regulatory sequences and promoter regions which when bound by transcription factors, permit expression of that particular gene. For example, steroid hormones once inside their target cell, bind specific receptors forming complexes which allow particular regions of DNA to be transcribed (see Chapter 4). Similarly, many of the immune responses are mediated via a cytosolic protein called nuclear factor kappa B (NFκB), which is itself activated by cytokine stimulation of the cell. Following the ingestion of certain drugs, for example barbiturates, enzyme systems in the liver are induced to accelerate hepatic detoxification, that is drug removal (see Chapter 6).

Living cells must adapt and respond to changes and challenges placed upon them by their environment. Metabolic control is concerned with mediating such responses. Without adaptation, life would not continue.

3.3   Glycolysis and the Krebs TCA cycle as models of control of metabolic pathways

Having now considered the principles of metabolic regulation, we can turn our attention to some real pathways and study how the theory is put into practice. Further details of the control reactions of glycolysis and gluconeogenesis are given in Section 6.4.2.

Glycolysis is the major route of carbohydrate metabolism in all cell types and the TCA is a 'focal point' allowing the integration of carbohydrate, amino acid and lipid metabolism. The two pathways are illustrated in Figures 3.9 and 3.14. These two well-known pathways exemplify many of the general principles of metabolic regulation described above.

3.3.1   Glycolysis

The three control steps in glycolysis are reactions catalysed by HK, PFK-1 and PK. All three reactions involve ADP/ATP cycling and are strongly exergonic suggesting they operate far from the true equilibrium position. Such reactions are physiologically difficult to reverse and so act as metabolic 'one-way streets'.

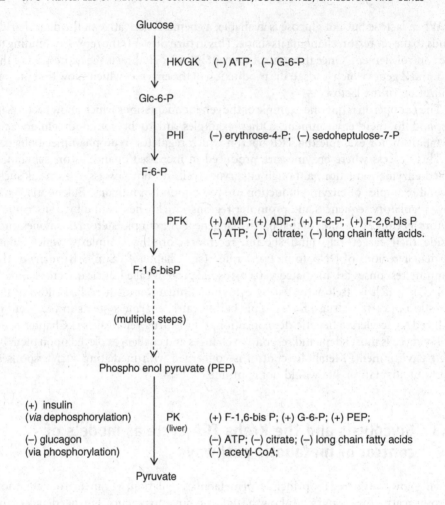

Glucose

HK/GK   (−) ATP;   (−) G-6-P

Glc-6-P

PHI   (−) erythrose-4-P; (−) sedoheptulose-7-P

F-6-P

PFK   (+) AMP; (+) ADP; (+) F-6-P; (+) F-2,6-bis P
(−) ATP; (−) citrate; (−) long chain fatty acids.

F-1,6-bisP

(multiple steps)

Phospho enol pyruvate (PEP)

(+) insulin
(*via* dephosphorylation)   PK   (+) F-1,6-bis P; (+) G-6-P; (+) PEP;
(liver)
(−) glucagon   (−) ATP; (−) citrate; (−) long chain fatty acids
(via phosphorylation)   (−) acetyl-CoA;

Pyruvate

Key: (+) = activation   (−) = inhibition
HK/GK = hexokinase/glucokinase; PHI = Phosphohexoisomerase;
PFK = phosphofructokinase; PK = pyruvate kinase; G-6-P = glucose-6-phosphate
F-6-P = fructose-6-phosphate; F-1,6bisP = fructose-1,6 bis phosphate

Figure 3.9 Hormone and metabolite-mediated control of glycolysis

$$\text{Glucose} + \text{ATP} \xrightarrow{\text{HK/GK}} \text{glucose-6-phosphate} + \text{ADP} \quad \Delta G' = -27\,\text{kJ/mol}$$

$$\text{Fructose-6-phosphate} + \text{ATP} \xrightarrow{\text{PFK}} \text{fructose-1, 6-bisphosphate} + \text{ADP}$$

$$\Delta G' = -21\,\text{kJ/mol}$$

$$\text{Phosphoenolpyruvate(PEP)} + \text{ADP} \xrightarrow{\text{PK}} \text{pyruvate} + \text{ATP} \quad \Delta G' = -17\,\text{kJ/mol}$$

(typical values for $\Delta G'$ in cardiac muscle)

3.3.1.1 The Hexokinase reaction

Four tissue-specific isoenzymes of HK (denoted types I to IV) have been described; type IV is usually known as GK and is found only in the liver. As noted previously, the hexokinases have a K_m for glucose of about 1 mmol/l whilst that for GK is about 10 mmol/l.

The hexokinases are inhibited by product (glucose-6-phosphate) and co-substrate (ATP). A high cytosolic concentration of ATP (strictly speaking a high ATP-to-ADP ratio) in peripheral tissues is a signal that there is sufficient energy available and so no further oxidation of glucose is necessary. Product inhibition by glucose-6-phosphate regulates the entry of glucose into the pathway. Glucose remains in the blood stream and is taken up by the liver where it may be converted into glycogen for storage. Hepatic GK is *not* subject to product inhibition so any Glc-6-P which cannot enter glycolysis can be converted into glucose-1-phosphate (Glc-1-P) the first step towards glycogen synthesis or may enter the pentose phosphate pathway.

3.3.1.2 The PFK-1 reaction

Refer again to Figure 3.4 before proceeding. This is the overall rate limiting reaction of glycolysis and is therefore the main control point. The net flux of substrate through this step is determined by the opposing effects of PFK-1 (the glycolytic enzyme) and FBP, which is important in gluconeogenesis. Both enzymes exhibit some activity at all time establishing what is referred to a 'futile cycle' because ATP is consumed in one direction but not generated in the reverse. Refer to Figure 3.7.

This apparently 'futile' process has benefits in maintaining the temperature in the flight muscles of bees, but the importance to humans, as described in Section 3.2.2.5, is that if both reactions are already operating either the forward or reverse reaction can very quickly speed-up if circumstances dictate; rather like a rolling start at the beginning of a race.

The mechanism of control of PFK-1 is mainly allosteric; inhibition by ATP, citrate, long chain fatty acids and activation by AMP and F2,6-bisP. With the exception of F2,6-bisP, the named regulators all indicate the fuel or energy status of the cell. Control by citrate helps to synchronise the rates of glycolysis and the TCA cycle.

PFK-1 is a classic example of a tetrameric allosteric enzyme. Each of the four subunits has two ATP binding sites; one is the active site where ATP is co-substrate and the other is an inhibitory allosteric site. ATP may bind to the substrate (active) site when the enzyme is in either the R (active) or T (inhibited) form. The other co-substrate, F-6-P binds only to the enzyme in the R state. AMP may also bind to the R form and in so doing stabilises the protein in that active conformation permitting ATP and F-6-P to bind.

ATP binds to the inhibitor site only when the protein is in the T form. In occupying the inhibitor site, ATP 'locks' the enzyme in the inactive conformation, preventing binding of F6P (Figure 3.10). Overall, the cytosolic ratio of ATP-to-AMP ratio determines the proportion of PFK-1 in the R conformation. Compare with the effect of the ATP-to-ADP ratio on HK above (Section 3.3.1.1)

Figure 3.4 also shows that fructose-2,6-bis phosphate (F2,6bisP) is an important allosteric regulator of PFK-1; this is partly because the turnover of F2,6bisP influenced by the hormone glucagon. F2,6bisP is synthesized from fructose-6-phosphate by PFK-2,

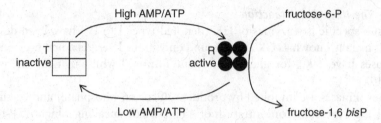

Figure 3.10 Phosphofructokinase

which is an interesting enzyme due to the fact that it has *both* kinase *and* phosphatase (FBP) activity;

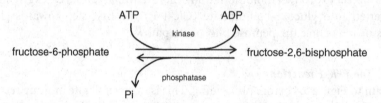

The switch in the action of the enzyme between its kinase and phosphatase activities is brought about by phosphorylation mediated by the serine/threonine protein kinase A (PKA), the same cAMP dependent enzyme which plays a role in the control of glycogen metabolism. In its kinase form, PFK-2 is dephosphorylated but phosphorylated in the phosphatase form.

Thus, in the presence of glucagon (a hormone associated with the fasting state), F2,6-bisP is dephosphorylated to F-6-P, the allosteric activation of PFK-1 is thereby lost and the rate of substrate flow through glycolysis is reduced (Figure 3.11). Instead, gluconeogenesis (synthesis of glucose from non-carbohydrate sources and needed for the central nervous system in particular) is facilitated and the liver becomes a net producer, not consumer, of glucose.

3.3.1.3 *The PK reaction*
The third and final control step is mediated by PK. This enzyme, like HK, exists as a number of isoenzymes in different tissues and, like the PFK reaction, is controlled by both the concentration of metabolites and covalent effects. Furthermore, PK also illustrates two other means of metabolic control, namely enzyme induction and feed-forward, regulation.

PK in the liver is regulated at the gene level with its synthesis being increased by carbohydrate feeding, via the actions of insulin, glucocorticoids or thyroxine (T4). Feed-forward allosteric control is mediated by F-1,6bisP and glucose-6-phosphate, whilst ATP citrate and acetyl-CoA inhibit the PK reaction. In the liver, covalent phosphorylation is brought about by the action of glucagon and PKA, reducing PK activity thus slowing glycolysis in favour of gluconeogenesis. The muscle isoenzyme of PK is not inactivated by glucagon signalling so glycolysis in that tissue can continue even if blood glucose concentration is low.

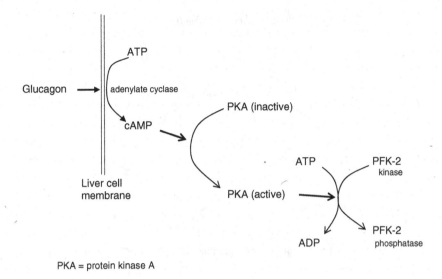

PKA = protein kinase A

Figure 3.11 cAMP activates the metabolic switch from PFK-2 kinase to phosphatase activity

3.3.2 Control of the Krebs' TCA cycle

The control mechanisms operating on the TCA cycle are similar to those described for glycolysis above, that is allosteric and covalent. As might be predicted, it is the concentration of acetyl-CoA and the ATP-to-ADP and NADH-to-NAD$^+$ ratios which are crucially important as these indicate energy status within each mitochondrion and implicitly therefore the energy status of the whole cell. High concentrations of ATP or NADH slow down the cycle, an effect which is partly mediated by covalent modification.

PDH is a multi-enzyme complex consisting of three separate enzyme 'units'; pyruvate decarboxylase, transacetylase and dihydrolipoyl dehydrogenase. Serine residues within the decarboxylase subunit are the target for a kinase which causes inhibition of the PDH; the inhibition can be rescued by a phosphatase. The PDH kinase (PDH-K) is itself activated, and the phosphatase reciprocally inhibited, by NADH and acetyl-CoA. Figure 3.12(a and b) show the role and control of PDH.

During periods of prolonged fasting or starvation, acetyl-CoA can be generated by catabolism of stored fat. Acetyl-CoA inhibits the PDH complex and allows pyruvate to be re-directed towards gluconeogenesis by pyruvate carboxylase (part of the pyruvate kinase bypass loop). Here is another example of reciprocal control, two enzymes being controlled by the same metabolic switch; pyruvate carboxylase is *activated* by a high NADH concentration, simultaneously with the *inactivation* of PDH. Glycolysis is suppressed in favour of gluconeogenesis, as stated above, to generate glucose for the brain and the red blood cells.

It is clear from Figure 3.13 that PDH controls the production of acetyl-CoA. The control of the TCA cycle (illustrated in Figure 3.14) itself is mediated mainly by the

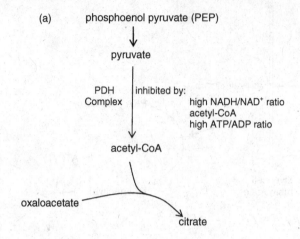

(a)   phosphoenol pyruvate (PEP)

pyruvate

PDH Complex   inhibited by:
high NADH/NAD⁺ ratio
acetyl-CoA
high ATP/ADP ratio

acetyl-CoA

oxaloacetate

citrate

Figure 3.12   (a) PDH complex

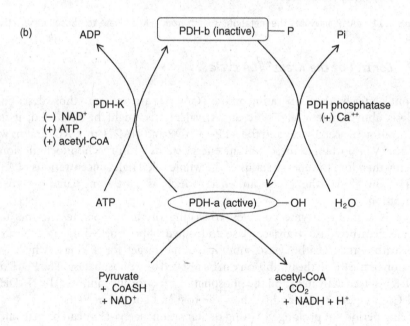

(b)   ADP   PDH-b (inactive) — P   Pi

PDH-K
(−) NAD⁺
(+) ATP,
(+) acetyl-CoA

PDH phosphatase
(+) Ca⁺⁺

ATP   PDH-a (active) —OH   H₂O

Pyruvate
+ CoASH
+ NAD⁺

acetyl-CoA
+ CO₂
+ NADH + H⁺

Key: PDH-K = PDH kinase (+) = activation; (−) = inhibition

Figure 3.12   (b) PDH complex: regulation

'pacemaker' enzyme citrate synthase, which is inhibited by citrate, succinyl-CoA (a substrate analogue) and a high ATP-to-ADP ratio. The concentrations of both citrate and succinyl-CoA reflect the overall rate of the cycle, and if sufficient ATP is available to the cell, acetyl-CoA can be channelled into other pathways, for example fat synthesis in

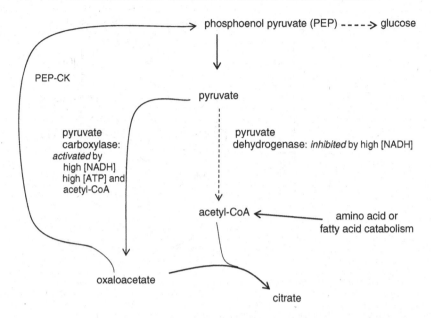

Key: PEP-CK = PEP carboxykinase: this enzyme begins the re-cycling of carbon
        atoms derived from amino acids (*via* pyruvate, 2-OG or OAA) or lactate (*via*
        pyruvate) to glucose (gluconeogenesis).

Figure 3.13 Reciprocal control of PC and PDH complex (compare with Figure 3.1)

the liver. Close consideration of the ICD and 2-oxoglutarate dehydrogenase (2-OGD) reactions show that they are similar to the PDH reaction; all are oxidative decarboxylations with similar coenzyme requirements, that is FAD and lipoate in addition to NAD^+, and both CD and 2-OGD are inhibited by ATP and NADH. The third control step is at 2-OGD. Succinyl-CoA and NADH both inhibit 2-OGD.

We should note at this point that the TCA cycle is more than just a means of producing NADH for oxidative phosphorylation. The pathway also provides a number of useful intermediates for other, often synthetic, pathways. For example, citrate is the starting substance for fat synthesis (Chapter 9); succinyl-CoA is required for haem production and 2-oxoglutarate and oxaloacetate in particular are involved with amino acid and pyrimidine metabolism. Pathways which have dual catabolic/anabolic functions are referred to as 'amphibolic'.

As is often the case, tissue-specific control mechanisms operate to optimise adaptation to particular conditions. For example, muscle contraction requires an increase in cytosolic calcium ion concentration (see Section 7.2.1, Figure 7.4). During exercise when energy generation needs to be increased, or from a more accurate metabolic point of view, when the ATP-to-ADP ratio falls rapidly, and the accompanying rise in $[Ca^{2+}]$ activate (i) glycogen phosphorylase which initates catabolism of

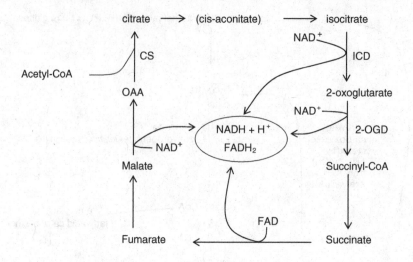

Cis-aconitate remains enzyme-bound.

The three key controlling enzymes of the TCA cycle.

CS   = citrate synthase
ICD   = isocitrate dehydrogenase
2-OGD = 2-oxoglutarate dehydrogenase

Figure 3.14   The TCA cycle

glycogen, (ii) PDH complex, (iii) ICD and (iv) 2-OGD complex). Thus, fuel in the form of glucose is made available from storage and its oxidation is potentiated for efficient ATP production; a perfect example of integration and regulation of metabolism.

Chapter summary

Integration and control of metabolic pathways is crucial for optimum utilization of substrates. Metabolism must adapt to changing physiological circumstances such as fed/fasting or resting/exercising. This is achieved mainly by changes in substrate concentrations, coenzyme ratios mediated by allosteric effects and hormonal control, usually via covalent modification of enzyme structure, to ensure that metabolism within different tissues responds in an appropriate manner to any given situation. Enzyme induction, that is enzyme synthesis on demand, is a relatively slow process which mediates control in a more long term sense than allosterism or covalent effects. Because substrates such as pyruvate, acetyl-CoA, oxaloacetate, 2-oxoglutarate and glucose-6-phosphate are at metabolic cross-roads, where streams of carbohydrate, lipid and amino acid metabolism converge, reactions involving these compounds are often tightly, but sensitively, controlled.

Problems and challenges

1. Control points often occur at or near to the start of a pathway or a branch points in a pathway. Why?

2. Verify the figure for the number of moles of ATP generated per mole of glucose. You will need to consider ATP generated from substrate level phosphorylation in glycolysis *and* from oxidative phosphorylation. Assume each NADH generates three ATP and each $FADH_2$ generates two ATP. (You may need to refer to the diagrams showing glycolysis and the TCA cycle).

3. What is the biological 'logic' in ATP, ADP, AMP, F-2,6 bisphosphate and citrate acting as regulators of such a key enzyme (PFK-1) in glycolysis?

4. Why is it desirable to have an enzyme in heart which works efficiently to remove lactate, that is an isoenzyme of lactate dehydrogenase with a low K_m for lactate?

5. What would be the effect of (a) a competitive inhibitor and (b) a non-competitive inhibitor on K_m and V_{max} of an enzyme catalysed reaction? Explain your answer.

6. Study the graph below which shows (a) the response of a typical allosteric enzyme in the absence of an inhibitor and (b) the same enzyme in the presence of an inhibitor.

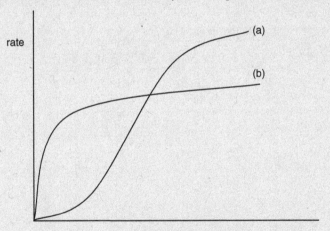

Substrate concentration

What conclusions can be drawn about the effect of the inhibitor on the enzyme?

7. Covalent modification of enzymes (molecular weight of several hundreds or thousands) by the incorporation of inorganic phosphate in the form of PO_3^- (formula weight = 85), seems to represent a small chemical change in the enzyme yet is an important control mechanism of enzyme activity. Explain how phosphorylation can exert its controlling effect on the activity of the enzyme.

8. In the fed state when the insulin:glucagon ratio is high, would you expect glycogen synthase, PK and PDH to be phosphorylated or de-phosphorylated? Give reasons to support your answer.

4

Biochemistry of intercellular communication; metabolic integration and coordination

---

Overview of the chapter

The metabolic activities of different tissues are controlled and coordinated partly by messenger-induced mechanisms. Messengers, often known generically as ligands because they act by binding to a receptor protein, are physiologically sub-divided into two major groups, hormones and neurotransmitters but they share many biochemical features. Hormonal stimulation of target cells requires ligand-receptor binding to initiate a transduction sequence. Transduction systems based on second messengers such as cAMP or inositol triphosphate (IP3), which act as relay mechanisms and are often responsible for controlling enzyme activity via phosphorylation/dephosphorylation cycles. Some hormones regulate cellular activity by changing DNA transcription and upregulating (inducing) the expression of enzymes or other proteins.

Key pathways

Hormone and neurotransmitter synthesis and catabolism; signal transduction cascades.

4.1 Introduction

'No man is an island entire of itself' . . . nor are cells and tissues!

(after John Donne, c.1570–1621; *Meditation* XV)

Functional specialization of tissues and organs allows for greater efficiency than in single-celled organisms where all reactions and processes occur in close proximity. An analogy is to think of a typical university. It would not be conducive to learning

---

Essential Physiological Biochemistry: An organ-based approach Stephen Reed
© 2009 John Wiley & Sons, Ltd

if a biochemistry lecture were to be held in the same lecture theatre at the same time as a presentation on, say, French literature of the seventeenth century. Multicellularity allows efficiency through metabolic and physiological specialization, but the disadvantage is ensuring *coordinated* activity; team work is vital for physiological success. Coordination of metabolic activity is achieved by an intricate system of communication where signalling molecules such as hormones (derived from the Greek for *'arouse to activity'* and sometimes called the 'first messengers'), growth factors, cytokines, and neurotransmitters are released by one cell and target (i) a distant cell (classic hormones), (ii) a neighbouring cell (local hormones and neurotransmitters) or (iii) the same cell (autocrine hormones), and initiate an appropriate metabolic and physiological response in that target. The word ligand describes a small molecule which binds to a larger molecule and in the context of cellular communication means the primary signal (hormone, neurotransmitter, growth factor, etc.).

Cell signalling and signal transduction are topics of great research interest, partly because defects of these processes are associated with diseases such as type 2 diabetes, cancer and obesity. In recognition of this is the fact that a number of Nobel Prizes for Medicine or Chemistry have been awarded to researchers of cell communication. This chapter describes the nature of the disparate signalling molecules and how they regulate the activity of their targets.

4.2 Physiological aspects

4.2.1 The classical endocrine system

The classical endocrine system is composed of a series of glands that secrete hormones directly into the blood where they are carried to act on cells in the body often quite distant from the place of secretion. Insulin, for example, secreted from β pancreatic islet cells has actions on fuel metabolism in most tissues of the body. Compare this with a radio or television broadcast originating in one place but arriving at multiple sites.

Many of the endocrine glands are listed in Table 4.1. This table includes not only the well-accepted hormone-secreting tissues but also those such as adipose tissue (see Chapter 9 for further details), which also have secretory activities.

Most of the classical endocrine glands are functionally hierarchical; a notable exception being pancreatic islets. A hormone secreted from one gland activates another gland to produce another hormone, which in turn activates the target tissue. Thus the concept of an 'axis' is formed (Figure 4.1a). Such axes constitute an 'amplification cascade' allowing the primary signal to be enhanced several fold. A major advantage of this arrangement is that control can operate at multiple levels, that is different cells or tissues, so making the system very responsive to changes in the physiological status of the whole organism. Negative feedback is the term used to describe how the rising

Table 4.1 Some classical hormones

Endocrine gland	Principal hormone secretions	Chemical nature of the hormone
Pituitary	ACTH	Peptides or proteins
	Antidiuretic hormone (ADH)	
	Luteinizing hormone (LH)	
	Follicle-stimulating hormone (FSH)	
	Growth hormone (GH)	
	Oxytocin	
	Prolactin	
	Thyroid-stimulating hormone (TSH)	
Thyroid	Thyroxine (T_4)	Amino acid derivatives
	tri-iodothyronine (T_3)	
Parathyroids	Parathyroid hormone (PTH)	Peptide
Pancreas	Glucagon (α cells)	Proteins
	Insulin (β cells)	
Adrenal cortex	Glucocorticoids	Steroids
	Mineralocorticoids	
	Sex hormones	
Adrenal medulla	Catecholamines	Amino acid derivatives
Gonads: ovaries & testes	Sex hormones	Steroids
Fetoplacental unit	Pregnancy hormones	Steroids
	Chorionic gonadotropin	Protein
Gut	Gastrin	Peptides
	Cholecystokinin	
	Pancreozymin	
Adipose tissue	Leptin adiponectin	Peptides

concentration of a hormone or other product of a tissue can inhibit the production of the signal which acts on that tissue or even on a tissue (gland) higher in the hierarchy. This is illustrated using the hypothalamic–pituitary–adrenal axis in Figure 4.1b. Positive feedback mechanisms where a signal from the target tissues increases the secretion of the original hormone are rarer. An example of positive feedback is increasing production of oxytocin by the pituitary in response to cervical stretch during childbirth.

Classical hormones, for example, insulin, thyroid hormones and cortisol, operate at distant and often multiple tissue sites and usually systemically on many different tissues. Paracrine and autocrine signals act locally on a particular target; paracrine acting on neighbouring cells and autocrine acting on the cell which produced the original signal. The signalling molecules diffuse through body fluids rather than being carried in the blood stream to reach their target. Talking face-to-face to a friend, or friends, or thinking out loud (talking to yourself) are useful analogies for the local, paracrine and autocrine respectively, hormone effects. Nitric oxide, eicosanoids and the neurotransmitter acetylcholine exemplify paracrine signals.

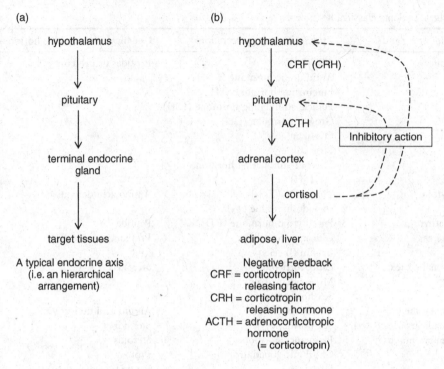

(a)

hypothalamus

pituitary

terminal endocrine
gland

target tissues

A typical endocrine axis
(i.e. an hierarchical
arrangement)

(b)

hypothalamus

CRF (CRH)

pituitary

ACTH

Inhibitory action

adrenal cortex

cortisol

adipose, liver

Negative Feedback
CRF = corticotropin
releasing factor
CRH = corticotropin
releasing hormone
ACTH = adrenocorticotropic
hormone
(= corticotropin)

(c)   representation of a ganglion.

An incoming signal (impulse) arrives via
one neurone ...

... and leaves the ganglion via
three neurones

Figure 4.1   Structural organization of a typical endocrine axis ((a) and (b)) and neural ganglia (c)

4.2.2   *The nervous system*

Communication between one neurone and another or between a neurone and an
effector tissue such as muscle is mediated via neurotransmitters. Although the message,
for example the sensation of pain, may be passed a long distance, neurotransmitters

are in effect local hormones acting on the next cell in the chain. To extend our analogy of social communication, the nervous system is hard wired, so, like a telephone conversation or a letter, the message is transmitted over a long distance but person-to-person (cell-to-cell) possibly passing through several switching centres or sorting offices *en route*.

Anatomically, the nervous system is divided into the central nervous system (CNS) consisting of the brain and the spinal cord and the peripheral nervous system comprised of neural cells forming a network throughout the body. The peripheral system is itself subdivided into two sections: the somatic system, where control of skeletal muscles allows movement and breathing, and the autonomic system which controls the actions of smooth muscle, cardiac muscle and glandular tissues. Further subdivision of the autonomic system based on anatomical and biochemical factors creates the sympathetic and parasympathetic nervous systems.

Delivery of a nerve impulse to its intended target requires that the message be passed from one cell to another at a synapse. Just as the physiological hierarchy of endocrine glands allows signal amplification, the close arrangement of synapses into ganglia (singular = ganglion) allows amplification and dispersion of a neural signal, as illustrated in Figure 4.1c.

4.3 Signalling molecules

4.3.1 Types of signalling molecules

In terms of their chemical structures, signalling molecules fall into five main categories: (i) peptides, (ii) steroids, (iii) amino acids and their derivatives, (iv) fatty acid derivatives, and (v) nucleotides.

Peptides form the largest group of signals and include most of the classical hormones, cytokines and growth factors. Peptide signals vary in size from small oligopeptides such as thyrotropin releasing hormone (TRH, three amino acids), oxytocin (nine amino acids) and the natriuretic factors (B-type natriuretic peptide, 32 amino acids), to fairly large peptides (parathyroid hormone, 84 amino acids) with molecular masses of up to 10 000 kDa. A few peptide hormones are only functional following glycosylation and others such as the gonadotrophins and thyroid stimulating hormone are dimeric.

Steroid hormones are produced by only two tissue types, the adrenal cortex and the gonads. A summary of the steroid hormones is given in Table 4.2. Steroid hormones are synthesized from cholesterol (Figure 4.2). This sterol lipid may itself be synthesized within the steroidogenic cell or it may be delivered to the cell by circulating lipoprotein complexes such as low density lipoprotein (LDL) or high density lipoprotein (HDL).

Amino acid derivatives include the thyroid hormones, catecholamines (e.g. adrenaline (epinephrine)) and dopamine, neurotransmitters such as γ-aminobutyric acid (GABA) and noradrenaline (norepinephrine). All of these signalling molecules retain

Table 4.2 Steroid hormones

Type	No. of carbon atoms	Examples	Origin
Oestrogens	18	Oestradiol, oestriol	Ovary, fetoplacental unit, adrenal cortex
Progestogens	21	Progesterone	Gonads and adrenal cortex
Androgens	19	Testosterone	Gonads and adrenal cortex
Glucocorticoids	21	Cortisol corticosterone	Adrenal cortex
Mineralocorticoids	21	Aldosterone	Adrenal cortex

some semblance of the structure of their parent amino acid whereas nitric oxide (NO), is synthesized from arginine but, being a simple diatomic gas, bears no similarity to its parent structure.

Fatty acid derivatives include a large and diverse group of compounds named eicosanoids, which includes thromboxanes, prostaglandins and leukotrienes, all of which are biochemically derived from arachidonic acid (a long-chain polyunsaturated fatty acid).

The billions of individual neurones within the nervous system communicate with each other and with the target tissues via chemical neurotransmitters. There is a bewildering array of chemicals which act as neurotransmitters or neuromodulators in peripheral or central nervous systems. These compounds fall into four major groups and some examples are shown in Table 4.3.

Somatic nerves originate in the CNS and terminate at the neuromuscular junction where acetylcholine is the transmitter. Nerves of the autonomic system also use acetylcholine as the neurotransmitter at the end of the preganglionic fibres within the ganglia. With few exceptions, the postganglionic sympathetic fibres secrete noradrenaline (norepinephrine) whilst postganglionic parasympathetic fibres secrete acetylcholine.

4.4 Synthesis of hormones

4.4.1 Peptides and proteins

As would be expected of active protein secreting cells, glandular epithelial tissue, the cytokine secreting cells of the immune system and the blood vessel endothelium, have an extensive internal structure consisting of rough endoplasmic reticulum and numerous mitochondria. Peptide hormones, growth factors and cytokines like all proteins are synthesized by DNA transcription and mRNA translation. The primary transcript of the mRNA may code for an inactive prohormone which requires careful proteolysis to produce the active hormone, as for example in the case of insulin. Adrenocorticotropic hormone (ACTH) is particularly interesting in this respect because

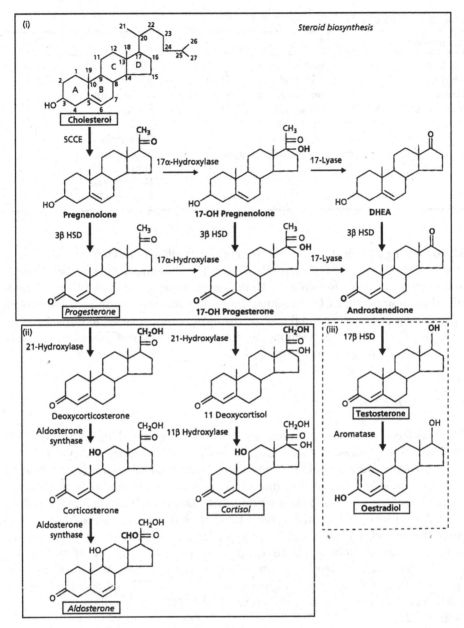

Figure 4.2 Steroidogenesis. Reproduced with permission from *Essential Endocrinology*, 4th edn, CGD Brook & NJ Marshall (eds). John Wiley & Sons, 2001

it is derived from a very large primary transcript called pro-opiomelanocortin (POMC), which following selective proteolysis yields not only ACTH but also β-endorphin and melanocyte stimulating hormone (MSH). Some peptide hormones undergo extensive post-translational processing in the Golgi network where oligosaccharide chains are

Table 4.3 Neurological signalling molecules

Amino acids	Amines	Peptides	Nucleotides
Aspartate	Acetylcholine	Enkephalin	Adenosine
Glycine	Adrenaline (epinephrine)	β-endorphin	ATP
Glutamate	Noradrenaline (norepinephrine)	Somatostatin	
γ-aminobutyrate (GABA)	Dopamine	Oxytocin	
	Histamine	Angiotensin II	
	Serotonin	Bradykinin	
	(5-hydroxytryptamine)	Neurotensin	
		Somatostatin	
		Neuropeptide Y	
		Substance P	

added forming glycoproteins or the formation of disulfide bonds to create the correct three-dimensional conformation of the protein. The individual components of dimeric hormones, (luteinizing hormone (LH), follicle-stimulating hormone (FSH), thyroid-stimulating hormone (TSH) and human chorionic gonadotropin (hCG)), are encoded by separate genes so the α and β peptide chains must be assembled to form the active hormone. The four dimeric hormones share a common α sub-unit but each possesses a functionally specific β subunit peptide.

4.4.2 Steroids

The pathways of steroid hormone production are outlined in Figure 4.2. The enzymes which operate the steroidogenic pathways are located either within the smooth endoplasmic reticulum or the inner mitochondrial membrane, indicating that there must be 'shuttling' of intermediates between the cytosol and the mitochondrion of the cell. The first reaction of the pathway is mediated by side-chain cleavage enzyme (SCCE), which is mitochondrial, but in order for the cholesterol to enter the mitochondrion, a transporter called steroid acute regulatory (StAR) protein is required. It is this transport process which is promoted by cyclic AMP (cAMP), generated by adenylyl cyclase following stimulation by ACTH acting on the adrenal cortex or the appropriate gonadotropin (LH or FSH) acting on ovaries and testes. The importance of adenylyl cyclase and cAMP are described more fully in Section 7.2.

With the exception of two dehydrogenases, all of the steroidogenic enzymes belong to the cytochrome P-450 (abbreviated as CYP) family of enzymes. The CYP enzymes are often involved with redox or hydroxylation reactions, and are also found in the liver where they are key players in biotransformation reactions (see Section 6.4). Different members of the CYP family are therefore involved with both synthesis in adrenal and gonads and hepatic inactivation of steroid hormones.

Testosterone offers an example of a hormone which is only weakly active in its secreted form and needs to be converted into an active form, 5α dihydrotestosterone (DHT). The enzyme responsible for the conversion, microsomal 5α reductase, exists in two isoenzymic forms and brings about the reduction the double bond in the 'A' ring of testosterone. Deficiency of 5α reductase, results in genotypically XY individuals being undervirilized during fetal and early life and so are often raised as females. Although it does have some androgenic activity, testosterone, like thyroxine described in the next section, may be seen as a prohormone.

4.4.3   Amino acid derived hormones

Two amino acids, tyrosine and arginine are of particular importance as precursors of signalling molecules. As outlined in Figure 4.3, tyrosine is the amino acid precursor of thyroid hormones tri-iodothyronine (T_3) and tetra-iodothyronine (T_4) and also of catecholamines adrenaline (epinephrine) and noradrenaline (norepinephrine).

4.4.3.1   Thyroid hormones
Three hormones are secreted by the thyroid; thyroxine (T_4) and tri-iodothyronine (T_3) are usually referred to as '*the*' thyroid hormones and calcitonin, a peptide, which under certain circumstances, affects calcium mobilization and is secreted from specialized so-called C cells. Only T_3 and T_4 will be discussed further at this point.

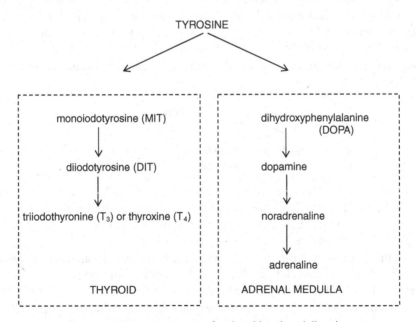

Figure 4.3 Tyrosine as a precursor for thyroid and medullary hormones

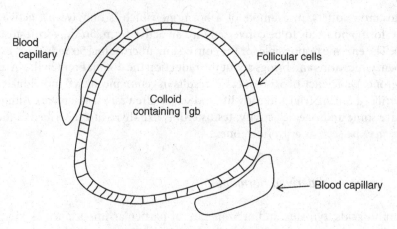

Figure 4.4 Thyroid follicle

Histologically, the thyroid gland consists of spherical follicles containing a 'scaffolding' protein called thyroglobulin (Tgb). This is a large, heavily glycosylated and tyrosine-rich protein composed of two identical subunits each with a molecular mass in excess of 300 000. Thyroglobulin synthesis occurs within the layer of epithelial cells which encloses each follicle but iodination of the Tgb occurs within the lumen of a follicle (Figure 4.4).

Thyroxine synthesis begins when iodide (I^-) is transferred from the blood stream to the thyroid follicle cell by an active ATP-driven membrane pump mechanism; this process is stimulated by cAMP following TSH stimulation of the gland. Iodide is transported through the follicular cell and secreted into the lumen of the follicle where it is oxidized to iodine and incorporated in to tyrosine residues by the enzyme thyroid peroxidase (TPO).

$$\text{Iodide} + H_2O_2 \xrightarrow{\text{TPO}} \text{iodine} + 2OH^-$$

$$\text{iodine} + \underset{\text{(part of Tgb)}}{\text{tyrosine}} \xrightarrow{\text{TPO}} \text{Tyr-I}$$

Iodination (sometimes referred to as 'organification' of Tgb) reactions form, as shown in Figure 4.5, monoiodotyrosine (MIT) and diiodotyrosine (DIT).

Transfer of the iodinated ring of the tyrosine residue of MIT or DIT to a neighbouring DIT forms T_3 or thyroxine, T_4 respectively (Figure 4.6).

When the thyroid gland is stimulated to secrete thyroxine, a small piece of iodinated Tgb is taken from the lumen into a follicular cell, where the hormones are released from the protein. Both T_3 and T_4 are secreted from the vesicle directly into the bloodstream but the plasma concentration of T_4 is substantially higher than that of T_3. In contrast, T_3 has a higher biological activity than T_4 and conversion of T_4 to T_3 occurs at the target site.

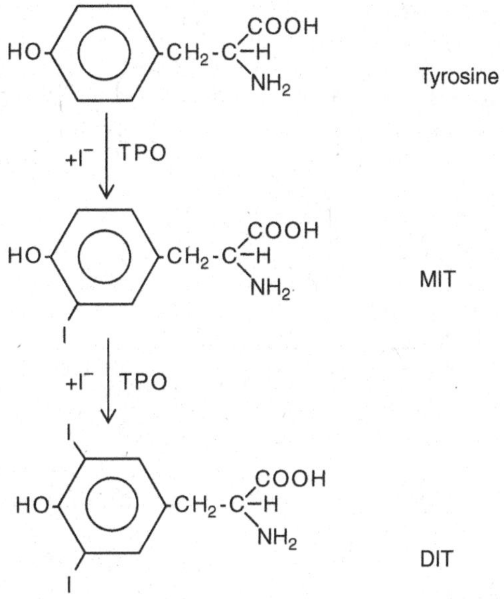

Thyroid peroxidase (TPO) oxidises iodide and incorporates iodine into tyrosine residues within Tgb.

Figure 4.5 Formation of MIT and DIT. Thyroid peroxidase (TPO) oxidizes iodide and incorporates iodine into tyrosine residues within Tgb

4.4.3.2 Catecholamines

Known most famously for their part in the 'fight or flight' response to a threat, challenge or anger, adrenaline (epinephrine) and dopamine from the adrenal medulla and noradrenaline (norepinephrine), mainly from neurones in the sympathetic nervous system are known collectively as catecholamines. Synthesis follows a relatively simple pathway starting with tyrosine (Figure 4.7).

The first step is catalysed by the tetrahydrobiopterin-dependent enzyme tyrosine hydroxylase (tyrosine 3-monooxygenase), which is regulated by end-product feedback is the rate controlling step in this pathway. A second hydroxylation reaction, that of dopamine to noradrenaline (norepinephrine) (dopamine β oxygenase) requires ascorbate (vitamin C). The final reaction is the conversion of noradrenaline (norepinephrine) to adrenaline (epinephrine). This is a methylation step catalysed by phenylethanolamine-N-methyl transferase (PNMT) in which S-adenosylmethionine (SAM) acts as the methyl group donor. Contrast this with catechol-O-methyl transferase (COMT) which takes part in catecholamine degradation (Section 4.6).

4.4.3.3 Nitric oxide

NO, molecular mass 30, is by far the smallest known signalling molecule. Physiologists had long known that acetylcholine stimulation initiates dilation of blood vessels and that this effect was lost if the smooth muscle layer surrounding the vessel had previously

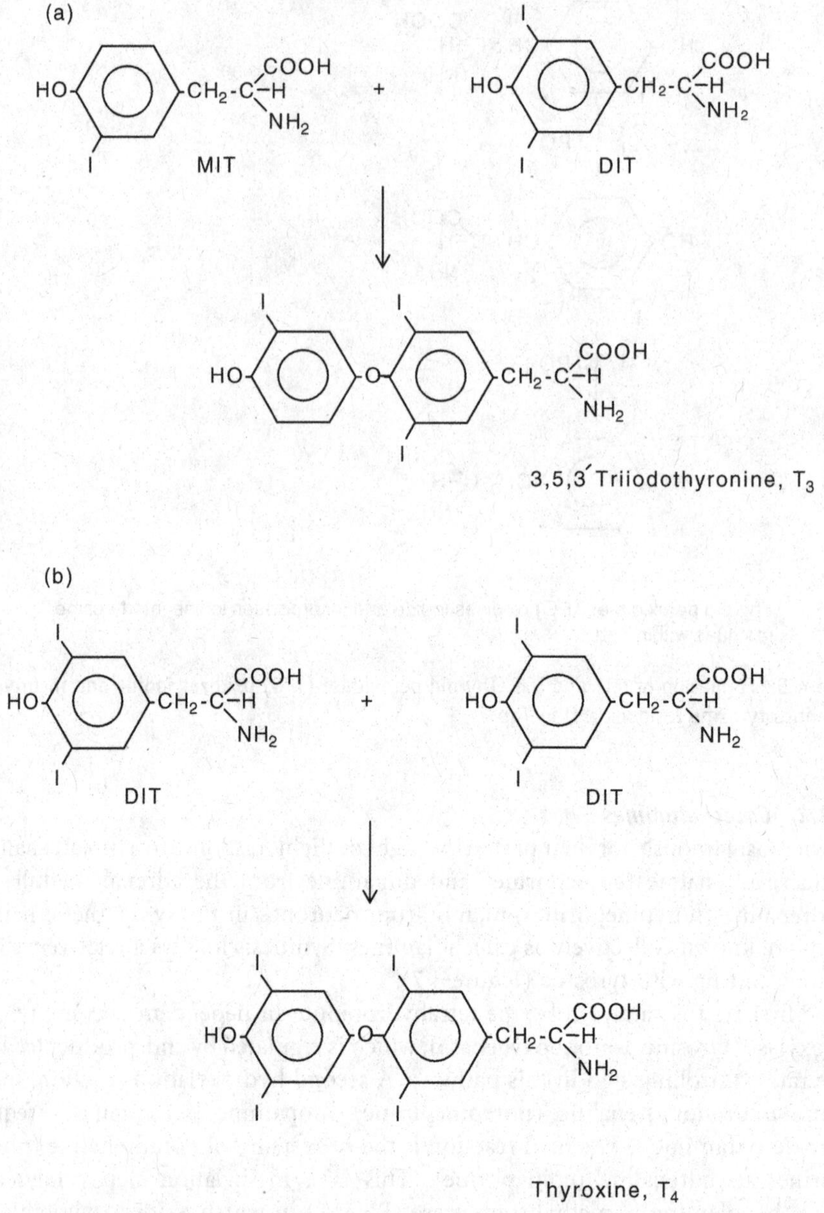

(a)

MIT

+

DIT

3,5,3′ Triiodothyronine, T$_3$

(b)

DIT

+

DIT

Thyroxine, T$_4$

Figure 4.6 Formation of (a) T$_3$ and (b) T$_4$ from MIT and DIT

been separated from the endothelium. The name 'endothelium-derived relaxation factor' (EDRF) was adopted to describe the molecule responsible for the vasodilation, but the exact identity and structure were unknown until the discovery of the role of NO in 1986.

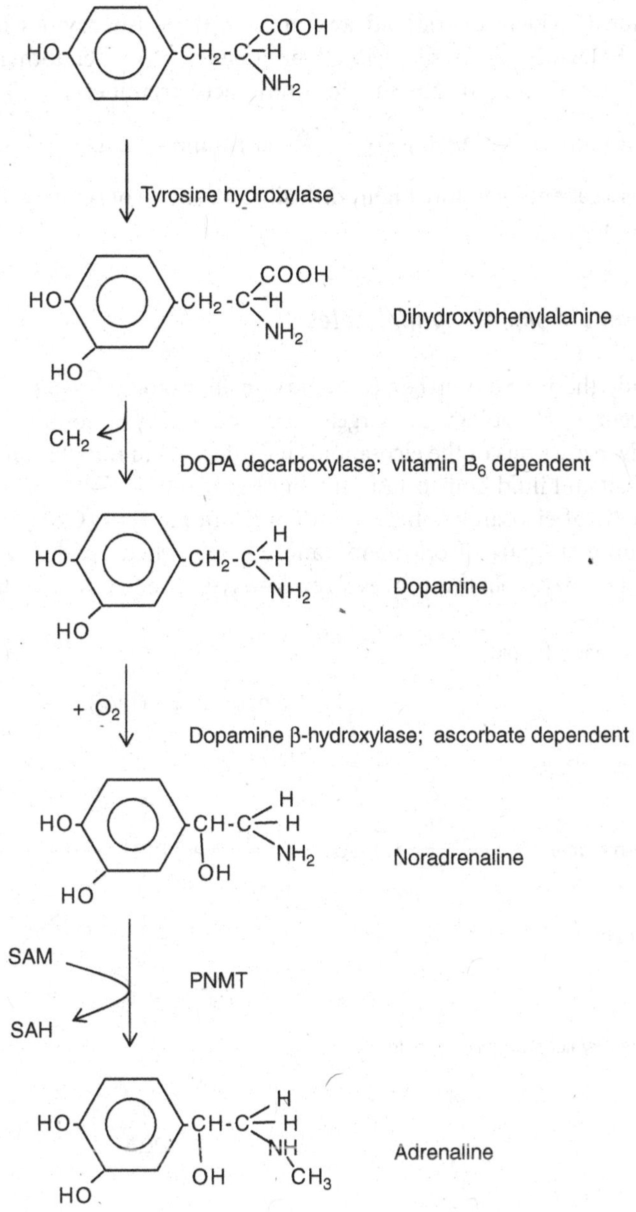

SAM = *S*-adenosyl methionine
SAH = *S*-adenosyl homocysteine

Figure 4.7 Catecholamine synthesis

Nitric oxide is widely distributed and at least three isoenzymes of nitric oxide synthase (NOS) have been described; iNOS (inducible), eNOS (endothelial) and nNOS (neuronal). The substrate for NOS is the amino acid arginine;

$$\text{arginine} + \text{NADPH} + O_2 \xrightarrow{\text{NOS}} \text{citrulline} + \text{NADP}^+ + \text{NO}$$

Because NO is a gas and therefore highly diffusible, its site(s) of action will be local to the site of production.

4.4.4   Eicosanoid signalling molecules

Like nitric oxide, the discovery of the eicosanoid signalling molecules was a significant event in twentieth century physiology, due largely to research led by Sir John Vane (Nobel Prize 1982). The diverse actions of the eicosanoids include roles in muscle contraction, blood coagulation, salt and fluid homeostasis, inflammatory responses and pain sensitivity.

The synthesis of eicosanoids begins with arachidonic acid (C20 : 4 n-6 fatty acid) which is component part of cell membranes. The synthetic pathway is outlined in Figure 4.8. A key enzyme in this process is cyclo-oxygenase (COX) which occurs in two

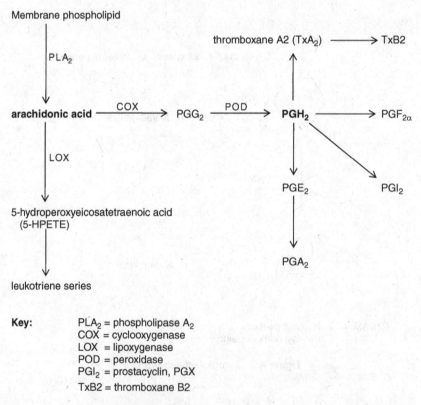

Figure 4.8   Eicosanoid synthesis

isoenzymic forms: Cyclo-oxygenase-1 (COX-1) is a constitutive enzyme, whereas cyclo-oxygenase-2 (COX-2) is inducible. A number of over-the-counter analgesics and non-steroidal anti-inflammatory drugs, for example paracetamol, ibuprofen and aspirin (acetylsalicylate) act by reducing the production of prostaglandins by the inhibition of COX-2. Arachidonic acid is derived from membrane phosphatidylcholine molecules by the action of phospholipase A_2 (PLA_2). This enzyme is inhibited by glucocorticoids, explaining the anti-inflammatory effects of cortisone.

4.4.5   Synthesis of neural signals

Synthesis of noradrenaline (norepinephrine) is shown in Figure 4.7. This follows the same route as synthesis of adrenaline (epinephrine) but terminates at noradrenaline (norepinephrine) because parasympathetic neurones lack the phenylethanolamine-N-methyl transferase required to form adrenaline (epinephrine). Acetylcholine is synthesized from acetyl-CoA and choline by the enzyme choline acetyltransferase (CAT). Choline is made available for this reaction by uptake, via specific high-affinity transporters, within the axonal membrane. Following their synthesis, noradrenaline (norepinephrine) or acetylcholine are stored within vesicles. Release from the vesicle occurs when the incoming nerve impulse causes an influx of calcium ions resulting in exocytosis of the neurotransmitter.

4.4.5.1   CNS signals
Serotonin (5-hydroxytrptamine, 5-HT) synthesis involves an hydroxylation reaction (catalysed by tryptophan mono-oxygenase) and a decarboxylation step, similar to that in adrenaline (epinephrine) synthesis.

Glutamate is a commonly occurring amino acid that acts as an excitatory transmitter in CNS. The molecule may be synthesized within the nerve ending either by transamination from 2-oxoglutarate (described in Section 6.3.1.1) or by deamination of glutamine (see Section 8.2.2). However, in common with other synaptic signals, there exists an efficient uptake mechanism in the axon to recycle glutamate that has been released.

In contrast to glutamate, γ-aminobutyric acid (γ-aminobutyrate, GABA) is an inhibitory compound. Metabolically derived from glutamate, GABA production illustrates a reaction, decarboxylation, common to the production of serotonin and the catecholamines (Figure 4.9).

4.5   Hormone and neurotransmitter storage, release and transport

Following their synthesis, hormones may be secreted immediately, for example the steroids, or stored within the tissue and only released 'on demand'. Proinsulin (pro = precursor) is synthesized as a single peptide chain and stored in vesicles within the

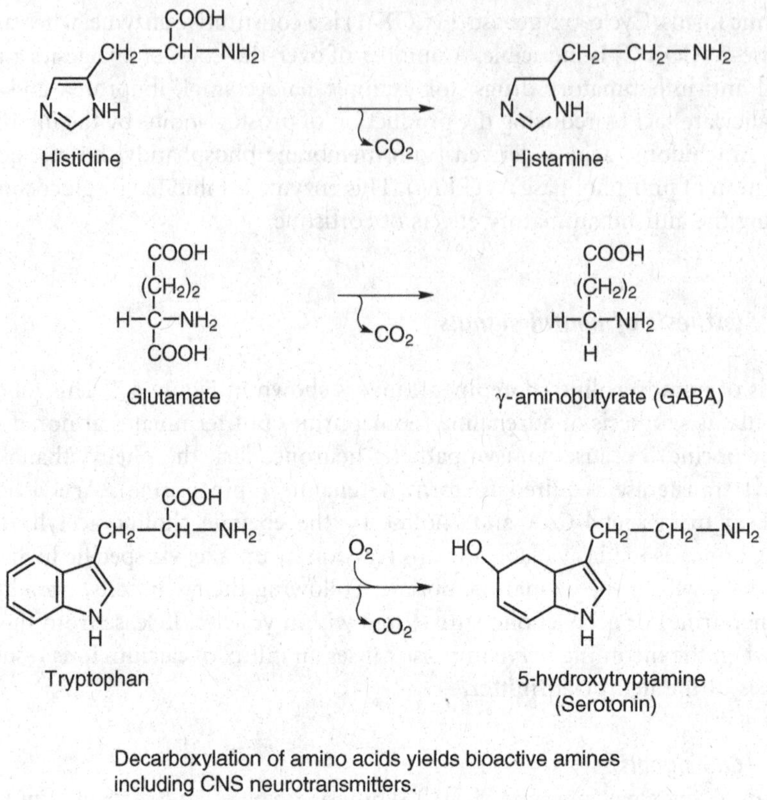

Decarboxylation of amino acids yields bioactive amines including CNS neurotransmitters.

Figure 4.9 Synthesis of CNS transmitter amines

β cells of the pancreatic islets in combination with atoms of zinc, but when required to regulate blood glucose concentration, the prohormone is cleaved and functional insulin is released into the circulation along with the C-peptide. This example of post-translational processing is mediated by peptidases which are contained in the vesicles along with the proinsulin. The fusion of the secretory vesicles with the cell membrane and activation of the peptidase prior to exocytosis of the insulin are prompted by an influx of calcium ions into the β-cell in response to the appropriate stimulus. Similarly, catecholamines are synthesized and held within the cell by attachment to proteins called chromogranins.

4.5.1 Hormone transport

Once released from the cell of origin, the signal ligand must travel to its site of action. For the classical endocrine hormones this means via the bloodstream. Given that blood plasma is approximately 94% water, the physical nature of the hormone is important. Peptides are hydrophilic and so circulate unbound to any other molecule whereas steroids and T_4 are complexed with specific binding proteins, for example cortisol-binding globulin (CBG) and sex hormone binding globulin (SHBG), and thyroid hormone binding globulin (TBG) respectively. Albumin may also act as non-specific hormone transport protein.

A dynamic equilibrium exists between the bound and the free fractions;

$$hormone–protein\ complex \rightleftharpoons protein + free\ hormone$$

The position of equilibrium lies strongly to the left. However, it is the free form of the hormone which is physiologically active with the bound fraction acting as 'reserve supply'. Typically therefore, the plasma total concentration of the hormone (bound + free fractions, measured typically in nmol/l) is of the order of 1000 times higher than the concentration of the free fraction alone (measured typically in pmol/l concentrations).

4.6 Hormone and neurotransmitter inactivation

The half-life of signalling molecules is necessarily short to prevent excessive stimulation of the target cell. Degradation of signals may be via specific (enzyme-based) and non-specific reactions and by elimination through the kidney. Measurement of the urinary excretion of hormones is often useful in diagnosing pathology such as Cushing's disease (inappropriate production of cortisol) or phaeochromocytoma (excessive production of catecholamines).

Peptide and protein hormones do not need to enter the target cell in order to change the biochemical activity of that cell but some hormone–receptor complexes are internalized and degraded by the cell. Peptides that remain in the plasma will also be destroyed and cleared from the circulation. Although peptide hormones are small enough to pass the renal glomerular barrier, intact peptides do not usually appear in the urine in appreciable amounts, implying there is degradation within the renal tubule.

In contrast, much is known about the catabolism of catecholamines. Adrenaline (epinephrine) released into the plasma to act as a classical hormone and noradrenaline (norepinephrine) from the parasympathetic nerves are substrates for two important enzymes: monoamine oxidase (MAO) found in the mitochondria of sympathetic neurones and the more widely distributed catechol-O-methyl transferase (COMT). Noradrenaline (norepinephrine) undergoes re-uptake from the synaptic cleft by high-affinity transporters and once within the neurone may be stored within vesicles for re-use or subjected to oxidative decarboxylation by MAO. Dopamine and serotonin are also substrates for MAO and are therefore catabolized in a similar fashion to adrenaline (epinephrine) and noradrenaline (norepinephrine), the final products being homo-vanillic acid (HVA) and 5-hydroxyindoleacetic acid (5HIAA) respectively.

Some of the catecholamine will enter the target cell rather than be recaptured by the neurone. Inactivation is brought about by the second enzyme, COMT which uses S-adenosyl methionine as a methyl donor as does PNMT (involved with catecholamine

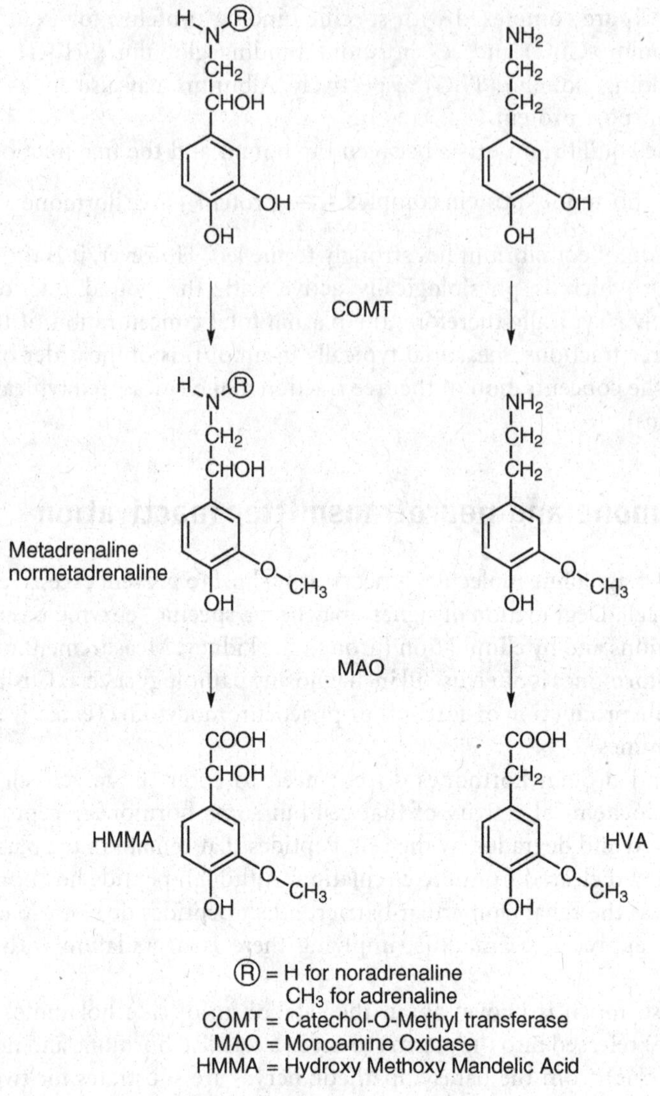

Figure 4.10 Inactivation of catecholamines

synthesis) to methylate the hydroxyl group at position 3 of the catechol ring; the product is normetadrenaline. Adrenaline (epinephrine) is metabolized in an identical fashion forming metadrenaline (Figure 4.10). Both normetadrenaline and metadrenaline are water soluble and appear in the urine but further modification may occur in the liver where conjugation with sulfate or glucuronide is possible. Steroid hormones are also inactivated in the liver, via CYP-450-dependent biotransformation reactions before being excreted in urine as sulfate or glucuronide derivatives.

4.7 Target tissue response to signals

Question: how does the target cell 'know' when a signal has arrived?

Question: why are only certain cells responsive to a particular signal?

Question: why does signalling sometimes 'go wrong' causing disease?

The answers to these questions can be given in one word: receptors.

If a hormone is likened to an e-mail message then a receptor is the recipient's mailbox address. There are many receptors in many tissues for many signals but irrespective of the particular situation, the concepts of intercellular communication are simple:

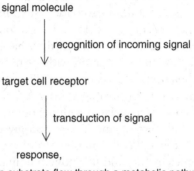

signal molecule

↓ recognition of incoming signal

target cell receptor

↓ transduction of signal

response,

e.g. change in substrate flow through a metabolic pathway, or

stimulation of cell division and growth of tissues, or

phagocytosis *etc*

4.7.1 Signal reception and transduction at the target tissue

The five 'S's of receptor biochemistry:

Structure

Specificity

Saturatability

Sensitivity

Signal transduction and amplification.

4.7.1.1 Structure

Receptors are proteins or glycoproteins found either on the surface of the target cell or located within the cell interior. The surface receptors engage peptide hormones which, being hydrophilic, do not traverse the fatty plasma membrane; intracellular receptors combine specifically with particular steroids or tri-iodothyronine, T_3.

Structurally, the peptide hormone receptors fall into one of three groups as illustrated in Figure 4.11. All three types have an extracellular domain (N-terminal) which may be glycosylated and is responsible for signal recognition (see 'Specificity' below), a transmembrane domain for anchorage and an intracellular domain (C-terminal) of variable length. Figure 4.11a is typical of a receptor tyrosine kinase (see later) associated with cytokines and growth factors. Figure 4.11b shows a 'seven pass' or serpentine receptor of the type used by classical peptide hormones and usually linked with G-proteins, whilst Figure 4.11c is a multimeric design (2α and 2β chains) with intracellular tyrosine kinase activity exemplified by the insulin receptor and insulin-like growth factor receptors.

Neurotransmitter receptors are also located in the surface membrane of the target. The nicotinic-type acetylcholine receptor is particularly interesting as it consists of five subunits; nicotinic receptors within the somatic system innervating skeletal muscle at the neuromuscular junction consist of 2α units and one each β, γ and ε whereas nicotinic receptors in ganglia and the adrenal medulla have 2α and 3β subunits. In both cases, the nicotinic receptor is an ion channel with the five subunits arranged in a pore-forming ring, so that when stimulated, cations are allowed to pass through the pore and enter the target cell. Many other neurotransmitter receptors also act as ion channels.

Not all types of receptor are associated with the outer membrane of the target cell. Receptors for vitamin D3, steroids and T_4 are non-glycosylated proteins which are located within the target cell, either free in the cytosol, (those which bind steroids), or are found within the nucleus (thyroid hormone binding receptors). These receptors vary in size from approximately 400 to nearly 1000 amino acids. All contain a DNA-binding domain, whose amino acid sequence is highly conserved in all types of intracellular receptor, and a specific ligand-binding domain. Intracellular receptors, once bound with their ligand, promote gene expression by acting as DNA transcription factors.

4.7.1.2 Specificity

Receptors exhibit 'structural complementarity' with their ligand in the same way that enzymes are complementary to their substrate. Often the actual binding of the hormone to its receptor involved just a small portion of both molecules. The peptide ACTH secreted by the pituitary gland contains 39 amino acids, but only about 12 of these near the N-terminal are required to engage the receptor. Furthermore, and as noted in Section 4.4.1, LH, FSH, TSH and hCG all share a common α subunit and their receptors recognize only the β unit.

Structural recognition by the receptor may not be absolute and the possibility of cross-reaction arises. Agonists are compounds which engage the receptor and bring about the usual physiological response whilst antagonists are molecules which act rather like competitive inhibitors of enzymes as they effectively block the receptor site making the cell unresponsive. These 'pseudo-signals' may be naturally occurring or arise as an intention of drug treatment. For example, naturally fair-skinned individuals who have an excess of ACTH circulating in their blood stream, as in Cushing's disease, may appear tanned because part of the ACTH molecule is sufficiently similar in

(a)   single pass

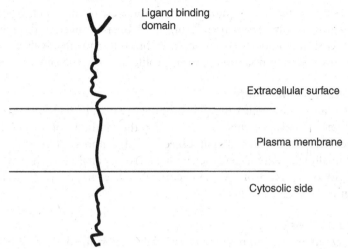

(b)   serpentine or 7-pass

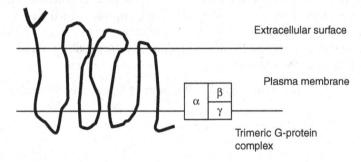

(c)   multimeric

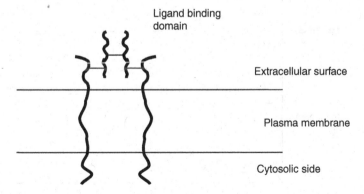

Figure 4.11   Generic structures of cell surface receptors (a) single pass (b) serpentine or 7-pass (c) multimeric

structure to MSH to engage the MSH receptor and initiate melanin production and deposition. The drug tamoxifen is an anti-oestrogen, which has been found to be beneficial in some women suffering from breast cancer. Although structurally only vaguely similar to oestrogen, tamoxifen binds to and effectively disables the oestrogen receptor resulting in a reduction in proliferation of the cancer cells.

4.7.1.3 'Saturatability'

An obvious consequence of signal binding to a defined region of the receptor is that the number of engagements is limited by the number of receptors. Once a signal concentration is reached that exceeds that number of receptors, the cell will be maximally stimulated. Here then we have another analogy with enzymes; the maximum velocity of the reaction (V_{max}) occurs when the enzyme is 'saturated' with substrate (Figure 4.12).

4.7.1.4 Sensitivity

The ability of a cell to respond to a signal is determined by the number of receptors it expresses and also the ease of engagement between the two molecules. A cell may be able to increase (upregulate) or decrease (downregulate) the number of receptors it expresses according to the prevailing physiological conditions. This is analogous to enzyme induction (see Section 3.2) as the control is mediated at the level of gene expression. A cell cannot however change the affinity (ease of binding) between the signal and the receptor. The affinity is defined numerically by K_d, which is a dissociation coefficient derived thus;

$$\text{unbound Signal} + \text{Receptor} \rightleftarrows \text{Signal–Receptor complex}$$
$$\quad\; [uS] \qquad\qquad\quad [R] \qquad\qquad\qquad\qquad\quad [SR]$$

$$K_d = \frac{[uS] \times [R]}{[SR]}$$

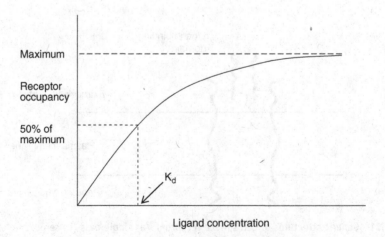

Figure 4.12 Saturation of receptor

If the affinity between the signal and the receptor is high, at any given total [S] most will be found as [SR], so [SR] $\gg$ [uS] and thus K_d is numerically small. Conversely, if the affinity of binding is low, [uS] $\gg$ [SR] and K_d is a large number. Notice again the conceptual similarity here with K_m for an enzyme–substrate combination (Section 2.3).

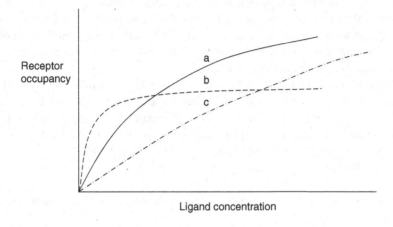

Figure caption on plot axes:
Receptor occupancy (y-axis); Ligand concentration (x-axis); lines labelled a, b, c.

Key

line a typical receptor characteristics

line b high affinity (low Kd), low capacity (low number)

line c low affinity (high Kd), high capacity

Figure 4.13 Receptor sensitivity

One could easily imagine a cell having subpopulations of high or low affinity receptors; low affinity for 'routine maintenance' of cell status and a small number of high affinity receptors for rapid responses in the face of a physiological challenge. Figure 4.13 shows how target cell sensitivity can be explained.

Thus, response sensitivity is determined by type and number of receptors. Several diseases, including nephrogenic diabetes insipidus, insulin resistance and other growth-related abnormalities are associated with defects in the processing of the incoming message. The generic term 'end organ failure' is used to describe pathologies in which the individual produces enough hormone but whose tissues are unable to respond resulting in symptoms of apparent hormone deficiency. The cause may be a genetically determined defect in the expression of receptors; as an analogy, the e-mail is sent but not received. Alternatively, there may exist a defect in the intracellular pathway which is normally modulated by signal-receptor binding; by analogy, the e-mail is received, but ignored!

4.7.1.5 *Signal transduction and amplification*
Knowledge of receptors explains how cells 'know' there is an incoming signal but does not explain the metabolic consequences of hormone activation. Transduction is the conversion of the external signal message, into a sequence of intracellular metabolic

events mediating changes in cell activity. Functionally, this can be viewed as occurring in two phases: (i) receptor-ligand binding and, for peptide signals the initial processing events which occur proximal to the cell membrane and (ii) 'downstream' events whereby the message is passed on to the 'metabolic machinery' within the cell. Because each step of the transduction process is enzymatic, and because each enzyme can act on many substrates, a cascade is initiated. This results in signal amplification further increasing the sensitivity of the response. The specificity of the cellular response to signal is therefore determined by both the possession of the correct receptor and the appropriate intracellular metabolic machinery.

Four transduction mechanisms can account for many peptide signal-stimulated responses, that is those associated with surface receptors. All four models assume that the receptor undergoes some degree of conformational change following signal engagement and leading to further protein, usually enzyme, changes within the cytosol of the target cell close to or within the plane of the membrane.

The four models are;

- Gated ion channels (e.g. neurotransmitter receptors described under 'Structure' above);

- Second messenger generation linked to membrane-bound G-protein;

- Receptor tyrosine kinase (RTK) activation;

- Receptor tyrosine phosphatase (RTPase) activation.

Gated ion channels are exemplified by the nicotinic receptors of the autonomic nervous system where the ligand is acetylcholine. The electrical impulse (nerve impulse) is propagated when sodium and potassium ions move across the neuronal membrane causing depolarization. The flow of ions is permitted by the ligand-induced opening of protein channels. When stimulation of the cell is removed, the channel closes and the membrane repolarizes as the ions move in the opposite direction and the cell regains its resting state. (Figure 4.14 ion gates).

4.7.2 G-proteins

G-proteins are so called because they bind a guanosine nucleotide, either GTP or GDP. Their transduction mechanism involves the production of a second messenger such as $3'5'$ cAMP, $3'$ $5'$ cyclic GMP (cGMP) or IP3 and diacylglycerol (DAG), derived from AMP, GMP and phosphatidyl inositol-3,5bisphosphate respectively (Figure 4.15). It is the second messenger that initiates the downstream amplification process phase of transduction.

G-proteins can be found membrane bound or free in the cytosol. The membrane bound proteins are trimeric complexes of α, β and γ subunits. The β and γ subunits may

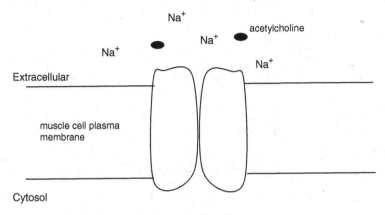

(a) receptor sites unoccupied and gate closed

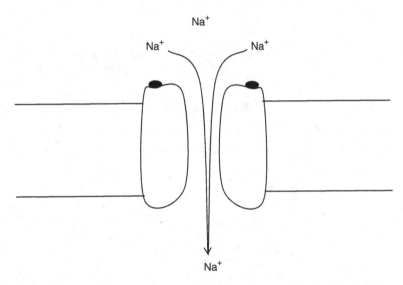

(b)   acetylcholine receptor sites occupied and gate is open allowing influx of ion

Figure 4.14 Gated ion channel (ligand activated, acetylcholine) at neuromuscular junction. (a) receptor sites unoccupied and gate closed (b) acetylcholine receptor sites occupied and gate is open allowing influx of ion

be considered as a single functional component, Gβγ, which is distinct from the α subunit, Gα. The Gα component has inherent GTPase activity. In the resting state, GDP is bound but on stimulation, this is exchanged for GTP. The GTPase activity converts the GTP into GDP and the G-protein returns to its resting state. The GTPase activity is therefore an auto-regulation mechanism to limit the transduction process and prevent over-stimulation of the cell.

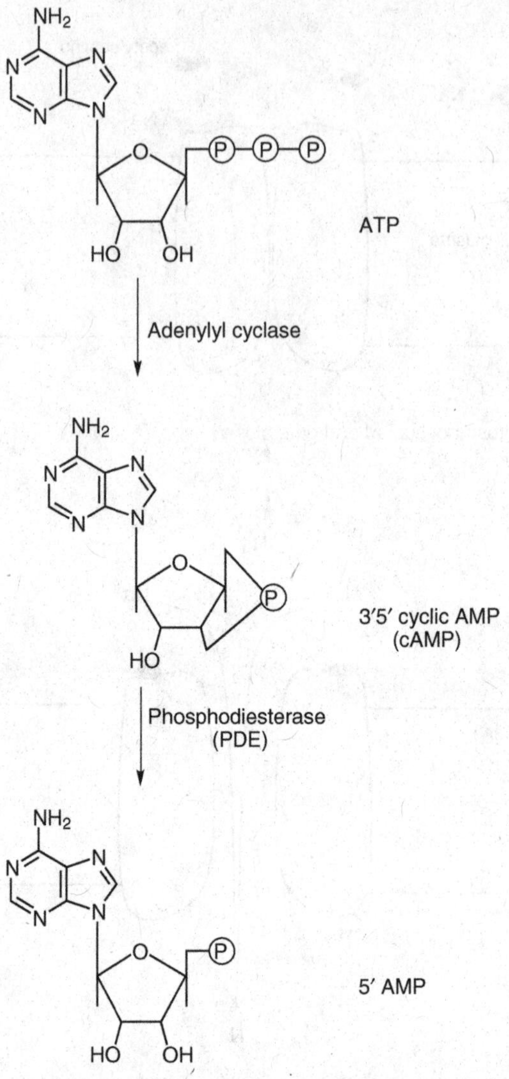

Figure 4.15 Formation and degradation of cAMP

These G-proteins act as a 'bridge', linking the receptor with an enzyme which begins the metabolic cascade. Functionally, the β/γ component engages with the signal-bound receptor and the α subunit, which carries the guanosine nucleotide, interacts with an effector enzyme which is responsible for the actual generation of the second messenger. The role and importance of G-proteins were described by Sutherland and colleagues (Nobel Prize in 1971).

Several different types of G-protein complex have been described. In general terms, if a cell becomes activated by ligand binding, the G-protein complex is said to be

Table 4.4 Gα sub-units associated with particular hormones

Sub-unit type	Hormone
$G_s\alpha$	ACTH (ac)
	glucagon (ac)
	TSH (ac and PLC)
	LH & FSH (ac and PLC)
	Parathyroid hormone (ac and PLC)
	Prostaglandin E2 (ac)
$G_i\alpha$	Calcitonin (PLC)
$G_q\alpha$	Thyrotrophin releasing hormone (ac and PLC)
	Oxytocin (PLC)

ac = adenylyl cyclase-linked receptor, cAMP as second messenger.
PLC = phospholipase C-linked receptor, IP_3 as second messenger.
Hormones often engage with different types of Gα sub-unit in a concentration-dependent fashion.

stimulatory (G_s) but inhibition may occur via engagement with G_i-containing complexes. Two other types are known; the rod cells in the retina of the eye contain a specific G-protein called transducin, designated G_t which conveys the message via cGMP as second messenger, whilst the G_q complexes are associated with transduction via phospholipase. Representative examples of G-proteins are shown in Table 4.4.

The sequence of events that follows hormone binding to its surface receptor is as illustrated in Figure 4.16. The hormone–receptor complex engages β/γ subunit of the GDP carrying G-protein complex. The G complex undergoes a conformational change, GTP replaces the GDP and the α subunit of the complex dissociates from the β/γ subunit/hormone-receptor assembly. The alteration in the structure of the α allows it to activate an inactive effector enzyme, well known examples of which are adenylyl cyclase (AC) and phospholipase C (PLC). As the inherent GTP'ase activity of the Gα subunit takes effect, the bound GTP is hydrolysed forming GDP. The Gα subunit can now no longer bind to the effector but instead reforms a complex with Gβγ and is ready for the next incoming signal-induced cycle.

At this point, the detailed events of signalling pathways begin to differ as the exact nature of the components of the cascade will vary from one cell-type to another. Furthermore, 'branching' of the downstream transduction may occur resulting in the ultimate activation or induction of several regulatory enzymes within different metabolic pathways. To continue the social communication analogy introduced earlier in this chapter, an e-mail message has been forwarded by the original recipient to a network of other address boxes.

Cyclic AMP diffuses away from the membrane and engages its own target which is an inactive protein kinase, called cAMP dependent protein kinase or protein kinase A (PKA). The inactive PKA is a tetramer of two catalytic subunits and two regulatory subunits. Binding of cAMP to the regulatory subunits causes structural changes and the two regulatory subunits dissociate from the two catalytic sub-units. The now activated protein kinase A, that is the C subunit dimer, initiates a downstream cascade by

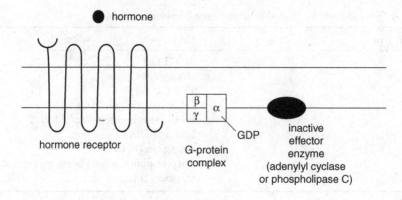

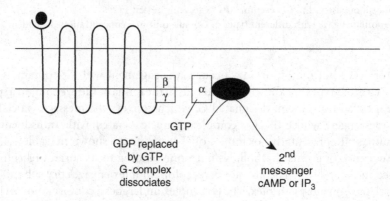

Activation of membrane-bound enzyme
generating 2nd messenger following hormone
binding to its receptor.
Adenylyl cyclase produces cAMP from ATP
PLC produces two 2nd messengers, IP$_3$ and DA

Figure 4.16 Signal transduction by G-proteins

phosphorylating and altering the activity of target enzymes on serine/threonine residues.

For example, Figure 4.17 shows how glycogen catabolism is initiated in the liver in response to glucagon stimulation. Glucagon stimulation of the liver cells initiates a cascade beginning with the production of cAMP and which ends with the activation, via phosphorylation, of glycogen phosphorylase. Inactive glycogen phosphorylase, an enzyme, is actually a substrate for protein kinase A. This type of covalent modification of an enzyme structure to bring about control of its activity is described in Section 3.2.2.3. Note that the cAMP-dependent cascade simultaneously *inactivates* glycogen synthase, illustrating reciprocal control of two enzymes with opposing roles.

In addition to its direct effect on enzyme activity in the cytosol of target cells, cAMP can also initiate gene expression without actually entering the nucleus. PKA may also

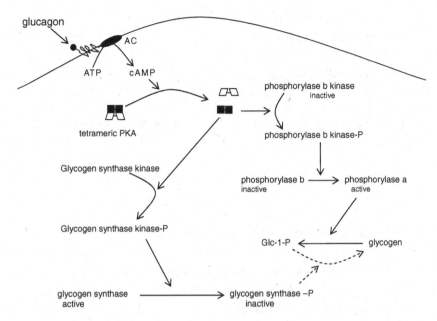

Figure 4.17 cAMP activation of glycogen phosphorylase in liver. Glucagon stimulation of liver cells activates glycogen degradation by glycogen phosphorylase and simultaneously inhibits (- - - -) glycogen synthesis. Both effects are mediated by phosphorylation of target enzymes, glycogen synthase kinase and phosphorylase b kinase. AC = adenylyl cyclase black squares = catalytic subunit dimer of PKA

phosphorylate a cytosolic protein called cAMP response element binding protein or more simply, CREB. CREB is a nuclear transcription factor, which engages with specific regions of DNA called cAMP response enhancer element (CRE) to permit gene expression. There is some similarity here with steroid and thyroid hormone receptors in that they too interact with DNA and initiate transcription.

PLC is a generic name for a family of isoforms of an enzyme which remain membrane bound as the presence of phospholipids is required for activity. Signal transduction via PLC as the effector is mediated by not one but two second messengers; inositol 1,4,5 triphosphate (IP3) and DAG, which are the products of hydrolysis of membrane phospholipid by PLC (Figure 4.18).

The DAG produced by PLC action diffuses within the membrane bilayer where it encounters and activates protein kinase C (PKC), which catalyses the phosphorylation of serine and threonine resides on in specific cytosolic proteins and enzymes. Calcium ions are also activators of PLC and it is inositol 1,4,5 P_3 which induces the increase of cytosolic calcium concentration by engaging receptors on endoplasmic reticulum. The rising calcium concentration also activates a number of cytosolic enzymes, often following binding to calmodulin . . . a *cal*cium-dependent enzyme-*modul*ating prot*ein*. For example, glycogen catabolism involves calcium–calmodulin (CaCM) regulation of glycogen phosphorylase.

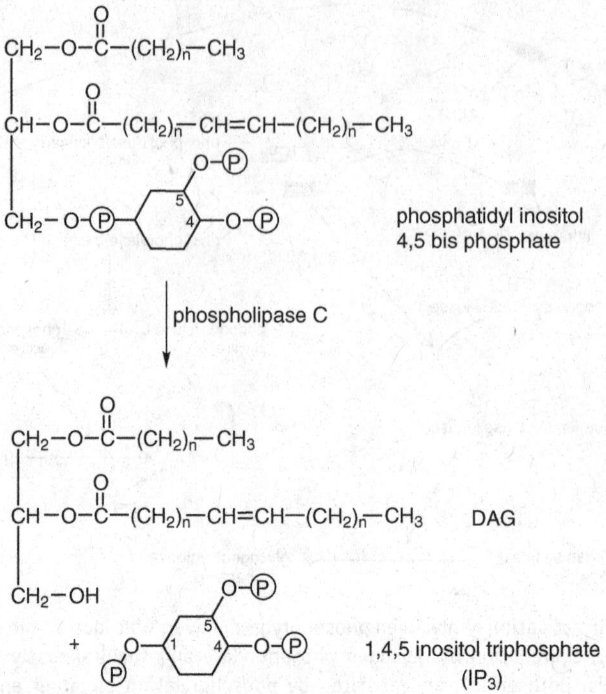

Figure 4.18 PLC and structures of IPs

Thus far, the discussion of G-proteins and effector enzymes has assumed that a ligand has engaged with its surface receptor. There is however, an important example of an alternative mechanism to activate an effector without the *direct* involvement of G-protein complex. NO is a local hormone, a neurotransmitter and part of the cell's armoury of oxidizing agents called free radicals.

NO is a gas that diffuses freely through cell membranes. Once inside its target cell, NO binds directly with and activates guanylyl cyclase, an enzyme that converts GTP into cGMP; cGMP is a second messenger for a number of tissue specific processes in neurones, smooth muscle and white blood cells.

Although NO does not itself use a G-protein for signalling, the mechanism of NO production in vascular endothelium is initiated by IP3 via a G-protein-linked acetylcholine receptor on the cell surface. The IP3 causes activation of nitric oxide synthase via calcium– calmodulin and the NO generated diffuses from the endothelial cell into the adjacent smooth muscle cell where cGMP is produced.

4.7.3 Receptor tyrosine kinases

Kinases are the class of enzymes which transfer the terminal phosphate group from ATP to a substrate. In particular, tyrosine kinases transfer the phosphate to a tyrosine residue which is part of the primary amino acid sequence of a target protein substrate. Figure 4.19 illustrates this. Note the single-letter abbreviation for tyrosine is Y so sometimes you may see symbols such Y-P or Yp to represent a phosphorylated tyrosine residue.

Signal transduction may involve two types of tyrosine kinase:

1. receptors with inherent tyrosine kinase activity (RTKs), for example insulin receptor;

2. receptors, which recruit cytoplasmic tyrosine kinases, for example growth hormone, leptin operating through Janus kinase (JAK).

RTK constitute a large group of surface proteins with dual recognition/kinase properties. The extracellular N-terminal of the protein binds with the ligand, often a growth factor peptide, whilst the intracellular portion of the protein has an active site designed to phosphorylate tyrosine residues within target proteins. Because of this dual functionality, there is no need for G-proteins or similar to 'bridge the gap' between signal reception and initiation of the cytosolic cascade.

Signal transduction from RTKs is complex, involving a number of proteins, found either bound to the inner face of the plasma membrane or free within the cytosol, which are recruited to a cascade pathway. These proteins have what appear (to the new student of cell signalling) to be rather idiosyncratic names; Grb, raf, ras, Shc, MEK and ERK. Although these terms do have a meaning, don't be too distracted by them when you first study signal transduction. Initially, look upon them simply as protein links in the chain which either take the signal from the cell surface to the cell nucleus where gene expression is regulated or they alter the activity of particular enzymes. As we have seen already with steroid and thyroid hormone receptors, eukaryotic gene expression may be 'switched on' when particular proteins bind their promoter regions. With TK-linked receptors, it is not the receptors themselves which engage the gene, but intracellular proteins that have been 'activated' by external signals, mediated by the TK receptor.

Ligand binding to the receptor domain of the RTK brings about phosphorylation of the intracellular C-terminal tail of the protein (Figure 4.20). Although the term *auto-phosphorylation* is often used in this context, it is likely that the mechanism is really one of cross-phosphorylation. Single-chain RTKs aggregate and so bring their active sites into close proximity to the tyrosine phosphorylation sites of their neighbours. For multi-chain RTKs, the distortion of the shape of the whole protein allows the intracellular domains to phosphorylate each other, as illustrated by the insulin receptor described later.

The phosphorylated tyrosine (Tyr-P) residues act as 'docking' sites for cytosolic proteins; the RTK becomes a 'molecular coat hook' (Figure 4.21) as it forms the focal point for protein aggregation. The cytosolic proteins, generically known as downstream adaptor proteins, which bind to the RTK, have Tyr-P recognition sites called SH2 or SH3 domains. The whole assembly forms a multicomponent complex which propagates transduction of the signal. Well-characterized and important adaptor proteins include the insulin receptor substrate (IRS) proteins represented in Figure 4.21.

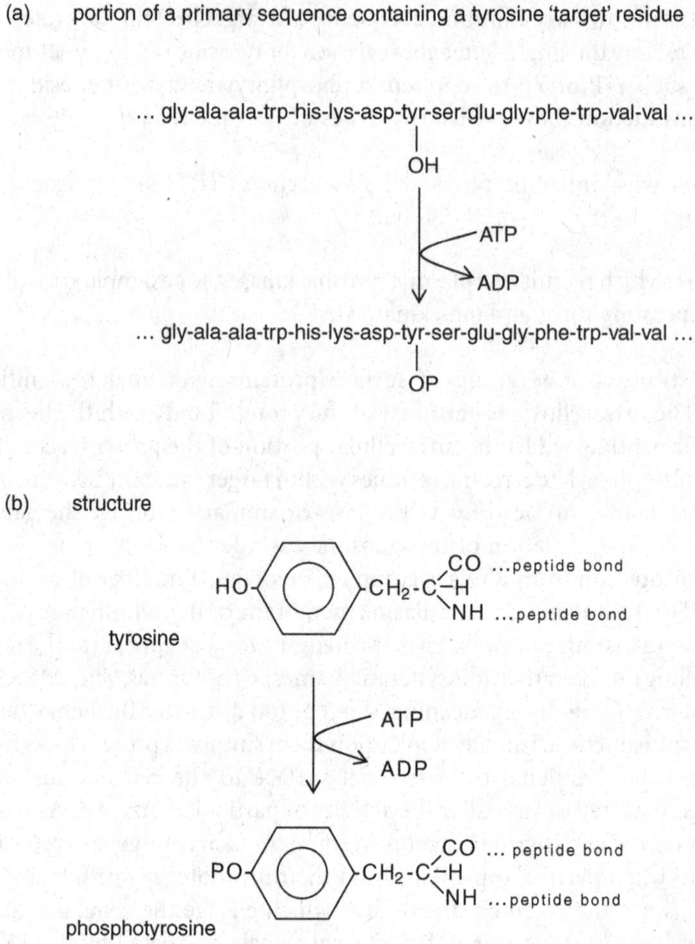

(a)   portion of a primary sequence containing a tyrosine 'target' residue

... gly-ala-ala-trp-his-lys-asp-tyr-ser-glu-gly-phe-trp-val-val ....
|
OH

ATP
ADP

... gly-ala-ala-trp-his-lys-asp-tyr-ser-glu-gly-phe-trp-val-val ....
|
OP

(b)   structure

HO—⬡—CH_2-C—H CO ...peptide bond
                    NH ...peptide bond

tyrosine

ATP
ADP

PO—⬡—CH_2-C—H CO ... peptide bond
                    NH ...peptide bond

phosphotyrosine

Figure 4.19 Tyrosine kinase (a) portion of a primary sequence containing a tyrosine 'target' residue (b) structure

A particularly interesting downstream protein is phosphatidyl inositol 3-kinase (PI3K), which is functionally associated with RTKs for insulin, prolactin, many growth factors and cytokines. So far, our discussion has focussed on intracellular transduction through *protein* phosphorylation; PI3K is involved with the generation of a series of phosphorylated *lipid* signals, but may also phosphorylate proteins, not at tyrosine, but at serine or threonine resides. Furthermore, G-linked receptors may be able to activate PI3K, so PI3K is very much at 'transduction cross roads'.

The term PI3K defines a family of isoenzymes that are important mediators of metabolic control, cytosolic trafficking and mitogenic regulation in many types of animal and plant cells. Enzymes that are so well conserved through species are often pivotal in cell physiology and PI3K is a focus of research interest in cancer biology and diabetes, two significant causes of morbidity and mortality.

PI3K isoforms are heterodimers made up of a large (110 kDa) catalytic unit and a smaller regulatory unit (50–85 kDa) which has the SH2 binding sites allowing interaction with an activated phosphoprotein. Structural variations of PI3K arise because there are two types of the larger subunit (p110α and p110β) and five types of the regulatory subunit (p50, p55α, p55γ, p85α and p85β). The products of PI3K activity are phosphoinositol-3-P, phosphoinositol-3,4-P_2, phosphoinositol-3,4,5-P_3. Phosphoinositol-3,4-P_2, phosphoinositol-3,4,5-P_3 bind to and activate a number of proteins including serine/threonine kinases such as protein kinase B (PKB also called AKT), protein kinase C, phosphoinositide dependent kinase (PDK) and glycogen

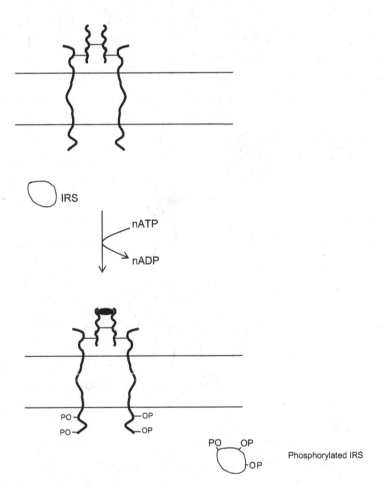

(a)   those receptors with inherent TK activity; insulin receptor

Figure 4.20 Tyrosine kinase-linked receptors (a) those with inherent TK activity; insulin receptor. The intracellular tail and cytoplasmic proteins become phosphorylated following hormone engagement with the ligand-binding domain

(b)   those receptors which recruit cytosolic TK;
       receptors for growth hormone, erythropoietin (EPO), Prolactin.

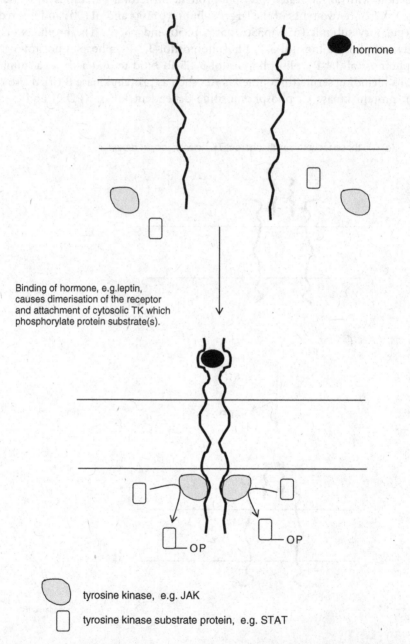

Binding of hormone, e.g.leptin,
causes dimerisation of the receptor
and attachment of cytosolic TK which
phosphorylate protein substrate(s).

tyrosine kinase, e.g. JAK

tyrosine kinase substrate protein, e.g. STAT

Figure 4.20 (b) those receptors which recruit cytosolic TK; receptors for growth hormone, erythropoietin (EPO), Prolactin. Binding of hormone, for example leptin, causes dimerisation of the receptor and attachment of cytosolic TK which phosphorylate protein substrate(s)

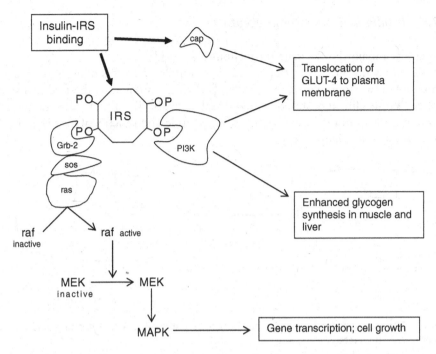

Figure 4.21 IRS signalling. Grb-2, sos, and ras are adaptor proteins which assemble onto the phosphorylated IRS. Raf is a membrane bound GTP-binding protein which when activated has Ser/Thr kinase activity

synthase kinase (GSK). The inositol phosphates are clearly therefore acting as classical second messengers.

4.7.4   JAK/STAT signalling

JAKs and signal transducers and activators of transcription (STATs) are functionally analogous with IRS and PI3K. JAKs are physically associated with a cell surface receptor (e.g. for leptin, erythropoietin (EPO), growth factors or cytokines); STATs are free monomeric proteins within the cytosol but following phosphorylation by a JAK, individual proteins dimerize and then move into the nucleus of the cell where they control gene expression.

The JAK family of proteins is derived from four genes; *JAK1*, *JAK2*, *JAK3* and *TYK2*. The JAK2 and TYK2 proteins are found widely in tissues; JAK3 expression is restricted to lineages of white blood cells. STAT proteins arise from expression of at least seven genes and additional isoforms may be generated by post-translational processing. Taken together, the variety of JAK and STAT proteins generate a large number of potential JAK/STAT interactions resulting in the regulation of a wide range of genes.

4.7.5   Insulin and the insulin receptor

Diabetes is a major health issue throughout the world.

Structurally insulin is a small peptide, with a molecular mass of around 5500 and composed of two subunits, denoted α and β chains. Insulin is synthesized as a single peptide, Proinsulin and stored within the pancreatic β-cells. At the moment of secretion, pro-insulin is cleaved, releasing C-peptide and functional insulin in to the blood circulation (Figure 4.22).

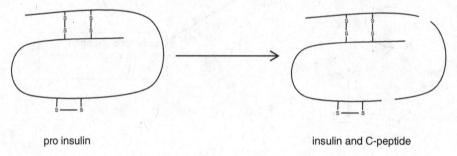

pro insulin   insulin and C-peptide

Figure 4.22 Insulin is produced from pro insulin by proteolysis. Active hormone and inactive C-peptide are co-secreted

Insulin is a potent anabolic hormone. Tissue responses to stimulation by insulin include:

- Nutrient homeostasis; cell uptake of glucose (especially important in adipose and muscle), amino acids (all cells) and fatty acids; stimulation of glycolysis but inhibition of gluconeogenesis (liver), synthesis of glycogen (liver and muscle), triglyceride (liver and adipose) and protein (all cells);

- DNA synthesis, gene expression and growth promotion (mitogenic effects) (all cells);

- Inhibition of apoptosis (all cells);

- Stimulation of the N^+/K^+ ATPase 'pump' (adipose and muscle).

A challenge posed to researchers was therefore to account for diverse physiological effects emanating from the same receptor/hormone interaction. Structurally, the insulin receptor (IR) is a tetrameric protein, composed of two smaller extracellular α units and two larger transmembrane β units (see Figure 4.20a).

Typically of receptor tyrosine kinases, binding of insulin to the extracellular domains of the IR causes autophosphorylation of specific tyrosine residues within the intracellular region of the β units. Some RTKs, as described above and as illustrated by JAKs described above and also shown in Figure 4.20, would at this point recruit adaptor proteins to bind directly to the phosphorylated intracellular chains. Insulin however is a little different because the tyrosine kinase region of the protein causes phosphorylation of a number of distinct intracellular proteins, known collectively as insulin receptor substrates (IRS). It is the IRS rather than the IR itself which acts as the docking protein for downstream adaptor proteins and enzymes. Specifically, the IRS family includes IRS-1, IRS-2, Shc, gab-1 and CAP; the possibility of multiple signal transduction pathways arises as shown in Figure 4.21.

Mitogenic effects are mediated through MEK/MAPK (mitogen-activated protein kinase) and changes in metabolic activity mainly via PI3K.

4.7.5.1   *IRS and glucose uptake (e.g. by muscle and adipose cells) and metabolism (by all cells)*

Glucose cannot easily traverse the outer plasma membranes of cells, thus its entry into cells is carrier mediated by so-called GLUT (glucose transporter) proteins. In the absence of insulin stimulation, GLUT 4 proteins are held within vesicles in the cytoplasm of skeletal and cardiac muscle cells and adipocytes. Engagement of the IR leads to mobilization of the GLUT 4 from the 'depot' to the plasma membrane where they are incorporated into that membrane. The process is mediated by two mechanisms, one is PI3K dependent and other is PI3K independent, operating through CAP signalling (Figure 4.21).

Once inside the cell, glucose disposal may be through oxidation (glycolysis and TCA cycle) or non-oxidative pathways, mainly glycogen synthesis. Insulin regulates key enzymes of glycolysis, notably GK/HK and glycogen synthesis (see below) and simultaneously suppresses the hepatic expression of phosphoenolpyruvate carboxykinase (PEP CK), one of the enzymes of gluconeogenesis. In short, glucose uptake and metabolism are stimulated and glucose release from the liver is suppressed; blood glucose concentration is restored almost to typical fasting values of around 5–6 mmol/l within 2 h after a meal.

4.7.5.2   *IRS and glycogen synthesis*

Under basal (i.e. in the absence of insulin) conditions, glycogen synthase, a key enzyme in glycogenesis (see Section 6.3.3) is kept in its inactive phosphorylated form by the action of glycogen synthase kinase (GSK-3). Following it's activation by PI3K, AKT (one of the protein 'links' in the insulin signal transduction chain) causes the *in*activation of GSK-3; glycogen synthase is therefore *not* phosphorylated. The *de*phosphorylated from of GS is the active form, so glycogen synthesis is promoted in the fed state when insulin is present (see Figure 4.23).

4.7.5.3   *IRS and protein synthesis*

Insulin promotes amino acid uptake and protein formation. AKT, noted above, is also implicated in mechanisms which regulate protein synthesis. Acting via GSK-3 again, under basal conditions, GSK-3 phosphorylates a key protein translation regulator (called eIF2B). Thus, if GSK-3 is inactivated, eIF2B is not phosphorylated and mRNA translation is permitted.

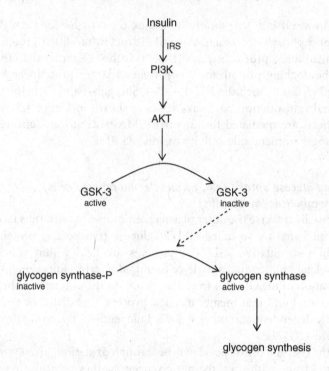

AKT inactivates GSK-3 and so relieves the inhibition of glycogen synthase.

Figure 4.23   Regulation of glycogen synthesis. AKT inactivates GSK-3 and so relieves the inhibition of glycogen synthase

4.7.5.4   *Insulin receptor and lipid metabolism*

Triglyceride and fatty acid synthesis are promoted by insulin stimulation of liver and adipose tissues by causing the phosphorylation of the first and controlling enzyme in the pathway; acetyl-CoA carboxylase (see Section 6.3.2). This enzyme catalyses the formation of malonyl-CoA and requires both allosteric activation by citrate and covalent modification for full activity.

In adipose tissue, insulin stimulation suppresses triglyceride hydrolysis (to free fatty acids and glycerol) by activating cAMP phosphodiesterase (cAMP PDE). Cyclic AMP, (3′,5′ cAMP), is required to stimulate hormone sensitive lipase (HSL), the enzyme which hydrolyses triglyceride within adipocytes; PDE converts active 3′,5′ cAMP to inactive 5′ AMP thus preventing the stimulation of HSL. The net effect of insulin on lipid metabolism is to promote storage.

4.7.5.5   *Mitogenic responses to insulin stimulation*

Growth promoting effects of insulin occur via interaction of the IR with Grb-2 or SHC adaptor proteins. The cascade from Grb-2 or SHC includes ras, raf, sos and MEK, culminating in activation of the gene transcription factor MAPK. Refer again to Figure 4.21.

4.7.5.6   *How are the IR cascades switched-off?*

Clearly, it is essential that cell can return to its basal state once insulin (or indeed any other messenger) secretion has declined. In the case of the IR cascade this occurs partly through endocytotic internalization and destruction of the insulin–IR complex and partly due to specific enzymes which degrade intracellular second messengers. For example, a phosphatase called SHIP (src homology containing inositol phosphatase activity) degrade inositol (1,4,5) P_3 thus breaking the link between a membrane effect and the downstream signalling. Experiments have shown that phosphorylation of serine in IRS allows the binding of a protein identified as 14-3-3, which inhibits insulin receptor-induced tyrosine phosphorylation of IRS. This might be significant if phosphorylation of the serine is brought about by PI3K, creating self-regulating mechanism; this is as yet uncertain.

Two tyrosine kinase-based mechanisms have been described: the IR and the JAK/ STAT cascades. It should not be assumed that these are either/or mechanisms as there may be parallel transduction (cross-talk) between the two pathways. For example, leptin, an appetite suppressor described more fully in Chapter 9 is a cytokine-like peptide produced by adipose tissue, which signals in the hypothalamus of the brain via JAK2/STAT3, but also influences the IRS/PI3K pathway.

In contrast to the actions of insulin described above, glucagon (the signal for 'fasting'), adrenaline (epinephrine) and cortisol (stress hormones), and also growth hormone, raise blood glucose concentration and mobilize the body's fuel reserves. Consequently, these hormones are sometimes referred to as 'insulin antagonists'. The actions of glucagon and adrenaline (epinephrine) on glycogen metabolism are described in Chapters 6 and 7 and details about fatty acid liberation from adipose tissue may be found in Chapter 9. Correct and appropriate fuel homeostasis is achieved by the complementary activities of the hormones mentioned. Excessive or deficient hormone production resulting in an imbalance of hormone action, or the inability of target organs to respond to a stimulus will lead to pathology, notably diabetes.

4.8   Diabetes mellitus (Greek: *'large volumes of sweet fluid'*)

The impact of diabetes mellitus (DM) on the health and wellbeing of societies is a growing concern in the twenty-first century, especially given its often close association with cardiovascular disease. The commonest cause of death in diabetic adults is coronary artery diseases such as myocardial infarction. The prevalence of DM has reached epidemic proportions with an estimated 200 million people affected worldwide (2008 figures) with a predicted rise to nearly half a billion cases before 2050 if current trends continue. The World Health Organization (WHO) has defined four major categories of diabetes based aetiology (causation) of hyperglycaemia as shown in Table 4.5.

This situation is largely due to the rise of type 2 DM which now accounts for 85–90% of all cases of DM; type 1 DM in which antibodies directed at the enzyme glutamic acid decarboxylase (GAD) attack and destroy the insulin-producing β cells

Table 4.5 Diabetes

WHO category	Aetiology	Notes
Type 1	Autoimmune disease resulting in an absolute deficiency of insulin.	Formerly referred to as juvenile onset diabetes, type I DM or insulin dependent diabetes mellitus (IDDM). Ketosis is common in poorly controlled subjects.
Type 2	Peripheral tissue resistance to the action of insulin	Includes those formerly classified as adult onset diabetes, type II DM or non-insulin dependent diabetes mellitus (NIDDM). Ketosis is rare.
	Insulin secretory defects	
Other specific causes	For example, genetic defects of insulin production or action at target site; endocrine disease; drug- or chemical-induced defects; viral infections	Many cases also previously classified as type II DM. endocrine disorders leading to the over production of insulin antagonists such as cortisol (Cushing's syndrome), thyroid hormones, growth hormone (acromegaly) or adrenaline (epinephrine) (phaeochromocytoma)
Gestational diabetes	Pregnancy	Usually transitory but many women have an increased long-term risk of diabetes.

of the pancreas is less common. Autoimmune destruction of endocrine tissues is not limited to the pancreas. The commonest form of primary hypothyroidism, sometimes called Hashimoto's thyroiditis is also an autoimmune condition as are some cases of Addison's disease of the adrenal cortex and hypoparathyroidism.

Although both main types of the diabetes have a genetic element, up to 10% of individuals in some ethnic groups being afflicted by the condition and development of type 2 DM in particular is influenced by environmental and personal habits, for example obesity and a sedentary lifestyle. Onset of symptoms may be acute (sudden) or slowly progressive and the WHO (1999) and more recently the American Diabetic Association (ADA, 2003) have devised classification schemes for various stages of DM according to plasma glucose concentrations (Table 4.6).

Untreated type 1 DM causes major and severe metabolic and physiological disruption. The lack of insulin leads to significant hyperglycaemia as cells cannot take up glucose from the blood; the body enters what is essentially a state of 'metabolic starvation' and low insulin concentration with high glucagon and adrenaline concentrations result in lipolysis of triglyceride in adipose tissue. The fatty acids released from their depot stores are transported by albumin and taken into cells where they undergo β-oxidation (see Section 7.5 for details) to provide acetyl-CoA and reduced coenzymes (NADH and $FADH_2$) to drive the TCA cycle and oxidative phosphorylation in tissues. However, the production of acetyl-CoA from fatty acids is greater than can be metabolized so much of the excess is converted in to so-called 'ketone bodies' acetoacetate, hydroxybutyrate and acetone (see Figure 4.21 and refer also to Figure 6.19 in Section 6.3), that is a state of ketosis arises, which if severe develops into

Table 4.6 Classification of diabetes

Classification	Source	Diagnostic Criteria. Plasma glucose concentration
Normal glucose regulation	WHO (1999)	FPG < 6.1 mmol/l and 2-h post glucose load <7.8 mmol/l
	ADA (2003)	FPG < 5.6 mmol/l
Impaired Fasting Glucose regulation (IFG)	WHO	FPG > 6.1 but < 7.0 mmol/l, and 2-h post glucose load < 7.8 mmol/l
	ADA	FPG > 5.6 but < 7.0 mmol/l
Impaired glucose tolerance (IGT)	WHO	FPG < 7.0 mmol/l and 2-h post glucose load > 7.8 but < 11.1 mmol/l
	ADA	No equivalent
Diabetes mellitus	WHO	FPG > 7.0 mmol/l or 2-h post glucose load > 11.1 mmol/l
	ADA	FPG > 7.0 mmol/l

Key to Table 4.6: FPG = fasting plasma glucose (NB: there is an approximate 15% difference between plasma and whole blood glucose values. The period of fasting is overnight, usually about 12 h in total). 2-h post glucose load: 75 g of glucose dissolved in approximately 300 ml of water and consumed within 5 min. This test has now replaced the original $2^{1}/_{2}$ h oral glucose tolerance test in which blood samples were taken every 30 min throughout the period of the test.
Note: Plasma glucose values were originally defined in units of mg/dl (= mg/100 ml); molecular weight of glucose = 180, so 100 mg/dl = ~5.6 mmol/l or 5 mmol/l = 90 mg/dl.

ketoacidosis. Acetoacetate and hydroxybutyrate lead to the metabolic acidosis and contribute to abnormalities in fluid and electrolyte balance that occur as the patient excretes large volumes of urine; an osmotic diuresis due to glucose in the glomerular fluid. Coma and death may follow if the subject is not treated with insulin injection.

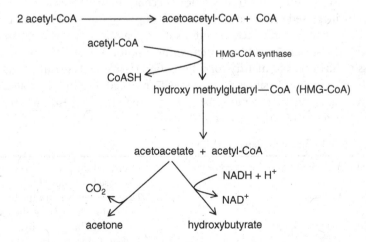

Figure 4.24 Ketogenesis: formation of acetoacetate, hydroxybutyrate and acetone

Peripheral tissue resistance to the action of insulin as evidenced by impaired glucose tolerance is a marker of possibly serious underlying metabolic disease leading to type 2 DM and cardiovascular disease (CVD). As peripheral tissues become less sensitive to the hormone; pancreatic insulin production increases to compensate, leading to a state of chronic hyperinsulinaemia. Eventually however, the pancreas fails to meet the increasing demand for insulin and the hyperinsulinaemia is not sufficient to maintain a normal blood glucose concentration. The plasma concentration of free fatty acids rises as the anti-lipolytic effect of insulin, mediated through hormone sensitive lipase in adipocytes, is lost. Clearly, if there is insulin resistance in the presence of hyperinsulinaemia, the molecular defect must lie in the hormone receptor or signal transduction mechanism. Table 4.7 lists some of the underlying biochemical defects known to contribute to insulin resistance.

The link between insulin resistance, hyperinsulinaemia, obesity and CVD is crucial. Albrink and her colleagues (1980) noticed coincidence of quantitative changes in plasma lipids and lipoproteins with obesity. Later, Reaven and his co-workers (1998) reported that hyperglycaemia, hyperlipidaemia and hypertension tended to associate together in subjects with coronary artery disease. This combination of findings became known as Reaven syndrome, Syndrome X or now more commonly as the metabolic syndrome and the coincident finding of these symptoms is interpreted as a 'high risk' factor for cardiovascular disease. Additional diagnostic criteria for the metabolic syndrome, and therefore CVD, include abdominal obesity (waist circumference of greater than 102 cm in men and greater than 88 cm in women), elevated plasma fasting triglyceride concentration (above 7 mmol/l), a low concentration of plasma HDL and evidence of renal glomerular damage.

As explained in Chapter 9, adipose tissue is not just a depot of triglycerides, but a metabolically important organ system which secretes a number of signalling molecules. During periods of weight gain leading to obesity, adipose tissue exhibits an inflammatory nature characterized by increased secretion of adipocytokines. Generalized insulin resistance due to a failure in the transduction mechanism may be exacerbated in obese individuals by increased secretion of leptin and adipocytokines, such as interleukin-6 and tumour necrosis factor, which inhibit insulin signalling in peripheral tissues. Also, a protein called C-reactive protein, an emerging risk marker of CVD is produced by adipose tissue during inflammatory conditions; CVD is just such an inflammatory state (see Chapter 5). Finally, obesity is associated with reduced insulin-stimulated tyrosine kinase activity and impaired binding between p85 of PI3K and IRS, whereas weight reduction helps to correct these defects.

Table 4.7 Insulin signalling defects

Decreased insulin binding to the receptor	Decreased numbers of receptors
	Decreased receptor affinity
	Defective synthesis of the receptor
	Autoantibodies against the receptor
Decreased signal transduction	Decreased kinase activity
	Post-receptor defects

Details of plasma lipoproteins and their metabolism are given in Section 5.5. Most of the cholesterol in the blood is carried as part of low density lipoprotein (LDL) or high density lipoprotein (HDL), whereas most triglyceride, in the fasting state, is carried by very low density lipoprotein (VLDL). The relative concentrations of these lipoproteins constitute the 'lipid profile' and determine CVD risk. Diabetics are more likely to show an unhealthy profile with elevated concentrations of LDL and triglyceride but reduced HDL concentration. This pattern can be partly explained by enhanced fatty acid liberation from adipocytes as a consequence of insulin resistance in that tissue and due to reduced removal from the circulation of triglycerides, which is also insulin dependent.

Most attention on insulin focuses on its actions in controlling fuel metabolism in liver, adipose tissue and muscle in particular, insulin also has direct actions on endothelial cells lining blood vessels and evidence indicates that hyperinsulinaemia promotes atherogenesis. Normally, insulin stimulates the production of nitric oxide by endothelial cells by activating nitric oxide synthase, (NOS) through the PI3K/AKT pathway. Nitric oxide helps to control blood pressure and also regulates blood clotting by its action on platelets. Insulin resistance will lead not only to a reduction in NO production but because the MAPK transduction pathway is functional, an increase in endothelin-1, a peptide which causes vasoconstriction, is promoted. Also, by impairing the effect of the Na^+/K^+ ATPase pump on vascular smooth muscle cells, insulin resistance makes those cells more sensitive to the actions of vasoconstrictors angiotensin and noradrenaline (norepinephrine).

In summary, results from several studies in to the effects of impaired insulin action can now be linked to CVD. Because of the numbers of people exhibiting sings of insulin resistance, there has been a lot of research effort has gone in to finding suitable therapeutic interventions. One group of drugs in particular, the thiazolidinediones (also known as glitazones) has attracted attention because they both increase insulin sensitivity and help to normalize plasma lipid concentrations.

Chapter summary

Intercellular communication is crucial for any multicellular organism to integrate its activities to achieve homeostasis. Chemical messengers emanating from the endocrine system, the nervous system or tissues such as white blood cells and the vasculature influence the actions of other cells. Structurally, the messengers fall into very few major groups, for example peptides, amines, steroids. Their turnover (synthesis, catabolism and clearance) is usually tightly regulated but often serious pathology can arise when the control is lost. Most signalling molecules recognize a receptor on or in the target cells and the effects are mediated by transduction mechanisms which may involve second messengers such as cAMP, cGMP or IP3, all of which are generated by G-protein involvement which influence the activity of enzyme-based pathways. Insulin, growth factors and cytokines do not use G-proteins but rely on tyrosine kinase activity to mediate their effects, whilst steroids, vitamin D3 and thyroid hormones exert their effects directly on DNA transcription. Diabetes is the commonest metabolic disease and one whose incidence is on the increase. Because of the diverse effects insulin has on overall fuel metabolism, most tissues of the body are affected by its inadequacy.

Case notes

1. **Hypoglycaemia**

    Hypoglycaemia (low blood glucose concentration) is not a disease but a symptom of an underlying pathology. The actual value of blood glucose concentration which constitutes 'low' is equivocal and variable, for example neonates are more tolerant of low values than children or adults. Noticeable effects of hypoglycaemia are due to neuroglycopaenia (low glucose concentration in the CNS) and tend to occur when there has been a rapid change in glucose concentration.

    a. Christine M developed glucose intolerance during her pregnancy, classified as so-called gestational diabetes. Her condition was well managed but for most of the pregnancy, her blood glucose concentration was a little higher than 'normal'. Her pregnancy was in all other respects uneventful and she delivered a son. *In utero* homeostasis is maintained by the mother so the baby had been subject to, and had become adapted to, the same elevated glucose concentration as had Christine. After birth, neonatal blood glucose concentration falls as the baby adapts to independent life, the lowest value, which may be as low as 2 mmol/l, is normally reached at about 3 h before the effects of glucagon and adrenergic stimulation begin to raise the concentration to normal.

    Christine's son appeared initially quite normal but later he became jittery and had an episode of fitting. Blood glucose concentration at the time of the fit was only 1.6 mmol/l, which accounted for the symptoms of neuroglycopaenia. Infants born to diabetic mothers are more prone to hypoglycaemia than those born to non-diabetics, possibly due to defects in the glucose sensor mechanisms, resulting in a much larger swing in glucose concentration in the pre- and post-natal periods. A significant proportion of women who exhibit gestational diabetes subsequently develop type 1 or type 2 diabetes within 10 years.

    b. Mr D., 48, consulted his general practitioner because he had suffered a number of recent incidents of dizziness, disorientation and what he described as mild panic attacks. The symptoms were unrelated to any particularly 'stressful' situation, but had occurred on two occasions soon after waking and once after exercise. Mr D.'s blood pressure was normal. Routine investigations revealed no abnormality.

    A diagnosis of insulinoma (insulin secreting tumour of the β-cells) was made when it was shown that Mr D. had elevated plasma insulin and C-peptide concentrations during a period of fasting; insulin is normally secreted in response to feeding. Subsequently, the site of the tumour was identified and its surgical removal proved effective in alleviating the symptoms. Insulinomas may have multiple locations making surgery difficult, but this type of tumour is benign.

    c. Adam Smith was a 24-year-old working in the finance industry in the City of London. Following a particularly long and hectic day at the bank, Adam went with colleagues to a bar for a social celebration during which, it was later estimated, he drank nearly 10 units of alcohol. As the evening progressed, he complained of feeling dizzy and confused, his speech became somewhat slurred and he began to perspire profusely. His friends assumed he was intoxicated and tried to take him out of the bar. At this point, Adam collapsed and was taken to the local Accident and Emergency department. The initial assumption of the medical staff was that he had simply drunk too much and should 'sober- up', however this view changed when a blood test revealed a low glucose concentration (2.3 mmol/l, compared with a typical value of approximately 5 mmol/l) and evidence of dehydration. Adam's condition improved following treatment with a dextrose-containing intravenous solution.

It transpired that Adam had eaten nothing within the 18 h prior to his arrival at A&E. This length of time of fasting would result in an almost complete exhaustion of hepatic glycogen, so normally the liver would maintain blood glucose concentration by the process of gluconeogenesis (described in Section 6.5), but metabolism of the 'binge' quantity of alcohol consumed upset the NAD^+/NADH ratio in the liver and as a result, hepatic glucose synthesis and release were severely impaired and Adam showed signs of acute neuroglycopaenia. Like the case of Mr D., this is an example of fasting hypoglycaemia but unlike Mr D. this has no real pathological basis.

2. **Addison's disease**
   Named after Thomas Addison who first described the clinical condition in the mid 1850s, Addison's disease is one of the commonest endocrinopathies. At one time, most cases of Addison's were due to infection, usually by tuberculosis, of the adrenal cortex but nowadays the likely cause is autoimmune destruction of the tissue, and may be associated with dysfunction of other endocrine glands.

   Mrs Peters was a 48-year-old who was referred to a specialist endocrinologist at her local hospital. She had experienced irregular menstrual periods (which she assumed signalled onset of early menopause), bouts of depression, weight loss and fatigue. A week before her first scheduled appointment at the endocrine clinic, she became unwell at home (diarrhoea and sickness) and was taken to A&E. The doctor who attended Mrs Peters noticed patches of skin pigmentation. Investigation revealed that she was hypotensive (low blood pressure), hypoglycaemic, hyponatraemic (plasma sodium 129 mmol/l, reference range 135–145 mmol/l) with a slightly raised plasma potassium (5.0 mmol/l, reference range 3.6–4.7 mmol/l) and raised plasma urea. Treatment with steroids was commenced.

   The biochemical picture is typical of an Addisonian crisis. Hypoglycaemia is due to lack of cortisol (an insulin antagonist; raises blood glucose during fasting); skin pigmentation is due excessive melanocyte stimulating hormone output from the pituitary arising from lack of negative feedback exerted by cortisol on ACTH production (ACTH and MSH are produced from the same pro-opiomelanocortin gene transcript). This indicates a primary cause, that is a defect within the adrenal cortex rather than due to a lack of pituitary ACTH. The sodium and potassium values are probably due to a lack of aldosterone although cortisol does have a very mild mineralocorticoid action.

3. **Thyrotoxicosis**
   Hyperthyroidism is a fairly common complaint which may occur at any age, but affects mainly females. There are several causes of increased thyroid hormone secretion but only rarely is the condition due to non-thyroidal (e.g. pituitary) illness.

   The patient, a 66-year-old woman, who was generally fit and well but had noticed episodes of restlessness, uncharacteristic irritability, muscle weakness and tremor, and weight loss.

Clinically she had a fixed stare and swelling at the front of the neck (goitre). The diagnosis of thyrotoxicosis was confirmed by the results of a blood test which showed an undetectable concentration of TSH but markedly raised T_3 result. The very low TSH is due to a normal negative feedback response to the raised tri-iodothyronine concentration.

This pattern of findings, especially the eye signs, indicates Graves' disease. This arises due to the presence of an antibody which mimics the action of TSH by binding to receptors on the thyroid. In contrast to case 2 above, here the antibodies are stimulatory rather then destructive and bring about inappropriate secretion of T_3 and T_4.

4. **Parkinson's disease**

This is a neurodegenerative condition in which a large proportion, up to 80%, of the cells located within the substantia nigra region of the brain are damaged or destroyed. The actual cause of the cell loss is unknown, but the possible involvement of unidentified toxins has been suggested. The onset of typical Parkinson's disease is typically after the age of about 60 years and the incidence rises to approximately 0.5% of all individuals over the age of 70. There is a rarer and more 'aggressive' form of the condition which may appear in much younger subjects.

Mrs Gordon had noticed that her husband, aged 68, had developed a slight tremor in his left-hand, that his movement had become slow and stiff (described technically as akinetic-rigidity) and he found increasing difficulty in rising from a chair or from bed. These symptoms grew more severe and, in time, the clarity of Mr Gordon's hand-writing began deteriorate and his walking became more shuffling than striding.

The pathology of Parkinson's disease is associated with a substantial reduction in neurotransmitters such as dopamine, 5-hydroxytryptamine, GABA, in the brain. Treatment is based on the use of replacement of dopamine or dopamine agonists which relieve the rigidity but do not affect a cure. Currently, there is no laboratory-based diagnostic test for PD and so diagnosis is based on clinical presentation alone.

5

Biochemistry of the blood and the vascular system

Overview of the chapter

Blood is mostly water in the form of plasma, but the cellular components may reasonably be seen to constitute a tissue even though they do not form a solid organ. Blood vessels regulate blood pressure in response to hormonal and neural signals and the endothelial cells that line the vessels synthesize a number of important local hormones such as nitric oxide and prostanoids. All blood cells are produced from the same origin before dividing into eryhtropoietic and leucopoietic lines. Haemoglobin within erythrocytes (red blood cells) is not only a transport protein for oxygen but also an effective pH buffer. White blood cells are involved with defence mechanisms through phagocytosis and secretion of chemical signals. Lipoprotein particles are transported through the blood plasma and are intimately implicated with one of the major pathologies of the twenty-first century, cardiovascular disease such as atherosclerosis.

Key pathways

Nitric oxide and eicosanoid synthesis; haem synthesis. The importance of the pentose phosphate pathway reduced glutathione in maintaining red cell integrity. The respiratory burst in phagocytes. Clotting and complement enzyme cascades. Metabolism of lipoproteins.

5.1 Introduction

Blood is a unique tissue in that it is (obviously) a fluid. A typical 70 kg adult male has about 5 l of whole blood coursing his vasculature at any time. The dynamics of the blood-vascular system are impressive: many millions of cells are released from the bone marrow each day; a red cell will travel over 250 miles (400 km) during its relatively short life span (120 days), and an endothelial cell lining the lumen of a blood vessel will

Essential Physiological Biochemistry: An organ-based approach Stephen Reed
© 2009 John Wiley & Sons, Ltd

survive the constant pounding of the blood surging through the vascular system and remain functional for many months or even years.

Blood is the transport medium of the body. Plasma, which accounts for approximately 60% of the total volume, carries a wide range of small and medium-sized metabolites; some are simply dissolved in solution (93% of the plasma is water), others are carried by specific carrier proteins. The chemical composition of the plasma is complex and reflects the chemical composition inside cells, which is why blood tests are so commonly used in diagnosis to 'see' the biochemical events occurring in tissues. The formed cellular elements of the blood perform several functions: defence against blood loss from bleeding (platelets, also called thrombocytes), defence against infection and immune surveillance (white cells, leucocytes), and gas transport and pH buffering (red cells, erythrocytes).

Haematopoiesis, the formation of the cellular components of the blood, the red and white blood cells and the platelets occurs in the bone marrow of the adult. Red cell production also takes place in the liver in the fetus from approximately 2 months of gestation until about 15 days after birth. Despite the great variation in structure and function of the blood cells, they all arise from the same haematopoietic stem cell (HCS) line (Figure 5.1). Development and maturation of the individual cell lineages is a complex process relying on appropriate stimulation and regulation by humeral (soluble hormone-like) factors and by direct cell-cell interactions. Blood cell proliferation and differentiation are responsive processes allowing the body to meet changing demands arising from, for example, anaemia or infection.

The humoral factors include numerous cytokines and growth factors, which act as classical endocrine, paracrine or autocrine regulators. Developing cells are sensitive to a range of such factors at all times during their maturation and, in the case of white blood cells, during their existence within the circulation. To ensure responsiveness during their development, cells must express on their surface a variety of receptors at different times.

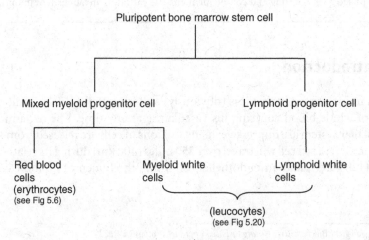

Figure 5.1 Blood cell maturation

5.2 The blood vascular system

5.2.1 The heart

The four chambers of the heart act as a 'double pump' creating dual, functionally separate circulations. First, deoxygenated blood from tissues enters the right side of the heart and is pumped into the pulmonary circulation where it exchanges carbon dioxide for oxygen in the lungs. Oxygenated blood returns to the left atrium and then passes to the left ventricle, from where it is pumped via the aorta through the systemic circulation to perfuse all tissues. Typically, in 1 min, the heart beats 70 times expelling around 5 l of blood, which is the total blood volume of a 70 kg adult male. In a life time, the heart pumps an estimated 250–300 million litres of blood around the body.

Histologically, the myocytes of the heart are branched, striated muscle cells. As would be expected, the cardiac cells contain numerous mitochondria, reflecting their dependence on aerobic metabolism, predominantly from fatty oxidation. Contraction of heart muscle is spontaneous, requiring no external neurological or hormonal signal but initiated by so-called 'pacemaker cells'. However, control of cardiac pumping allowing adaptation to circumstances is achieved via neural pathways. Noradrenaline (norepinephrine) released from postganglionic sympathetic fibres, increases both the rate and contractility of the heart whilst acetylcholine from the vagus nerve (part of the parasympathetic system) can slow the heart rate to as few as 35 beats per minute.

In addition to its pump function, the heart is also a secretory organ. Cardiac cells produce two small peptides, the natriuretic factors, which oppose the vasoconstrictive actions of noradrenaline (norepinephrine) from the sympathetic nervous system and of the peptide angiotensin II. By causing vasodilation and natriuresis (increased excretion of sodium in the urine), atrial natriuretic peptide (ANP) secreted from the atria and B-type natriuretic peptide (BNP) secreted by both atria and probably more significantly, from the ventricles, reduce blood pressure. The stimulus to secretion of natriuretic peptides is wall stretch of the chambers of the heart, indicating volume and pressure overload of the vascular system. A third member of the natriuretic peptide family, CNP, is secreted by endothelial cells.

Both ANP and BNP are synthesized as prohormones within the cardiac myocytes. Prior to its release, pro-ANP is stored in secretory granules and cleaved at the time of secretion by a protease called corin into an active peptide of 28 amino acids and an inactive N-terminal fraction of 98 amino acids. Similarly, the active BNP peptide consisting of 32 amino acids is co-secreted along with its inactive N-terminal fragment. Cleavage of the BNP prohormone occurs at the time of release and so both the N-terminal portion (1–76) of the prohormone and the active BNP (residues 77–108) can be found in plasma. If BNP and N-terminal BNP are present in abnormally high concentrations, doctors may make a diagnosis of heart failure, a weakening of the pumping mechanism, not to be confused with a heart attack.

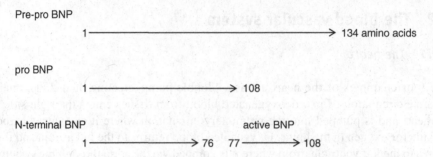

Structurally, the three NPs are alike in that they contain a 17-amino acid 'loop' with a conserved sequence (see also Figure 8.8).

5.2.2   Blood vessels

Blood is pumped away from the heart through arteries; it permeates the tissues through networks of very small capillaries where nutrient delivery, gas exchange and waste removal occur and is finally returned to the heart via the veins. The structures of the arteries and veins differ in important ways. First, the veins have one-way valves which prevent the back-flow of blood and second, the walls of the arteries are much thicker, due largely to the layer of smooth muscle cells. Both types of vessel are lined on their inner surface with endothelial cells. Refer to Figure 5.2.

5.2.2.1   Endothelial cells (the endothelium)
The endothelium is quite remarkable. Only one cell thick with a relatively slow turnover, it is able to resist the pressure of the blood flowing within the vessels and to maintain a smooth surface permitting non-turbulent blood flow. Endothelial cell precursors arise in the bone marrow and express on their luminal surface a range of functionally important proteins including many which belong to the CD group of antigens and others such as adhesion molecules. The endothelium forms a physical barrier between the bloodstream and the tissues, but for white cells to counter infections in tissues it is clearly necessary for them to leave the circulation and enter the 'crime scene' of damaged tissue. In order to allow firstly arrest and then extravasation (movement out of the vascular system) of white cells, the endothelium expresses a number of proteins known collectively as adhesion molecules. For example, the selectins are calcium-dependent carbohydrate binding proteins, expressed not only on endothelium (E-selectin) but also on platelets (P-selectin). The carbohydrate recognition domain (CRD) of selectins is able to bind specifically with particular sugars found in glycoproteins which are part of the exterior surface of leukocytes. Blood cells are normally carried rapidly with the flow of blood, but when a cell has been 'captured' by selectin, it begins to roll slowly over the surface of the endothelium. A rolling white cell is brought to a halt when integrin proteins (another type of adhesion molecule) on its surface bind with recognition proteins on the endothelium; this is followed by extravasation by binding to intercellular adhesion molecule 1 or

(a) an artery

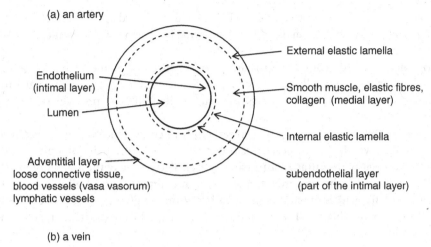

External elastic lamella

Endothelium
(intimal layer)

Smooth muscle, elastic fibres,
collagen (medial layer)

Lumen

Internal elastic lamella

Adventitial layer
loose connective tissue,
blood vessels (vasa vasorum)
lymphatic vessels

subendothelial layer
(part of the intimal layer)

(b) a vein

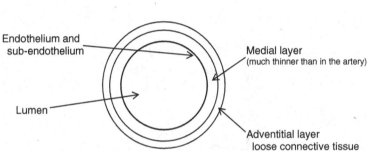

Endothelium and
sub-endothelium

Medial layer
(much thinner than in the artery)

Lumen

Adventitial layer
loose connective tissue

Figure 5.2 Cross section through blood vessels (a) an artery. Sizes of arteries vary from 25 mm diameter with 2 mm wall thickness in the aorta to 20 µm diameter with a 15 µm wall thickness in the smallest arterioles. (b) a vein. Note there are no elastic lamellae in veins so tension is maintained by elastic fibres which are arranged within the media. Typically, veins are approximately 5 mm in diameter but with a very thin wall, for example 0.5 mm. The vena cava is the largest vein at 30 mm diameter

2 (ICAM-1, ICAM-2) and vascular cell adhesion molecule (VCAM). Platelets are attracted to damaged endothelium where they adhere to prevent blood loss in a similar fashion to white blood cells, i.e. via adhesion molecule interactions, to form a clot (thrombus).

Physiologically, the endothelial cells serve more than just a barrier function. They are responsive to a number of chemical messengers such as cytokines and the expression of the adhesion molecules described above is increased (upregulated) by for example interleukin-1 or tumour necrosis factor when there is damage to the vessel or in the presence of infection. Furthermore, the endothelium secretes locally acting messengers such as prostaglandins, nitric oxide (NO) and peptides CNP and endothelin-1, which help to maintain vascular homeostasis by influencing important processes such as blood pressure, coagulation and cell proliferation. Much of our basic understanding of the structure and roles of prostaglandins (PG) can be attributed to the research work of Sir John Vane and associates in the 1970s. Somewhat later, Moncada identified NO as a vasoactive agent secreted from the endothelium.

Originally isolated from the prostate gland (hence the name *prosta-gland*-in), prostaglandins are chemically simple fatty acid-like molecules (Figure 5.3). Structural differences give rise to several series of structurally different prostaglandins, for example PGE, PGF and thromboxane.

Prostaglandins are a subgroup of a larger family of compounds known collectively as eicosanoids, which are synthesized from arachidonic acid (arachidonate); this is a 20-carbon omega-6 unsaturated fatty acid (C20:4). The source of the arachidonic acid for PG synthesis is the cell membrane. Most membrane phospholipids have an unsaturated fatty acid as arachidonate at carbon 2 on the glycerol 'backbone' to help maintain membrane fluidity. The arachidonic acid released from the membrane by the

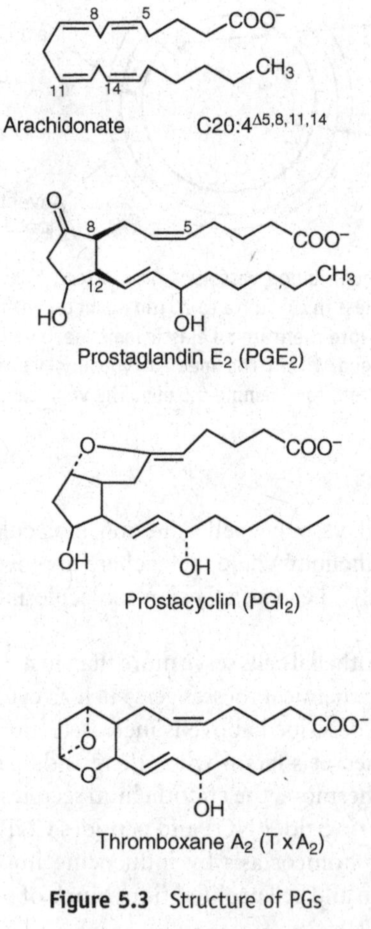

Arachidonate $C20:4^{\Delta 5,8,11,14}$

Prostaglandin E_2 (PGE$_2$)

Prostacyclin (PGI$_2$)

Thromboxane A_2 (T x A$_2$)

Figure 5.3 Structure of PGs

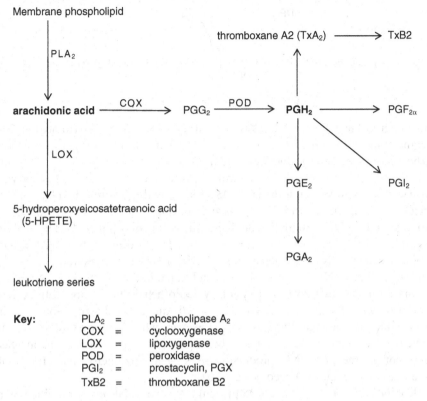

Figure 5.4 Eicosanoid synthesis

action of the enzyme phospholipase A2, is the substrate for cyclo-oxygenase (COX). Endoperoxides (PGG_2 and PGH_2) synthesized by COX are subsequently converted into prostacyclin by a specific synthase in endothelial cells or into thromboxane within platelets. A subscript numeral, for example PGE_2, indicates the number of double bonds present in each molecule (Figure 5.4).

The various prostaglandins have a diversity of actions in different tissues. Importantly in the blood vascular system, prostaglandins are involved with clotting and prevention of blood loss following injury. Prostacyclin (PGI_2) has anticoagulant properties and along with nitric oxide (see next section) is a vasodilator whilst thromboxane (TxA_2) derived mainly from platelets is a procoagulant. Blood coagulation is outlined in Section 5.4.

NO (molecular weight $= 30$) is small but plays a big role in physiological regulation, not least in the vasculature where its effects were first seen (see Chapter 4). Endothelium-derived relaxation factor (EDRF) was discovered its ability to cause dilatation of vessels by relaxing the arterial muscle layer. Only much later was EDRF discovered to be a gas, nitric oxide. More recent interest in NO is based on the evidence that it is anti-atherogenic. The pathogenesis of atherosclerosis is complex but many of the known effects of NO can be implicated in this common and serious condition.

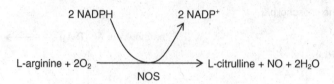

Figure 5.5 Nitiric oxide synthase reaction

As shown in Figure 5.5, nitric oxide is synthesized from the amino acid arginine under the influence of nitric oxide synthase (NOS). This enzyme exists in several tissues as either a constitutive or inducible protein (iNOS).

The term NOS is used to denote a family of three related but distinct isoenzymes: neuronal NOS (nNOS); endothelial NOS (eNOS, endothelium and platelets) and inducible NOS (iNOS, endothelium, vascular smooth muscle and macrophage). In addition to reduced nicotinamide adenine dinucleotide phosphate (NADPH) shown in Figure 5.5, NOS enzymes also require flavin adenine dinucleotide (FAD), flavin mononucleotide (FMN) and tetrahydrobiopterin (BH4) as coenzymes.

Although all three NOS isoforms have been located in the vasculature, nNOS is the least important in regulating vessel physiology. Endothelial NOS is constitutive (always active), producing NO in nanomolar (10^{-9} mol/l) quantities, having 'housekeeping' roles maintaining vascular homeostasis. However, when iNOS expression is up-regulated by the effects of, for example, cytokines or bacterial toxins causing increased transcription of the gene, NO production rises in to the micromolar (10^{-6} mol/l) concentration range and NO becomes cytotoxic.

Unlike iNOS and nNOS, the eNOS protein is post-translationally modified by the attachment of fatty acids, myristate or palmitate. This modification is important because the fatty acids help to attach the enzyme, in an inactive form, to the inner face of plasma membrane of endothelial cells or platelets. Several mechanisms serve to release eNOS from its membrane bound state and thus activate the enzyme.

Being a gas, NO is freely diffusible and penetrates cell membranes easily. NO is produced by and acts within the endothelium and platelets but is also a paracrine hormone targeting vascular smooth muscle cells (VSMC) and white blood cells.

A number of the actions of NO in target cells may be explained by its binding to the haem-containing protein soluble guanylyl cyclase (sGC), an enzyme which generates a second messenger, cyclic GMP (cGMP) from GTP.

NO-induced production of cGMP is fully or partly responsible for:

- vasorelaxation (the EDRF effect) in VSMC;

- inhibition of proliferation of VSMC and endothelial cells;

- inhibition of platelet adhesion and aggregation on endothelium;

- inhibition of endothelial apopotosis (at low NO concentration).

In its turn, cGMP regulates the cytosolic Ca^{2+} concentration by controlling the entry of Ca^{2+} via ion-specific channels or by preventing the release of Ca^{2+} into the cytosol from reservoirs such as the sarcoplasmic reticulum in muscle cells or the endoplasmic reticulum in many other cell types.

Examples of cellular effects by NO which are *not* related to cGMP include:

- Inhibition of leucocyte adhesion to endothelium. By acting on the gene transcription factor called nuclear factor kappa B (NFκB), NO limits the expression by endothelial cells of monocyte chemotactic protein-1 (MCP-1) and VCAM-1. As both MCP-1 and VCAM-1 are involved with pro-inflammatory responses, NO is in this manner an anti-inflammatory agent;

- Antioxidant effects. Superoxide ($O_2^{\bullet -}$) is a free radical which contributes to 'oxidative stress', a process which causes significant damage to the molecular structure of cells. NO induces the synthesis of endothelial cell superoxide dismutase (ecSOD) the enzyme which forms hydrogen peroxide H_2O_2 from $O_2^{\bullet -}$. A positive feedback loop is set up as H_2O_2 induces eNOS which produces more NO and so on. This is a protective action as superoxide which is potentially damaging to cells is 'buffered' by NO;

- Pro-oxidant effects. The antioxidant effect described above occurs when NO concentrations are low. If the concentration of NO rises, rather than reducing the damaging effects of $O_2^{\bullet -}$ the two may combine to form an even more powerful oxidant called peroxynitrite ($ONOO^-$);

- Mitochondrial function. NO is able to react with transition metals such as iron, including those contained within haem groups. Even at low NO concentrations there is competition between oxygen and NO for reversible binding to cytochrome c oxidase. If mitochondrial O_2 is low respiration slows, which may confer anti-apoptotic benefit to the cell. As NO concentration rises and peroxynitrite is formed, electron transport is irreversibly inhibited, there is increased production of superoxide and other reactive oxygen species and apoptosis occurs.

5.2.2.2   *Vascular smooth muscle cells (VSMC)*

These cells, located in the medial layer of the vessel wall, are normally relatively non-proliferative, but they secrete proteins which make up the extracellular matrix of the vessel wall. However, the narrowing of the arteries, a critical, often fatal, consequence of atherosclerosis is due to proliferation of VSMC and their migration into the intimal layer of the vessel wall.

Vascular smooth muscle cells are a major contributor to the maintenance of blood pressure. If blood pressure falls, blood flow becomes sluggish and tissues may become starved of oxygen (hypoxia) and other essential nutrients, whilst high blood pressure (hypertension) causes damage to a number of tissues including the heart itself, the brain and the kidneys. Vessel tone due to VSMC contraction is regulated by a number of external factors, notably adrenaline (epinephrine), angiotensin II and, as described above, NO. An imbalance of each of these has been implicated in impaired regulation of blood pressure leading to damaging hypertension.

Hypertension is a well known contributor to a number of diseases, but hypotension can be equally dangerous as tissues rely upon pressure to maintain a constant delivery of nutrients via the micro-circulation of fluid around individual cells. As pressure and volume are interdependent, blood pressure is controlled by either changing the internal diameter of the vessels (vasoconstriction or vasodilation) or by increasing the fluid volume contained within the vasculature.

- **Direct neural control**
  Adrenaline (epinephrine)-producing (adrenergic) and acetylcholine secreting (cholinergic) neurones of the autonomic nervous system have direct and complimentary effects on the tone of blood vessels.

- **Renin–angiotensin–aldosterone (RAA) system (see also Chapter 8)**
  Specialized cells, called the juxtaglomerular apparatus (JGA), within the renal cortex are able to detect a fall in blood pressure and respond by secreting a proteolytic enzyme called renin (not to be confused with rennin). The substrate for renin, a liver-derived peptide called angiotensinogen, circulates in the plasma. Renin removes two amino acids from the N-terminal to produce angiotensin I, which is itself a substrate for angiotensin-converting enzyme (ACE). ACE removes two more amino acids to produce angiotensin II.

Angiotensin II has two effects: first, as a vasoconstrictor acting via receptors on vascular smooth muscle cells, and second, it stimulates the adrenal cortex gland to produce aldosterone (a mineralocorticoid steroid hormone, see Chapter 4). Aldosterone promotes the reabsorption of sodium from the renal tubule into the bloodstream and the resulting increase in osmolality (osmotic potential) of the blood causes water reabsorption in the nephrons. The outcome is an increase in blood volume and, therefore, pressure which inhibits (by negative feedback) further renin secretion from the JGA.

5.3 Circulating blood cells

5.3.1 Red cells and haemoglobin

Erythrocytes are unique among human cells in that they do not contain a nucleus or any other major subcellular organelle. The absence of such intracellular structures means that the volume of a red blood cell (typically 90 fl; femtolitres 10^{-12} l) is able to maximize its carriage of haemoglobin and the essential enzymes required to maintain the red cell during its brief lifespan of approximately 4 months. It would be wrong however to consider erythrocytes to lack any subcellular or molecular organization. Their intricate cytoskeletal architecture consisting of, among others, α and β spectrin filaments, ankyrin, actin and protein 4.1, maintains the flexible biconcave disc shape of the mature cell. The flexibility of the membrane allows the cells (approximately 7 μm in diameter) to contort in order to flow through even the narrowest capillaries (4 μm diameter).

Red blood cells are amongst the most numerous of the human cell lines; an average healthy 70 kg male having a total of approximately 25×10^{12} cells in his 5 l of blood. A typical red cell contains in excess of 600 million haemoglobin molecules which equates to a total of about 300 g of haemoglobin, an amount that is far greater than for any other protein in the body. The lack of a nucleus clearly indicates that red cells cannot divide and at the end of their life, 'worn out' RBCs are removed by the cells of the reticuloendothelial system. Approximately 2% (5×10^{11}) of the red cell number are removed and replaced by new ones *each day*. Haem synthesis is outlined later in this chapter and its catabolism is discussed in Chapter 6.

5.3.1.1 *Erythropoiesis*

The early cells of the erythrocyte development sequence are 'typical' cells in that they have a complex internal arrangement of organelles. The proerythroblast (= pronormo-blast), the first identifiable form of the red cell lineage within the bone marrow, has a large nucleus with nucleoli and a dark staining cytoplasm, characteristic of cells which undertake extensive protein synthesis. These progenitor cells also contain mitochondria which play an important role in the synthesis of haemoglobin as well as energy (ATP) production. As the cells mature, they reduce in size (from about 12 µm to about 7 µm diameter) and lose their internal organelles. Karyohexis, the extrusion of the nucleus, occurs during the late normoblast stage which is just when the enzymes that bring about haemoglobin synthesis are at their peak. The reticulocyte, characterized by stippling within the cytoplasm due to remnants of RNA, is the penultimate form before the mature erythrocyte is formed.

Erythropoiesis is very carefully regulated, partly by the action of a number of humoral growth factors, including most notably the glycoprotein hormone erythro-poietin (EPO). EPO has 165 amino acids in its primary structure suggesting a molecular weight of less than 20 000. In fact, the hormone has a molecular weight in excess of 30 000. The difference between the two figures reflects the extensive glycosylation of the protein.

Erythropoietin is synthesized mainly in the peritubular cells of the renal cortex. These cells contain a protein called hypoxia-inducible factor 1 (HIF-1) and are able to 'detect' (exactly how is not known) the oxygen content of the blood perfusing the kidney. HIF-1 enhances the transcription of the EPO gene so generating more, or less, EPO as physiological circumstances demand. The targets for EPO are erythrocyte progenitor colony-forming units (CFU_E) and proerythroblasts (Figure 5.6); the later (more mature) cells in the sequence lack the EPO receptor.

In common with other protein hormones, especially those associated with cell growth and differentiation, EPO signals through cell surface tyrosine kinase-linked receptors (EPO-R), illustrated in Figure 5.7. Hormone binding to the EPO-receptor (EPO-R) initiates an intracellular cascade of tyrosine phosphorylations, which arise from the activation of EPO-R bound JAK2 (see Section 4.7 for further details) Cellular growth and division are stimulated by an intracellular cascade comprising proteins such the Ras/Raf/MAPK combination and STAT5.

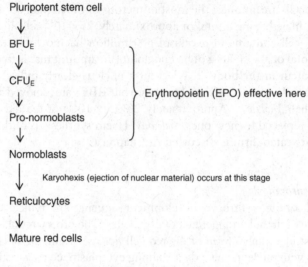

Abbreviations: BFU$_E$ = burst forming units, erythrocyte lineage
CFU = colony forming units

Figure 5.6 Erythropoiesis

Signalling by EPO also rescues erythroid progenitors from cell death (apoptosis) so increasing the number of RBC forming cells. Furthermore, EPO stimulates the production of haem and the globin chains required to synthesize haemoglobin. In short, EPO increases the number of erythrocytes and their haemoglobin content, not surprisingly therefore erythropoietin can be used to treat anaemia (literally '*without blood*'). Genetic engineering techniques allow fairly large quantities of the hormone to be manufactured. The synthesized protein is also used by dishonest sportsmen or women to boost their athletic performance.

There are many causes of the clinical condition referred to as anaemia. One particular type, whose cause can be traced to a genuine metabolic defect is megalo-blastic anaemia and is due to a deficiency of the vitamins B$_{12}$ (cobalamin) and/or folate. These vitamins are required for normal cell division in all tissues, but the rapid production of red cells makes them more susceptible to deficiency. In megaloblastic anaemia the blood haemoglobin concentration falls the synthesis of haem is not impaired. Examination of the blood reveals the appearance of larger then normal cells called macrocytes and megaloblasts are found in the bone marrow.

Very small amounts of cobalamin are required each day (<5 µg) and the diet normally provides plenty more than the minimum, so dietary B$_{12}$ deficiency is uncommon, except in very strict vegetarians. Pernicious anaemia arises when a defect in the stomach results in too little secretion of a protein called intrinsic factor, without which, cobalamin cannot be absorbed in the ileum of the small intestine.

Like many vitamins, cobalamin is functionally active as a derived coenzyme, coenzyme B$_{12}$. Structurally, this is composed of a corrin ring; a haem-like porphyrin ring containing cobalt (Co^{3+}) at the centre held by four coordination bonds. The fifth

(a)   unstimulated cell

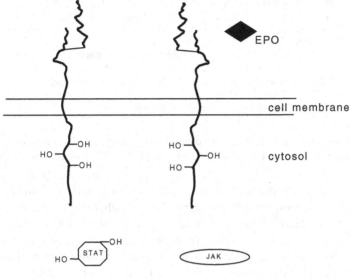

(b)   stimulated cell

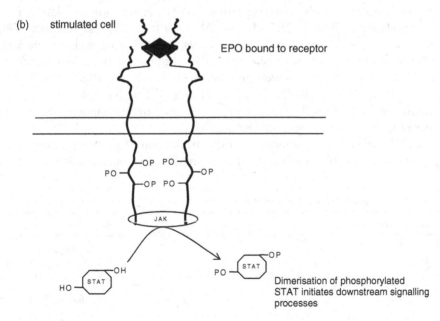

Figure 5.7 EPO-R with JAK/STAT and cascade (see also Figure 4.20). (a) unstimulated cell (b) stimulated cell

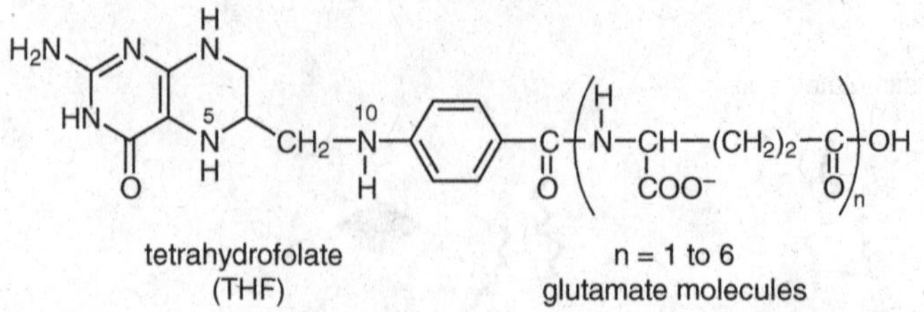

tetrahydrofolate
(THF)

n = 1 to 6
glutamate molecules

The N^5 and N^{10} atoms which carry one-carbon groups are indicated

Figure 5.8 Structure of folate

coordination position of the cobalt is taken up by a ribonucleotide-derived moiety and the sixth coordination position is occupied with *either* deoxyadenosine *or* a methyl group. Deoxyadenosylcobalamin is the active form of the vitamin in the metabolism of methylmalonyl-CoA (an intermediate in the catabolism of branched-chain amino acids and fatty acids with an odd number of carbon atoms), whereas methylcobalamin is involved with folate metabolism. It is this link with folate that explains the importance of B_{12} in red cell formation. In short, vitamin B_{12} helps to recycle tetrahydrofolate (THF) but to understand the connection between the two vitamins, we need first to study the chemical structure and biochemical roles of folate.

The chemical structure of folate (or folic acid) is shown in Figure 5.8. In humans, folate usually occurs as polyglutamate derivatives. The active form of folate is THF, sometimes shown as FH_4) is derived from folate via two reductase reactions. THF functions as a carrier of 'one-carbon' groups in varying oxidation states (Table 5.1).

These one-carbon groups, which are required for the synthesis of purines, thymidine nucleotides and for the interconversion some amino acids, are attached to THF at nitrogen-5 (N^5), nitrogen-10 (N^{10}) or *both* N^5 *and* N^{10}. Active forms of folate are derived metabolically from THF so a deficiency of the parent compound will affect a number of pathways which use any form of THF.

Folic acid derivatives are important in the production of purines and deoxythymidine monophosphate (dTMP, a pyrimidine) required for nucleic acid synthesis in

Table 5.1 Metabolically active THF derivatives

One-carbon unit	Name
Methyl —CH_3	N^5-methyl THF
Methylene —CH_2—	N^5,N^{10}-methylene THF
Formate	N^5-formyl THF
-C=O \ H	N^{10}-formyl THF
Formimino —CH=NH	N^5-formimino THF
Methenyl	N^5,N^{10}-methenyl THF
-C= \ H	

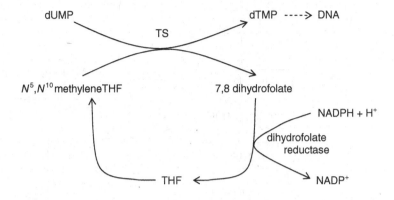

TS = thymidylate synthetase

Figure 5.9 Dihydrofolate reductase; a key enzyme in pyrimidine synthesis

dividing cells and thus deficiency of folate is associated with impaired formation of all nucleated cells. Because of their rapid rate of synthesis, red cell progenitors are affected adversely by the group of chemotherapeutics known as antimetabolites, for example methotrexate and trimethoprim used for the treatment of some forms of cancer and bacterial infections respectively. These drugs are inhibitors of dihydrofolate reductase, the enzyme which recycles dihydrofolate (DHF) to tetrahydrofolate (Figure 5.9), which, as indicated in Figure 5.10 is the parent compound of the numerous active forms of the vitamin.

It is the role of N^5-methyl THF which is key to understanding the involvement of cobalamin in megaloblastic anaemia. The metabolic requirement for N^5-methyl THF is to maintain a supply of the amino acid methionine, the precursor of S-adenosyl methionine (SAM), which is required for a number of methylation reactions. The transfer of the methyl group from N^5-methyl THF to homocysteine is cobalamin-dependent, so in B_{12} deficiency states, the production of SAM is reduced. Furthermore, the reaction which brings about the formation of N^5-methyl THF from N^5,N^{10}-methylene THF is irreversible and controlled by feedback inhibition by SAM. Thus, if B_{12} is unavailable, SAM concentration falls and N^5-methyl THF accumulates and THF cannot be re-formed. The accumulation of N^5-methyl THF is sometimes referred to as the 'methyl trap' because a functional deficiency of folate is created.

5.3.1.2 *Red cell antigens and the blood groups*
Karl Landstiener was the first to demonstrate the existence of different blood group antigens in 1900. We now recognize 29 different blood group systems comprising over 250 antigens, but the most well known are the ABO (also known as ABH) and rhesus (Rh) systems.

The ABO antigens are oligosaccharides attached to membrane proteins or lipids. The H antigen (carried on blood group O cells) is the biochemical

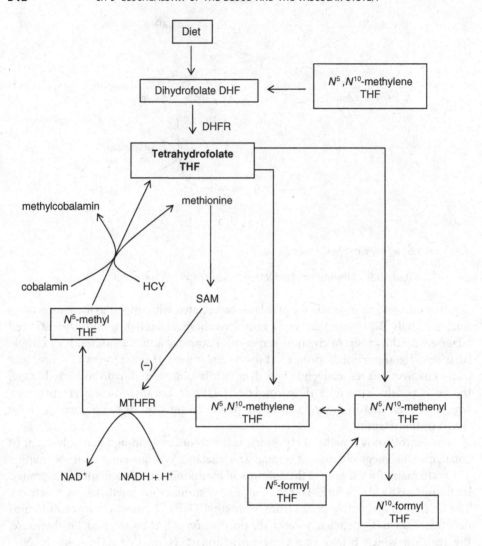

Figure 5.10 The interconversion of folate metabolites. N^5,N^{10}-methenyl THF is required for purine synthesis. N^5,N^{10}-methylene THF and N^{10}-formyl THF are required for deoxythymidylate (pyrimidine) synthesis

precursor of the other two antigens (A and B). The antigens are synthesized by the progressive addition of specific monosaccharides by the action of glycosyl-transferase enzymes. The exact blood group is determined by the terminal monosaccharide.

The glycosyltransferase reaction can be represented thus:

UDP-monosaccharide + oligosaccharide

oligosaccharide-monosaccharide + UDP

i.e. a trisaccharide is formed of a glycosidic bond between a monosaccharide and a disaccharide. The reaction is driven forward by prior 'activation' of the monosaccharide by addition of uridyl diphosphate (UDP).

Group O (H antigen)

fucose α1-2 galactose

| β1-4

N-acetyl glucosamine

|

R

Group B (B antigen)

galactose

| α1-3

fucose α1-2 galactose

| β1-4

N-acetyl glucosamine

|

R

Group A (A antigen)

N-acetyl galactosamine

|

galactose

| α1-3

fucose α1-2 galactose

| β1-4

N-acetyl glucosamine

|

R

5.3.1.3 Haemoglobin (Hb)

Haemoglobin is one of the most well studied and best understood proteins thanks largely to the early work of Max Perutz, John Kendrew and colleagues. Haemoglobin is now often used as a model allosteric protein and to illustrate the impact of protein structural alterations in disease.

There are a number of haemoglobin variants but all are tetrameric proteins and each sub-unit carries a haem group responsible for oxygen binding. Haemoglobin A_1 (HbA$_1$) is the commonest form and comprises two $\{\alpha\beta\}$ chain dimers. Variants, not all of which are pathological, arise due to changes in the primary sequence of the non-α chains, for example sickle cell haemoglobin (HbS) has a single amino acid substitution (valine to glutamate) at position 6 of the normal β chain whereas the physiologically normal HbF (fetal haemoglobin) has γ rather than β chains associated with α peptides.

Although haemoglobin is not a catalytic protein, it shares important features in common with enzymes, for example ligand binding, allosterism and 'inhibition'. Before continuing, the reader should ensure familiarity with the concepts of allosterism as described in Section 3.2.

The primary ligand for haemoglobin is oxygen which is loosely held by coordination bonding with iron of the haem prosthetic group. The binding and release dynamics of oxygen to haemoglobin are shown in Figure 5.11.

Normally, we measure substances in blood in concentration units such as mmol/l but the amounts of gases are always expressed in terms of their pressure; P_{O_2} is the partial pressure of oxygen in kiloPascals (kPa) or in older units, mmHg. The oxygen dissociation curve (ODC) shows how at very low (below physiological values) of P_{O_2}, oxygen binding to haemoglobin is 'difficult'. That is to say that a relatively large change in P_{O_2} causes very little O_2 uptake by haemoglobin. Once one of the subunits has taken up oxygen, conformational changes in adjoined subunits allow more rapid binding.

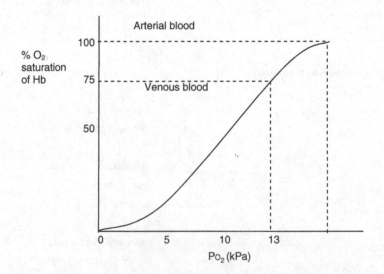

Figure 5.11 Oxygen dissociation curve

The curve flattens at very high P_{O_2} as the whole protein reaches 'saturation point'. At rest, haemoglobin in the blood which enters the general circulation from the left side of the heart is about 98% saturated with oxygen and the venous return is approximately 75% saturated with oxygen.

An important allosteric modulator of oxygen binding to haemoglobin is 2,3 bis-phosphoglycerate (2,3 BPG), which is produced via the Rapoport–Leubering shunt from glycolysis (Figure 5.12). As blood P_{O_2} falls, bis-phosphoglycerate mutase (BPM)

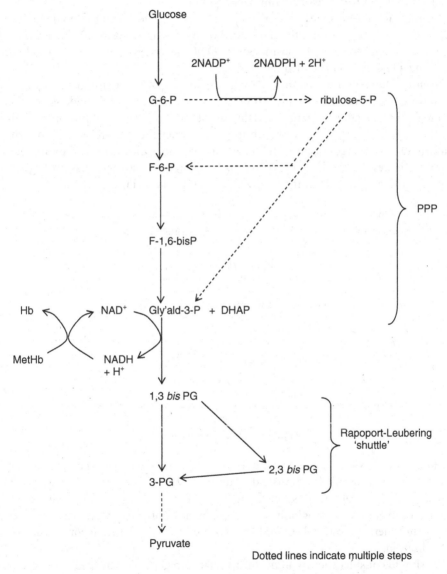

Figure 5.12   Links between glycolysis, pentose phosphate pathway and Rapoport shuttle

competes with phosphoglycerokinase for the available 1,3 BPG. The binding of 2,3 BPG causes the haemoglobin molecule to undergo a slight conformational change brought about as the BPG sits into a cleft between the two β chains of the haemoglobin molecule. This allosteric binding stabilizes the T configuration of haemoglobin thereby reducing the affinity of the protein for oxygen, so the ODC shifts to the right, that is oxygen is given up to the tissues more easily.

The *in situ* generation of 2,3 BPG is accelerated by low atmospheric Po_2, for example at high altitude. Lower Po_2 at altitude could compromise tissue oxygenation so to avoid cellular hypoxia, O_2 is released from haemoglobin.

Another, and more acute, 'trigger' to the release of oxygen from haemoglobin is the pH of the blood surrounding the red cell. Protons arising from metabolic activity of tissues cause a right-shift in the position of the ODC and oxygen is released when the pH falls. This phenomenon is called the Bohr effect.

Haemoglobin is more than just an oxygen delivery system; it also helps to regulate the pH of the blood. Like many proteins at a normal arterial blood pH of 7.40, haemoglobin is negatively charged and thus able to buffer protons. This property is due partly to the histidine residues within the primary sequence of the protein. Importantly, conformational changes occur in the haemoglobin molecule as it gives up oxygen to respiring tissues. Partially deoxygenated haemoglobin is a stronger base (i.e. greater proton binding ability) than the fully oxygenated protein (Figure 5.13).

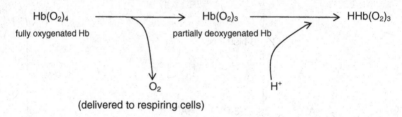

The source of the proton shown above is water;

$$CO_2 + H_2O \xrightarrow{\text{CA (CD)}} H_2CO_3 \longrightarrow HCO_3^- + H^+$$

Figure 5.13 Oxygen release

The carbon dioxide diffuses into the red cell and is hydrated by the enzyme carbonic anhydrase (CA, also called carbonate dehydratase, CD) to form carbonic acid, H_2CO_3 which, being a weak acid, spontaneously dissociates forming bicarbonate ions and a proton. Deoxyhaemoglobin ($Hb(O_2)_3$) is not only a stronger base than fully oxygenated haemoglobin ($Hb(O_2)_4$) but is able to bind CO_2 for transport to the lungs for expiration.

We should consider the fate of the bicarbonate ion generated by the dissociation of the H_2CO_3. The bicarbonate diffuses out of the red cell and into the plasma where it too buffers acids produced as a result of cellular metabolism. The export from the red cell of the HCO_3^- ion requires the import of another anion in order to maintain electrical neutrality, so a chloride ion enters the erythrocyte in a process called the chloride shift (Figure 5.14).

It will be apparent that in the absence of a buffer the production of carbon dioxide and metabolic acids would result a change in pH with all of the implied problems that

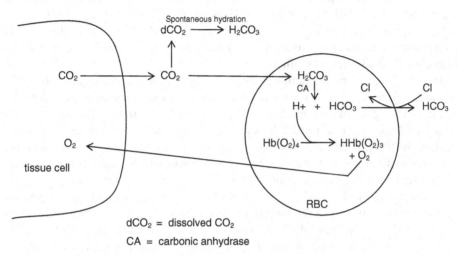

(a) in a tissue capillary

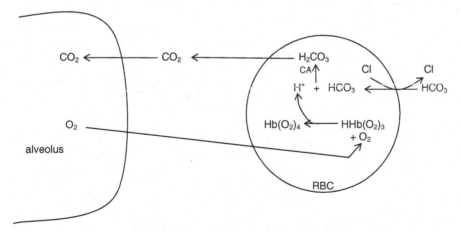

(b) in a pulmonary capillary

Figure 5.14 Chloride shift in a red cell (a) in a tissue capillary (b) in a pulmonary capillary

would cause to enzyme activity and metabolic control. Haemoglobin and the red cell are able to provide increased buffering capacity in exactly the region that it is required, that is tissue capilliaries.

Carbon monoxide, CO, is a very toxic gas because it is in effect, a competitive inhibitor of haemoglobin in that it can bind strongly to the haem group and thus prevent the binding of oxygen. As a result, the ability to switch between the R and T forms of the protein is lost and the oxygen dissociation curve shows hyperbolic rather than sigmoid kinetics indicating a loss of cooperativity between peptide subunits. Figure 5.15 shows the typical effect of a competitive inhibitor on an allosteric protein.

Circulating RBCs lack the enzymes to produce either haem groups or globin chains so the synthesis of haemoglobin is a defining feature of early RBC maturation. Globin chain production is like the formation of any other protein, but α and β chain synthesis is closely linked with haem synthesis.

Haem synthesis is a good example of a pathway that is partly compartmentalized. The pathway (Figure 5.16) occurs in all cell types for the production of respiratory cytochromes and begins within mitochondria but the majority of the reactions occur in the cytosol cell. Because mature red cells have no subcellular organelles, haem synthesis occurs only in early RBC progenitor cells. Although this is a relatively simple pathway, there are a number of well-known enzyme defects that cause a group of diseases called the porphyrias.

The haem molecule would be incomplete without iron so this must be delivered to the progenitor red cells. Iron is toxic so it is carried in the plasma bound to a specific protein named transferrin (Tf). Uptake of iron is via a Tf receptor, of which there are approximately 300 000 per cell. The whole iron/Tf complex is taken into the cell by endocytosis where the iron is released and made available for incorporation into the porphyrin ring by ferrochelatase.

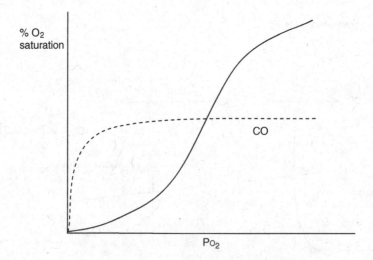

Figure 5.15 Effect of CO binding to haemoglobin

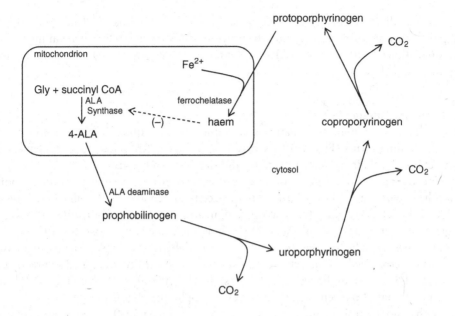

Figure 5.16 Haem synthesis. NB: PBG is converted into two 'forms' of uroporhyrinogen, the series I and III isomers but only the series III compounds are physiologically important

In the liver (the other major site of haem synthesis), regulation of haem biosynthesis is a classic example of end-product negative feedback; haem inhibits the transcription of the *ALAS* gene and thus production of the enzyme at the rate controlling step in the whole process, ALA synthase is suppressed. The control mechanism in the developing erythroblasts is linked to the concentration of iron in the cytosol. The translation of the *ALAS* mRNA is inhibited by a repressor protein which binds to the 5′ end of the primary gene transcript. Iron can interact with the repressor protein to relieve the inhibitory effect. If iron concentration exceeds the availability of protoporphyrin groups, the synthetic pathway is given a 'kick-start' to ensure that supply of protoporphyrin matches demand. As haem production rises, the 'de-repression' of the iron is lost and *ALAS* mRNA remains untranslated. Here we have an example of metabolic control operating at the level of the message, between that of the gene and of the protein. Furthermore, it could be argued that the ultimate control of protoporphyrin synthesis is determined by the expression of Tf receptors on the surface of erythroblasts, as it is this which determines cytosolic iron concentration giving an entirely different level of control.

Any inability to form haem in appropriate amounts will release the feedback control and the pathway will accelerate until the enzyme 'block' point is reached, resulting in the accumulation of the precursors (ALA and PBG) and/or intermediates (porphyrinogens) up to the block point.

5.3.1.4 RBC metabolism

The metabolic capabilities of mature circulating RBCs are quite different from that of the developing forms (Figure 5.1). Once the developing RBCs lose their nucleus and mitochondria, their range of metabolic abilities is severely curtailed.

Due to their lack of mitochondria, mature erythrocytes are entirely dependent upon glycolysis to produce ATP and NADH to meet cellular demands. Ironically in a cell type which is so rich in oxygen, glycolysis operates in an anaerobic environment so RBCs are significant net producers of lactate, which is an important substrate for hepatic gluconeogenesis. The NADH produced by glyceraldehyde-3-phosphate dehydrogenase cannot be reoxidized via mitochondrial shuttles but is usefully consumed indirectly to maintain haem-iron in the reduced (Fe^{2+}) state. Haem groups which contain Fe^{3+} are non-functional as oxygen carries (e.g. methaemoglobin, MetHb).

Methaemoglobin is formed in all red cells on a daily basis. Oxygen, being strongly electronegative, can abstract an electron from a donor to form a free radical ion called superoxide, $O_2^{\bullet-}$; Fe^{2+} is just such a electron donor:

$$Hb(Fe^{2+}) + O_2 \rightarrow MetHb(Fe^{3+}) + O_2^{\bullet-}$$

5.3.1.5 Free radical generation; superoxide dismutase and glutathione

Oxygen is vital for human life but it can also be a 'toxic hazard'. The high electronegativity of oxygen means that it can very easily become reduced forming reactive oxygen species (ROS, e.g. H_2O_2), many of which are free radicals, compounds with a single unpaired electron, for example superoxide, $O_2^{\bullet-}$. Oxygen free radicals are produced whenever molecular oxygen is being transferred within tissues. On occasions, free radical generation is intentional and the biochemical benefits of free radical generation are described in Sections 5.3.2.1, but here we will consider the potentially deleterious effects of these compounds.

Superoxide has a chemical half-life measured in microseconds, but in even this short time serious damage can be caused to all types of biological macromolecules. Peroxidation of membrane lipids could cause haemolysis but the oxidation of ferrous (Fe^{2+}) to ferric (Fe^{3+}) iron in haemoglobin due to free radical action is a more immediate cause for concern within the red cell (Figure 5.17).

To prevent membrane damage, the red cell has an armoury of antioxidants, notably a manganese or selenium containing enzyme called superoxide dismutase (SOD) and the sulfydryl compound glutathione (GSH), a tripeptide of γ-glutamate, cysteine and glycine.

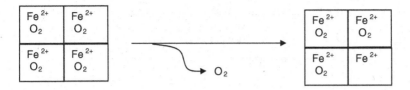

Fully oxygenated haemoglobin releases one atom of O_2 and the haem group-iron remains in the reduced state.

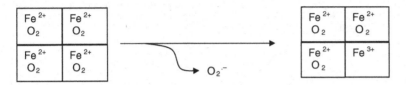

Figure 5.17 Superoxide generation during oxygen release from haemoglobin

SOD removes superoxide (see (a) below), hopefully before it causes extensive cellular damage, and the glutathione acts as 'suicide substrate' by allowing itself to be oxidized (see (b) below) so protecting membrane lipids and preventing crosslinking (via disulfide bridges) within haemoglobin chains due to oxidation of cysteine residues.

$$\text{(a)} \quad 2H^+ + \underset{\text{superoxide}}{O_2^{\bullet-}} \quad \rightarrow \quad \underset{\text{hydrogen peroxide}}{H_2O_2}$$

(b) 2 Gly-Cys-γGlu + $O_2^{\bullet-}$ →   Gly-Cys-γGlu   + 2H₂O
         | |
         SH   Gly-Cys-γGlu
   reduced glutathione (GSH)   oxidised glutathione (GSSG)

a.   the SOD reaction

b.   the oxidation of glutathione (GSH) by superoxide/GSH peroxidase.

Glutathione, which is synthesized by two ATP molecules requiring synthetase enzymes (γ glutamate cysteine synthetase and glutathione synthetase), is present at a concentration of about 2 mM in a red cell. To be an effective 'redox buffer', the ratio of GSH to GSSG must be kept high. This is achieved by the reduction of GSH by NADPH and glutathione reductase:

$$GSSG + NADPH + H^+ \xrightarrow{\text{Glutathione reductase}} 2GSH + NADP^+$$

Refer also to Figure 5.18.

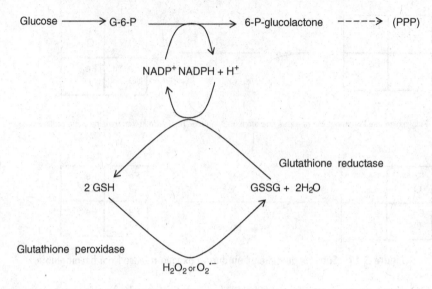

Figure 5.18 Roles of NADPH and glutathione in neutralizing oxygen free radicals PPP = pentose phosphate pathway

Thus a constant supply of the reduced coenzyme (NADPH) is required, hence the importance of the pentose phosphate pathway (PPP) as described in Section 5.3.1.6.

The enzymes GSH peroxidase, GSH reductase and SOD collaborate to ensure that the red cell is protected from the effects of methaemoglobin and superoxide. Disposal of hydrogen peroxide may occur by catalase, in a reaction which is also dependent on NADPH. This vital coenzyme is provided via the PPP. Although the PPP operates in all cell types for the provision of pentose sugars and nucleotides, its role in the RBC is more directed to cell survival than cell division.

The generation of superoxide creates two metabolic problems: (i) how to regenerate useful $Hb(Fe^{2+})$ and (ii) how to dispose of potentially lethal superoxide. Needless to say, Nature has found ways to do both!

i. The reduction of Fe^{3+} requires an electron transfer process involving $FADH_2$ and cytochrome b5 reductase;

Glyaldehyde-3-P

| NAD⁺ | FADH₂ | cyt. b5 (ox) | Hb(Fe²⁺) |

NAD⁺   FADH₂   cyt. b5 (ox)   Hb(Fe²⁺)

NADH + H⁺   FAD   cyt b5 (red)   MetHb(Fe³⁺)

1,3 BPG

The scheme shown above, sometimes called the diaphorase reaction, is an important mechanism in the RBC for the reduction of methaemoglobin. A mechanistically simpler reaction is catalysed by methaemoglobin reductase:

$$MetHb + NADH + H^+ + \tfrac{1}{2}O_2 \rightarrow Hb + NAD^+ + H_2O$$

ii. Disposal of superoxide occurs via SOD.

$$2O_2^{\bullet -} + 2H^+ \rightarrow H_2O_2 + O_2$$

The H_2O_2 so produced is also a reactive oxidizing species capable of causing cellular damage, so it too needs to be removed by catalase at the expense of glutathione (GSH), thus;

$$HOOH + 2GSH \rightarrow GSSG + 2H_2O$$

The other feature of RBC glycolysis which is significant is that at two points, substrates are diverted away from the main pathway to produce important metabolites, namely

a. NADPH (via the PPP) and

b. 2,3 bis-phosphoglycerate (via the Rapoport–Leubering shunt, Figure 5.12), which promotes oxygen release from haemoglobin.

5.3.1.6 Pentose phosphate pathway (= hexose monophosphate shunt); generation of NADPH; G6PDH deficiency

The pentose phosphate pathway (PPP) operates in all cells and tissues where it provides ribose for the synthesis of nucleotides. This pathway is especially important in adipose tissue and in the liver where the NADPH generated is used in fatty acid synthesis (see Section 6.3) and the steroidogenic glands (for cholesterol synthesis, see Section 6.3.2.2) In fact the key control enzyme, glucose-6-phosphate dehydrogenase (G6PD), is highly conserved throughout evolution is referred to a 'housekeeping' enzyme. Given the ubiquitous occurrence of the PPP why is it appropriate to discuss it in a chapter on blood cells? The answer is first that the PPP has particular roles in red cells and phagocytic white cells. Maintaining the correct 'redox' balance is important in erythrocytes where the risk of damage from reactive oxygen species and free radicals is high because of their high oxygen content. Second, the provision of NADPH in phagocytes such as neutrophils and macrophages is essential for bacterial killing, a specialized function of those types of cells.

Mechanistically, the PPP may be seen as a by-pass sequence beginning and ending with glycolysis. As we saw in Chapter 1, glucose-6-phosphate (G6P) is at a metabolic crossroads. G6P is formed from glucose in the first step of glycolysis; two red cell enzymes may then 'compete' for G6P as a substrate; phosphohexose isomerase (PHI, in glycolysis) or the $NADP^+$-linked G6PD, which begins the pentose phosphate pathway and is a key control point (Figure 5.19).

The PPP has two chemical phases; an initial irreversible oxidative phase when G6P is decarboxylated and oxidized to form ribulose-5-phosphate (Ru-5-P) followed by a more complicated but reversible non-oxidative phase involving interconversions of phosphorylated monosaccharides with four, five, six or seven carbon atoms. The

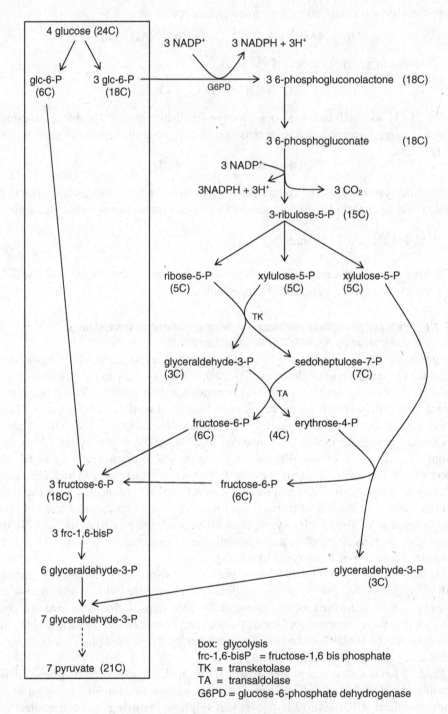

Figure 5.19 Pentose phosphate pathway

non-oxidative phase of the PPP produces ribose-5-phosphate which is the starting material for nucleotide synthesis, but of course the *mature* red cell does not need to produce either DNA or RNA. However, all of the blood cell progenitors that are nucleated and divide rapidly to maintain the continual supply of new cells for the circulation do require a source of ribose and deoxyribose. The reversible reactions also allow a route for the re-entry of carbon atoms into glycolysis at the level of fructose-6-phosphate (F6P) or glyceraldehyde-3-phosphate. This last point means that ATP generation by glycolysis is not compromised a very important consideration in cells which rely on substrate level phosphorylation for their energy supply.

The prime purpose of the PPP in erythrocytes is the generation of reduced $NADP^+$, which is used to recycle oxidized glutathione an important 'redox' regulator within the cell. Defects in NADPH production, for example in G6PD deficiency, have potentially serious effects not only on the red cell due to chemical damage brought about by oxygen free radicals, but could compromise oxygen delivery to all other tissues. G6PD deficiency is an inherited condition, with at least 300 known mutations of the gene, occurring most commonly in racial groups who are subject to malaria. The malarial parasite (*Plasmodium falciparum*), which develops partly within the red cells, is susceptible to free radical attack. The lack of G6PD and therefore NADPH is seen to be advantageous as the small amount of superoxide produced normally is sufficient to destroy the parasite without causing too much damage to the host red cell.

Genetically-determined deficiency of G6PD is the most common cause of haemolysis arising from enzyme defects. Mutated glycolytic enzymes such as hexokinase, phosphofructokinase, aldolase and pyruvate kinase can also bring about haemolysis but the occurrence of these defects are much rarer than for G6PD deficiency (see Case Notes at the end of this chapter).

5.3.2 White blood cells

There are several types of white blood cell (leucocytes), which differ structurally and functionally from each other as well as from RBCs. Generally, leucocytes are larger than RBCs, are more spherical in shape and have a more extensive intracellular architecture with typical eukaryotic organelles. The white blood cells (WBCs) contribute to the body's immune and defence mechanisms as they are able to engulf and destroy invading organisms (phagocytosis), secrete immunoglobulins, engage in allergic reactions and influence blood vessel activities. In order to fulfil their varied defensive functions, leucocytes need to interact appropriately with their surroundings. This is achieved largely by the differential expression on their surface of proteins many of which have been classified as 'cluster of differentiation' (CD) antigens. Specific cell–cell or cell–matrix contact is determined by possession of the correct CD proteins, acting as surface receptors bringing about a biochemical change within the cell.

Ontologically, WBCs are divided into two categories: (i) the myeloid series and (ii) the lymphoid series, both derive from the same HSC which produced the

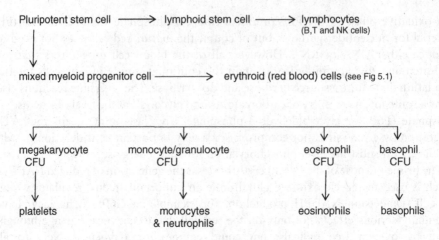

Figure 5.20 Leucopoiesis CFU = colony forming unit

erythroid cells and as will be described later, the platelet forming megakaryocytes. Figure 5.20 outlines leucopoiesis.

Like red cell development, the maturation and functioning of white cells are controlled by a number of growth factors (colony-stimulating factors, CSFs) and cytokines, especially some of the interleukins. Unlike red cells however, mature circulating white cells are continually susceptible to cytokine regulation to bring about functional adaptation. Granulocyte colony-stimulating factor (G-CSF), for example influences the differentiation of early myeloid progenitor cells into neutrophils or monocytes but also, when circumstances dictate, stimulate mature neutrophils to engulf and destroy invading organisms. White cells also need to 'sense' the presence of other leucocytes and endothelial cells and protein ligands. All such interactive functions are mediated via cell surface receptors identified under the CD system and many such surface receptors are linked with cytosolic tyrosine kinases.

So-called, B lymphocytes, which constitute about 20% of the total lymphocyte population in the blood, have only one major function which is to synthesize and secrete immunoglobulin antibodies. Following an antigenic challenge, B cells develop into plasma cells whose role is to produce specific immunoglobulin molecules in what is known as the humoral arm of the immune response.

During their early development in the bone marrow, immature B cells express on their surface IgM and/or IgD immunoglobulin (sIgM or sIgD respectively). These molecules will act as antigen recognition sites but do not contribute to the metabolic signalling process. Expressed alongside the sIg is a heterodimer of immunoglobulin light chain-like molecules, denoted Igα and Igβ. The sIg and Igα/Igβ complex is known as the B-cell receptor (BCR).

Thymus-derived, or T, lymphocytes also have an immunoglobulin-like surface receptor linked with accessory proteins to form the T-cell receptor (TCR). There are similarities between the BCR and TCR in that the recognition part of the T receptor complex is also a dimer. In most T cells, the dimer consists of α and β individual peptide chains; a very few T-cell lines exhibit alternative peptides, γ and δ. As in the BCR, the recognition unit is not directly involved with signalling; this responsibility is undertaken by another dimeric protein, CD3/ζ (ζ is pronounced 'zeta').

5.3.2.1 *The respiratory burst in neutrophils and monocytes*

One of the biologically harmful effects of free radicals was discussed in Section 5.3.1.5. Here the beneficial role of free radicals and ROS as part of immune defence is considered.

The respiratory burst causes a localized and intense production of ROS (e.g. hydrogen peroxide) and free radicals such as hydroxyl radicals ($OH^{\bullet-}$) and hypochlorous radicals ($OCl^{\bullet-}$) a sort of 'leucocyte bleach' (Figure 5.21). The conventional view of the process is that these chemicals are required to destroy, by oxidation, invading bacteria and fungi. This view, and an alternative, are discussed more fully below.

The production of the free radicals is controlled by a complex multi-meric protein called phagocyte NADPH oxidase (Phox). The 'core' component of Phox is a haemcontaining flavocytochrome b_{558}, which is itself a heterodimer of glycoprotein 91 ($gp91^{phox}$) and protein $p22^{phox}$. The numbers indicate the approximate molecular weight of the two components. Like the respiratory proteins of the mitochondria, cytochrome b_{558} is an electron-transporting protein. Additionally, four accessory proteins are required for activation and full function of cytochrome b_{558}; $p67^{phox}$ is an activator of b_{558}, $p47^{phox}$ is an adapter molecule which allows $p67^{phox}$ to attach to $p22^{phox}$, $p40^{phox}$ whose function is thought to help translocate $p67^{phox}$ to the phagosome membrane and p21, a guanine nucleotide binding protein (see Figure 5.22).

When neutrophils encounter bacteria, possibly coated with opsonin proteins of the complement system, the invaders are engulfed by the phagocytes and taken into the cells by endocytosis. A small part of the neutrophil membrane is used to create a phagosome – that is, a vacuole enclosing the bacterial cells. Within a matter of a few

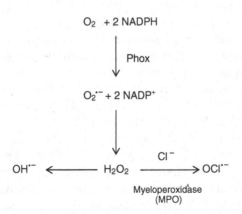

Figure 5.21 generation of free radicals and reactive oxygen species in a phagocyte. Note here the dependence on the PPP to produce NADPH

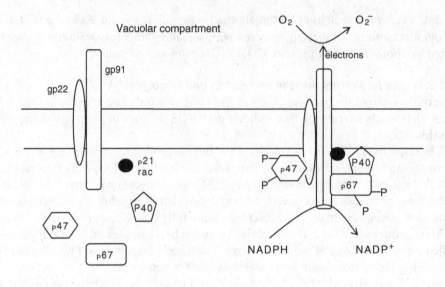

Figure 5.22 NADPH oxidase complex. gp22 and gp91 form a heterodimer; p40, p47 and p67 are accessory proteins. Phosphorylation of the accessory proteins causes aggregation of the components and activation of the oxidase activity

seconds, the component parts of NADPH oxidase are assembled in the phagosome membrane and electrons supplied by NADPH are transported from the cytosol into the vacuole. The electrons are used to reduce molecular oxygen to superoxide which then accepts hydrogen ions to become hydrogen peroxide. The consumption of protons at this step raises the intravacuolar pH to approximately 7.9, significantly higher than the cytosolic pH. Highly reactive hydroxyl radicals ($OH^\bullet$) are formed from hydrogen peroxide whilst myeloperoxidase produces $OCl^{\bullet-}$. Thus, an armoury of ROS and free radicals is generated within the phagosome. Coincident with the onset of NADPH oxidase activity is fusion of cytoplasmic granules with the phagosome and discharge into the vacuole of bacteriolytic proteins.

As stated above, conventional wisdom was that the ROS and radicals were themselves sufficient for pathogen killing within the vacuole. Experimental evidence that this assumption was valid under typical physiological conditions within the cell and phagosome was, however, lacking. Moreover, evidence began to accumulate that cells genetically engineered to be deficient in proteolytic enzymes normally found sequestered within cytoplasmic granules, showed that, even in the presence of complete activity of NADPH oxidase, bacterial killing was often ineffective.

An alternative model to explain the destruction of pathogens within phagosomes has been proposed based on the effects of NADPH oxidase activity on ion distribution. Transfer of electrons into the vacuole and the simultaneous consumption of protons creates an electrochemical gradient across the phagosome membrane, which may be *conceptually* compared with the proton gradient generated within mitochondria for ATP production. To help dissipate the charge differential, potassium ions, which are abundant in the cytosol of all cells, pass into the phagosome, and simultaneously chloride ions, derived from the cytoplasmic granules along with the proteolytic enzymes, exit the vacuole. In short, the electrochemical environment within the phagocytic vacuole is dramatically changed.

The final piece in this fairly complex puzzle is the fact that the bacteriolytic enzymes such as neutrophil elastase and cathepsin G contained within granules and released into the phagosome, are activated by potassium but inhibited by chloride. Thus it seems that the function of NADPH oxidase, at least in part, is to create the appropriate electrochemical environment within the phagosome to allow enzymatic killing of invading pathogens. In addition, the creation of the physiologically atypical conditions within the phagosome would help to explain why neutrophil elastase and cathepsin are not active in the resting white cell. For more information on this interesting and original theory, carry out a Web search for research literature by A.W. Segal and colleagues.

5.3.3 Physiology of platelets

Platelets (also called thrombocytes), the smallest of the formed elements of the blood, are formed by fragmentation of megakaryocytes. Each megakaryocyte (*mega* = 'big', *karyo* = 'nucleus') can produce approximately 3000 individual platelets and a total of about 2×10^{11} platelets are formed each day.

In common with RBCs platelets lack a nucleus, but unlike erythrocytes platelets have a complex internal organelle structure including mitochondria, lysosomes, peroxisomes and secretory granules. They are able to produce, store and secrete a wide range of chemical factors (e.g. nucleotides, serotonin, platelet-derived growth factor, fibrinogen and β-thromboglobulin). Like WBCs, platelets express on their surface adhesion molecules, especially integrins and selectins. All of these chemicals are involved with haemostasis.

5.4 Coagulation and complement: two of the body's defence mechanisms

The prevention of excessive blood loss through breaches of the vascular barrier is important to maintain oxygen delivery and blood volume. A fall in blood volume would cause a drop in blood pressure and the metabolism of all of the major organs would be badly affected. Failures of the haemostatic mechanisms can lead to haemorrhage (bleeding disorders) or thrombosis (hypercoagulation disorders).

Haemostasis is achieved by the interplay between platelets, proteins within the plasma (the clotting factors) and endothelial cells. The endothelial cells lining the vasculature normally present a non-thrombogenic surface, and secrete chemicals (prostacyclin, PGI_2, and NO) which inhibit platelet adhesion and aggregation. If the endothelial layer is broken and blood leaks from the vessels, contact between collagen in connective tissue and platelets initiates clotting at the site of injury. Normally, as discussed in Section 5.2.2.1, endothelial cells are antithrombotic, following damage they change their phenotype to become procoagulant. The integrins and selectins allow platelet attachment to damaged endothelium, to other platelets and to fibrinogen, effectively forming a plug to stem bleeding. Subsequent fibrin deposition completes the process which prevents blood loss.

The protein-based clotting process is a classic example of an enzyme cascade (see Figure 5.23). The clotting factors (which are designated with a Roman numeral, I to XIII) are synthesized in the liver and circulate in the blood as inactive precursors, strictly, proenzymes. Most of the clotting factors are serine protease enzymes, that is they are enzymes which cleave other proteins (substrates) by a mechanism which involves a serine residue at the active site.

Following injury, a thrombus temporarily plugs the leak and stops the loss of blood until repair can be effected, but if blood clots were not removed from the luminal surface of the vessels, blood flow to tissues would become compromised. Fibrin strands are degraded by plasmin (a serine protease) forming fibrin degradation products (Figure 5.24). The plasmin itself circulates as an inactive precursor, plasminogen, which has to be activated. One activator is tissue plasminogen activator (tPA) produced by endothelial cells. Tissue PA (also a serine protease enzyme) binds to fibrin in the clot and so brings about the conversion of plasminogen *only* at the sites it is needed.

The complement system which functions as part of the immune response is composed of about twenty proteins which circulate in the blood stream as inactive precursors. The complement cascade is functionally divided into two 'arms' called the classical and alternative pathways, reflecting their different initiating events but which converge at C3. A simplified scheme is shown in Figure 5.25.

The cascade consists of a number of steps which involve protein structure modification by proteolysis or through conformational change and aggregation.

5.5 Blood as a transport medium

Blood plasma, which is approximately 93% water, contains very many soluble compounds ranging in size from small ions to large proteins. Several compounds of physiological importance are not water soluble, so a means to overcome their insolubility must be sought. The answer lies in proteins.

Haemoglobin, described in Section 5.3.1.3, is the most well known but it is just one of a number of carrier proteins present in blood. Albumin is quantitatively the most abundant protein in plasma. It is synthesized in the liver and circulates with a half life of about 3 weeks before being degraded or eliminated. Albumin has two very important functions to fulfil. First, it makes a significant contribution to the oncotic pressure of the blood and so influences the distribution of fluid between the intracellular and

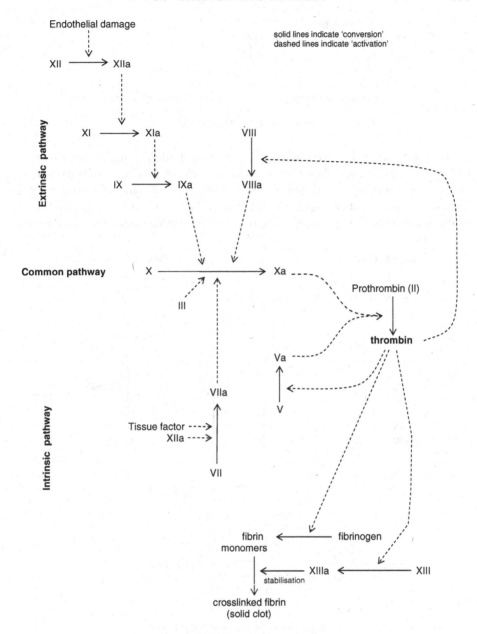

Figure 5.23 Coagulation cascade

extracellular compartments. Too much (overhydration) or too little fluid (dehydration) inside cells can have disastrous consequences on cell function.

Second, albumin is a non-specific carrier protein. A wide range of chemically disparate compounds are bound loosely to albumin for transport through the blood stream. Important examples include calcium, bilirubin, drugs and free fatty acids.

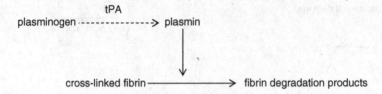

Figure 5.24 Fibrinolysis

Of the plasma total concentration of calcium (around 2.5 mmol/l), approximately half is bound to albumin. The unbound fraction is physiologically active in roles such as clotting, in regulating neuromuscular membrane potential and of course for bone formation. There exists an equilibrium between the bound and free fractions, so the albumin can be seen as a 'buffer' able to release or take up calcium as circumstances

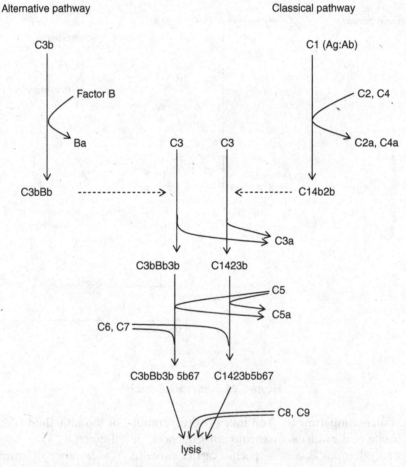

Figure 5.25 Complement cascade. The classical pathway requires antigen:antibody (Ag:Ab) interaction to activate C1, the alternative pathway is antigen independent

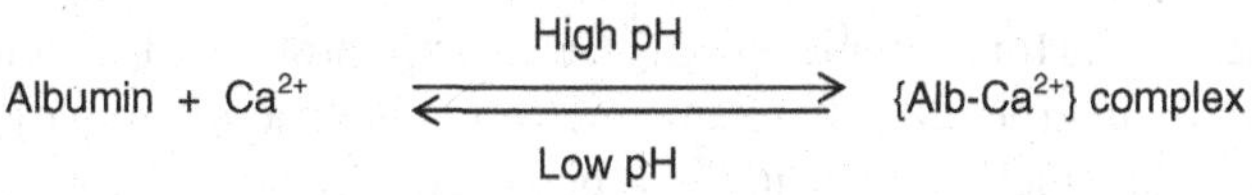

High pH

Albumin + Ca^{2+} $\rightleftharpoons$ {Alb-Ca^{2+}} complex

Low pH

Figure 5.26 pH dependence of calcium binding to plasma albumin

dictate. The binding of calcium to albumin is pH dependent and sudden changes in blood pH can alter the equilibrium sufficiently to cause physiologically significant changes in the concentration of free (i.e. not protein-bound and ionized) calcium in plasma (Figure 5.26).

Bilirubin is the waste product derived from haem catabolism. In order to be eliminated from the body, mainly via the gut, bilirubin must be processed through the liver (see Section 6.4). Bilirubin is, however, insoluble in water, so to reach the liver from the spleen where a substantial amount of red cell destruction occurs, bilirubin must first be bound to albumin. As blood perfuses the liver, bilirubin is transported into the hepatocyte where it is conjugated with glucuronic acid prior to excretion.

Muscles, including the heart, prefer to utilize free fatty acids as their energy-generating fuel. Fatty acids, which are hydrophobic, are derived from either the diet or the storage adipose tissue and are carried by albumin from the depot site to the muscles. An even greater problem is faced in transporting cholesterol, cholesterol esters and triacylglycerols (triglycerides) around the body. This challenge is met by specific carriers called lipoprotein particles.

Lipoproteins (Table 5.2) are macromolecular aggregates with varying proportions of triglycerides and cholesterol (with some phosphoacylglycerols) and apoproteins. The apoproteins act as recognition 'flags' for receptor binding, for example apo B and apo E,

Table 5.2 Lipoprotein particles

Lipoprotein class	Lipid components	Main apoprotein components	Enzymes present	Role
High density Lipoprotein (HDL)	Phospholipids Cholesterol, some TG	Apo A$_I$ and/or apo A$_{II}$; Apo C$_{II}$	Paraoxonase LCAT^a	Cholesterol scavenger. Anti-inflammatory
LDL	Cholesterol Some TG	Apo B100 Apo E		Cholesterol delivery
VLDL	Mostly TG synthesized in the liver; Some cholesterol	Apo B48 Apo E		TG delivery to adipose tissue
Chylomicrons	TG derived from the diet	Apo B48		TG delivery to adipose tissue.

^{a}LCAT: lecithin-cholesterol acyl transferase.
HDL, high density lipoprotein; LDL, low-density lipoprotein; TG: triglyceride; VLDL, very low-density lipoprotein.

or as enzyme activators for example apo C_{II}. Certain lipoproteins also include functional enzymes within their structure. These enzymes often have a role to play in lipoprotein metabolism, most of which occurs within the plasma whilst the lipoproteins are in transit, before the lipoprotein particle is removed from the circulation, usually, by receptor-mediated endocytosis.

High-density lipoproteins (HDL) and very low-density lipoproteins (VLDL) are synthesized in the liver. LDL is produced in the blood stream as VLDL particles are partially delipidated by lipoprotein lipase, a triglyceride hydrolysing enzyme located on the luminal surface of vessels in sites such adipose tissue.

It is important to realize that lipoproteins are dynamic particles, continually exchanging lipid and or protein with other lipoproteins or with cells (Figure 5.27).

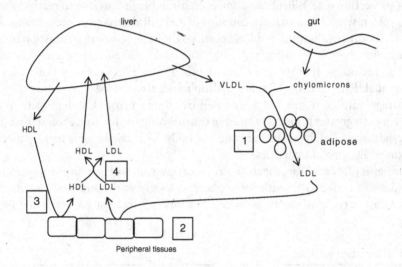

1	endothelial lipoprotein lipase hydrolyses triglyceride and the released fatty acids enter adipocytes. Partial de-lipidation of VLDL generates LDL.
2	LDL particles are transported through the plasma and taken into peripheral cells by apoB receptor
3	HDL particles will remove cholesterol from cells and the component enzyme LCAT esterifies cholesterol whilst it is part of the HDL
4	Cholesterol ester transfer protein (CETP) exchanges cholesterol ester (from HDL) with triglyceride (to HDL) and both lipoproteins may be cleared by the liver.

Figure 5.27 Lipoprotein metabolism. Lipid exchange between lipoprotein particles and cells.

Lipid exchanges between lipoproteins and cells is associated with loss of apoproteins also so the chemical nature of lipoproteins changes significantly whilst being transported within plasma. An understanding of their metabolism gives useful insights to the processes of atherosclerosis, a significant pathology associated with the vasculature and one which is responsible for a considerable amount of mortality. The cause of atherosclerosis is complex involving genetic and lifestyle factors but the development of lesions (injury) to the wall of especially the vessels of the heart and brain is well known and shows the typical signs of an inflammatory condition.

At the biochemical level a number of important events occur. Initial damage to the vessel wall elicits secretion of cytokines, which attract white blood cells and cause them to adhere (via selectin attachment) and penetrate (via integrin interactions) the endothelial layer. Low density lipoproteins (LDL) accumulate in the area of the lesion and the lipids they carry become partially oxidized (oxLDL), probably by the action of reactive oxygen species and free radicals such as superoxide produced by leucocytes. Phagocytosis of oxLDL by macrophages via a cell surface receptor SR-B1, causes the cells to develop into lipid-laden 'foam cells'. An atheroma (lipid-filled swelling) begins to grow in the intima of the artery wall (Figure 5.28).

Concomitantly, the coagulation cascade begins as platelets are exposed to the subendothelial connective tissue. A clot of platelets and fibrin deposits forms on the luminal face of the vessel reducing the diameter of the artery and so restricting the supply of blood to tissues 'downstream'. The lipid-rich swelling and the clot form a plaque which may continue to enlarge, further restricting blood supply to tissues (a condition called ischaemia) resulting in tissue hypoxia.

There may be few or no symptoms if the plaque is 'stable', but if the surface in contact with the flowing blood is weak and tends to break away (friable plaque) the body perceives this as a new injury and the whole process of damage limitation involving clotting begins again, resulting in enlargement of the plaque. Eventually the plaque may completely occlude the vessel; the death of tissue cells due loss of blood flow is called an infarction.

A high plasma concentration of LDL (usually measured as LDL-cholesterol) is a risk factor for the development of atheroma whereas a high concentration of HDL is an 'anti-risk' factor for cardiovascular disease (CVD). Fundamental discoveries relating to cholesterol metabolism and the importance of the LDL receptor made by Nobel laureates Joseph Goldstein and Michael Brown led to an understanding of the role of LDL in atherosclerosis. The impact of HDL in reducing CVD risk is often explained by the removal of excess cholesterol from tissues and its return to the liver, a process known as reverse cholesterol transport. However, evidence from research by Gillian Cockerill and others shows that HDL has a fundamental anti-inflammatory role to play in cardioprotection.

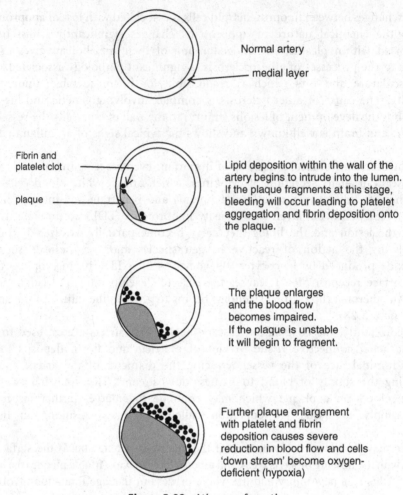

Normal artery

medial layer

Fibrin and platelet clot

plaque

Lipid deposition within the wall of the artery begins to intrude into the lumen. If the plaque fragments at this stage, bleeding will occur leading to platelet aggregation and fibrin deposition onto the plaque.

The plaque enlarges and the blood flow becomes impaired. If the plaque is unstable it will begin to fragment.

Further plaque enlargement with platelet and fibrin deposition causes severe reduction in blood flow and cells 'down stream' become oxygen-deficient (hypoxia)

Figure 5.28 Atheroma formation

Chapter summary

Blood is the stream of life, carrying essential nutrients to cells and removing metabolic waste products. Red cells have no internal organelles so cannot generate energy from oxidative phosphorylation and thus rely on glycolysis for ATP production. Although not tissue-specific, the pentose phosphate pathway is a major metabolic route within red cells because generation of NADPH is vital in preventing free radical damage to the cell. The same pathway and coenzyme are important to phagocytic white cells as part of the respiratory burst involved with bacterial killing. Vascular cell biochemistry is often overlooked but as discussed in this chapter, endothelial cells play a key role in maintaining the health of the veins and arteries and indeed damage to these cells or changes in their biology may result in pathology, notably, atherosclerosis. The liquid component of blood is the body's main transport medium but proteins suspended in the plasma have numerous functions including defence against blood loss following injury, immune defence against microbes and carrier proteins for many water insoluble compounds.

Case notes

1. **G6PD deficiency**

   An elderly African female living in the UK visited her general practitioner (GP) complaining of shortness of breath and extreme lethargy. Clinically, the GP noted she had som eyellowing of the eyes, had a weak and rapid pulse with heavy and fast breathing. Routine blood tests revealed the following:

Haemoglobin	4.8 g/l	(ref: 11.5–15.5 g/l)
Red cell count	1.2×10^{12}/l	(ref: $5–11 \times 10^{12}$/l)
Mean cell Hb conc (MCHC)	36.5 g/dl	(ref: 27–32 g/dl)
Haematocrit (PCV)[a]	0.12	(ref: 0.35–0.45)
Mean cell volume (MCV)	91 fl	(ref: 80–98 fl)
White cell count and platelet count:	normal	
Sickle test for HbS:	negative	
Serum ferritin	4700 µg/l	(ref: 10–120 µg/l)
Serum bilirubin	366 µmol/l	(ref: <20 µmol/l)
Serum enzyme activities:	lactate dehydrogenase (LD) and aspartate transaminase (AST) were both elevated but alanine transaminase (ALT) result was normal.	

   [a]PCV = packed cell volume is a measure of the relative volume of red cells to plasma in whole blood; this is a ratio so has no unit.
   See also Chapter 6 for details of jaundice and bilirubin production.

   Clearly this patient has both clinical and haematological symptoms of severe anaemia. The cause is too few red cells; low RBC count and PCV but the erythrocytes which are present contain a higher than usual concentration of haemoglobin (MCHC result). Iron deficiency and vitamin B_{12} deficiency can be ruled out by the high serum ferritin and normal MCV results respectively. The negative HbS screen rules out sickle cell anaemia which is fairly common in Africans.

   The jaundice *could* be due to liver damage. Hepatocytes contain AST, ALT and LD; red cells also contain AST and LD but do not contain significant amounts of ALT. These data suggest increased red cell destruction rather than liver cell damage and the patient was diagnosed with haemolytic anaemia.

   As a follow-up test, the activity of red cell G6PD was measured.

   G6PD   1.5 u/g Hb   (ref: 6–12 u/g Hb)

The lack of G6PD activity means that the production of NADPH is compromised, oxidized glutathione (GSSG) cannot be reduced (to GSH) and the red cells show morphological abnormalities leading to lysis.

Relative G6PD deficiency is one of the commonest genetic diseases, affecting an estimated 400 million people of ethnic groups originating mainly in Africa, Asia, and the Middle East. There are numerous variant mutations of the G6PD gene but for convenience these are normally grouped into four major categories. In all cases, there is some residual G6PD activity and no major alterations within the gene have ever been noted. Importantly, the G6PD gene is on the X chromosome so the clinical effects are seen more acutely in males and women may be asymptomatic carriers. Occasionally, and as in this case, acute and potentially life-threatening haemolytic crises are brought about in susceptible individuals by exposure to a 'trigger', the most well known of which is divicine, a pyridine derivative of vicine, a β-glucoside found in fava beans, hence the name favism used to describe this condition.

2. **Chronic granulomatous disease (CGD)**

Peter was the second born, but only son, of three children. He appeared fairly well for the first year but was prone to developing infections and had had a few bouts of diarrhoea. At 18 months Peter showed lymphadenopathy (enlarged lymph glands) and hepatoslenomegaly (enlarged liver and spleen) and a severe chest infection was diagnosed as pneumonia. There was no family history of relevance to Peter's condition.

Following further episodes of infection Peter was referred for immunological investigations. The key finding was of a poor respiratory burst response of leukocytes to a challenge indicating a defect in the function of the NADPH oxidase complex confirming a diagnosis of chronic granulomatous disease. As was the case with Peter, symptoms usually appear during the first 2 years. Phagocytic cells are able to ingest micro-organisms but due to a defect in NADPH oxidase, the production of reactive oxygen species and free radicals is compromised. The micro-organisms are not destroyed and granulomas (swellings due to macrophage accumulation) occur. Recurrent infections occur, often with organisms which are not normally considered to be highly virulent.

Most cases of CGD, including Peter, are X-linked recessive traits in which the gp91 component of NADPH oxidase is affected. Autosomal recessive defects in p22, p47 and p67 also occur and these may present with symptoms in female children.

Treatment relies on the use of broad spectrum antibacterial and antifungal drugs; bone marrow transplant or gene therapy are possible options. The prognosis for a child with CGD is not good and many will not live beyond their mid-teens.

3. **Atherosclerosis**

Mr Leane is a 61-year-old retired civil servant who lives alone in a rural setting following the death of his wife 2 years ago. Mr Leane walks his two dogs at least twice a day; he drinks wine in moderation. Mr Leane's elder brother died of a stroke 5 years ago. Clinical examination shows that he is not overweight, has a normal blood pressure but has some slightly yellow patches in the skin.

Mr Grostmann is 60 years old, works long hours each week in running his own 'one-man business; he often drives long distances from his home to attend meetings with clients. Mr Grostmann has been divorced twice and now lives alone. Any free time he has is usually spent watching TV; he admits to consuming a bottle of whisky every week and often eats convenience foods. Examination by his doctor revealed a slightly high blood pressure and calculations based on his height and weight showed a moderate degree of obesity (body mass index, BMI 27 kg/m^2, normal 25).

Both subjects under went a routine health check-up. Analysis of blood samples collected after a 16 h fast gave the following results:

	Mr Leane	Mr Grostmann	
Total cholesterol	7.5 mmol/l	6.0 mmol/l	(ref: target = 5.2 mmol/l)
HDL-cholesterol	0.8 mmol/l	1.1 mmol/l	(ref: >1 mmol/l)
Triglycerides	2.1 mmol/l	1.8 mmol/l	(ref: <2 mmol/l)
Glucose	5.8 mmol/l	4.8 mmol/l	(3.5–5.5 mmol/l)
LDL-chol (by calculation)	6.5 mmol/l	4.5 mmol/l	(ref: <5)
total chol : HDL ratio	9.4	5.5	(ref <5)
Haematology results	No abnormalities	No abnormalities	

Of the two subjects, Mr Leene is at the greater risk for cardiovascular disease (CVD). Despite his healthier lifestyle Mr Leene has a family history of vascular disease (brother who died of a stroke), clinical signs of lipid deposits (yellow patches in skin) and a very poor lipid profile. Lipid-lowering drug intervention is required in this subject.

Mr Grostmann, is "at risk" but a change in eating habits may be sufficient to normalize his lipid profile. His CVD risk would be improved further by radical reassessment of his lifestyle and work–rest balance.

A number of chemical markers carried in the bloodstream are available to detect an acute myocardial infarction, or heart attack). These include cardiac muscle enzymes such as creatine kinase, but more reliable clinical information is given by cardiac structural proteins such as the troponins (see chapter for a description of muscle structure). Where the occlusion (blockage) of the artery is due to coagulation, 'clot busting' drugs such as urokinase or streptokinase can be administered to allow the flow of blood (reperfusion) to the damaged muscle. Whilst the myocardial cells experience anaerobic conditions, they begin to adapt their metabolism and when oxygen is re-introduced, there is an increased production of free radicals and ROS causing severe damage to the cells; this is called reperfusion injury.

6

Biochemistry of the liver

Overview of the chapter

The liver is a glandular organ with multiple metabolic functions. The organ has a rich blood supply delivering oxygen, nutrients and waste products for processing. Metabolically, the liver plays a central role being responsible for fuel homeostasis, biosyntheses, detoxification and has important storage functions.

Key pathways

In addition to the common pathways, glycolysis and the TCA cycle, the liver is involved with the pentose phosphate pathway; regulation of blood glucose concentration via glycogen turnover and gluconeogenesis; interconversion of monosaccharides; lipid syntheses lipoprotein formation; ketogenesis; bile acid and bile salt formation; phase I and phase II reactions for detoxification of waste compounds; haem synthesis and degradation; synthesis of non-essential amino acids and urea synthesis.

6.1 Introduction

The liver is the body's largest organ and is central to metabolic processes. The liver performs a wider range of biochemical functions than any other organ. For this reason, the liver is one of the tissues most commonly used by biochemists studying metabolic pathways and their control mechanisms. For simplicity, the various functions to be described in this chapter have been arranged under a number of subheadings.

6.2 Physiology of the liver

The biliary system consists of the liver, the gall bladder and its associated drainage ducts which combine with the pancreatic duct to form the common bile duct. Situated in the upper right quadrant of the abdomen this is a highly vascular organ being perfused by

Essential Physiological Biochemistry: An organ-based approach   Stephen Reed
© 2009 John Wiley & Sons, Ltd

the hepatic artery (oxygen rich) and hepatic portal vein (carrying nutrients from the gut after a meal). The liver is the heaviest organ in the human body weighing approximately 1.5 kg in a typical 70 kg adult. Macroscopically, the gross structure of the liver shows two main lobes (left and right), each of which has its own bile duct drainage system. The histology of the liver consists mainly of hepatocytes (the parenchyma) arranged into lobules which are roughly hexagonal plates of cells enclosing a central vein. In humans, the outline of the lobules is not distinct but the boundaries can be imagined as lines running from a portal area at each corner. Each portal area contains a branch of the hepatic artery, the hepatic portal vein and a bile ductule, (Figure 6.1). Blood flows through sinusoids from the periphery of each lobule towards the central vein (a tributary of the hepatic vein) whilst bile flows in the opposite direction along canaliculi. Lining the blood capillaries in the liver are to be found phagocytic Kupffer cells, which participate in the cleansing processes performed by the organ.

The subcellular structure of hepatocytes exemplifies a typical animal cell with a large central nucleus and a cytosol rich in mitochondria (up to 1000 per cell), extensive smooth and rough endoplasmic reticulum (ER) and Golgi apparatus. The ER and Golgi network suggest a secretory function and it is legitimate to classify the liver as gland. In summary, all of these cytological features indicate metabolically very active cells.

6.3 Synthetic functions

Because of its wide range and diversity of synthetic functions, the liver may be considered to be the body's factory, taking raw materials and producing many compounds some of which are exported for use in other tissues. This section describes the syntheses of proteins, lipids and lipoproteins, ketones, urea and haem. Not all of these pathways are unique to the liver, but all illustrate important points of hepatic function.

6.3.1 Amino acid and protein synthesis

In addition to their well known role in protein structure, amino acids also act as precursors to a number of other important biological molecules. For example, the synthesis of haem (see also Section 5.3.1), which occurs in, among other tissues, the liver begins with glycine and succinyl-CoA. The amino acid tyrosine which may be produced in the liver from metabolism of phenylalanine is the precursor of thyroid hormones, melanin, adrenaline (epinephrine), noradrenaline (norepinephrine) and dopamine. The biosynthesis of some of these signalling molecules is described in Section 4.4.

Humans have a limited capacity to synthesize amino acids *de novo*, but extensive interconversions can occur. Those amino acids which cannot be formed within the body and must be supplied by the diet are called 'essential'. Members of this group, which includes the branched chain amino acids leucine and valine, and also methionine and phenylalanine, are all dietary requirements. Such essential amino acids may be chemically converted, mainly in the liver, into the 'non-essential' amino acids. The term 'non-essential' does not equate with 'not biochemically important' but simply means they are not strict dietary components.

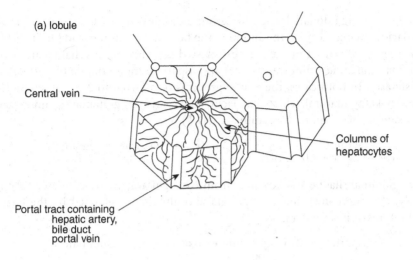

(a) lobule

Central vein

Columns of hepatocytes

Portal tract containing hepatic artery, bile duct portal vein

(b) detail of section between portal tract and central vein

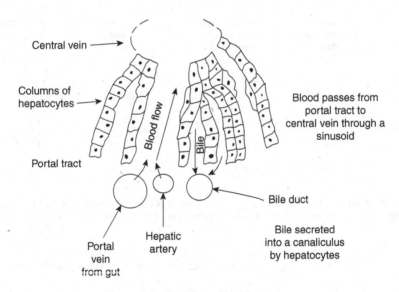

Central vein

Columns of hepatocytes

Portal tract

Blood flow

Bile

Blood passes from portal tract to central vein through a sinusoid

Bile duct

Bile secreted into a canaliculus by hepatocytes

Portal vein from gut

Hepatic artery

Figure 6.1 Structure of the liver

6.3.1.1 *Transamination*

A group of enzymes which is particularly important in amino acid metabolism in the liver (and also in muscle) is the transaminases, (also called aminotransferases). These are vitamin B_6 (pyridoxine) dependent enzymes which transfer an amino group from an amino acid to an oxo (keto) acid, thus:

$$\text{amino acid}_1 \; + \; \text{oxoacid}_1 \quad \leftrightarrow \text{oxoacid}_2 + \text{amino acid}_2$$

(amino group donor) (amino group acceptor)

Amino acid$_1$, having donated its amino group becomes an oxoacid (oxoacid$_2$ above) and oxoacid$_1$ having accepted the amino group becomes an amino acid (amino acid$_2$).

As examples, two enzymes that will be discussed again later in this chapter are alanine transaminase (alanine aminotransferase) and aspartate transaminase (aspartate aminotransferase). In both cases, the amino group is transferred to 2-oxoglutarate (also known as α-ketoglutarate), which is oxoacid$_1$ above, forming glutamate as amino acid$_2$.

For example, the alanine transaminase (ALT) reaction is:

$$\underset{\text{(amino group donor)}}{\text{alanine}} + \underset{\text{(amino group acceptor)}}{\text{2-oxoglutarate}} \leftrightarrow \underset{\text{(oxoacid}_2)}{\text{pyruvate}} + \underset{\text{(amino acid}_2)}{\text{glutamate}}$$

In effect, glutamate has been synthesized from 2-oxoglutarate, at the expense of alanine. Similarly, the interconversion of aspartate and glutamate catalysed by the enzyme aspartate transaminase (AST):

$$\text{aspartate} + \underset{\text{(2-OG)}}{\text{2-oxoglutarate}} \leftrightarrow \underset{\text{(OAA)}}{\text{oxaloacetate}} + \text{glutamate}$$

Figures 6.2a and b illustrate important examples of transamination reactions.

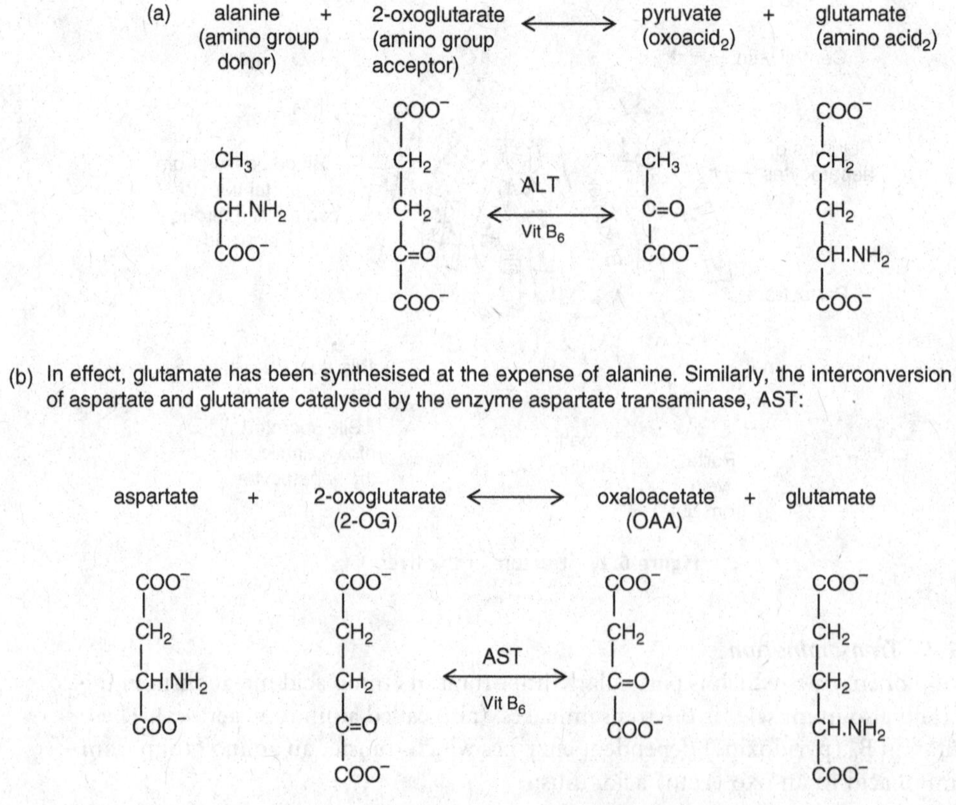

Figure 6.2 Transaminase reactions

Although aspartate and alanine have been identifited in the description above as the 'amino group donor', the K_{eq} for transaminase reactions is close to 1.0 so both reactions are, as indicated by the double-headed arrows, are easily reversible. Glutamate could just as easily act as the amino group donor if either reaction proceeds from right to left, resulting in the formation of alanine or aspartate respectively.

The role of pyridoxal phosphate (vitamin B_6), which is enzyme bound, is to 'hold' the amino group once it leaves the donor and before being incorporated into the acceptor (Figure 6.3).

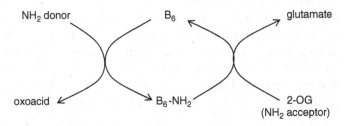

Figure 6.3 Role of vitamin B_6 in transaminase reactions

6.3.1.2 Phenylketonuria (PKU)

An example of an amino acid interconversion that has pathological significance is the hydroxylation of phenylalanine to form tyrosine in the liver (Figure 6.4).

Phenylalanine hydroxylase (PH) which requires tetrahydrobiopterin (BH_4) as a cofactor, is defective in cases of phenylketonuria (PKU). This is a rare (prevalence $\sim 1/$ 15 000 in the United Kingdom) genetic condition characterized by fair complexion, learning difficulties and mental impairment. If PH is either not present in the hepatocytes or is unable to bind BH_4 and is therefore non-functional, phenylalanine accumulates within the cells. Enzymes in 'minor pathways' which are normally not very active metabolize phenylalanine ultimately to phenylpyruvate (i.e. a phenylketone). To use the 'traffic flow' analogy introduced in Chapter 1, the main road is blocked so vehicles are forced along side roads. Phenylpyruvate is excreted in the urine (*phenylketone-uria*), where it may be detected but a confirmatory blood test is required for a reliable diagnosis of PKU to be made.

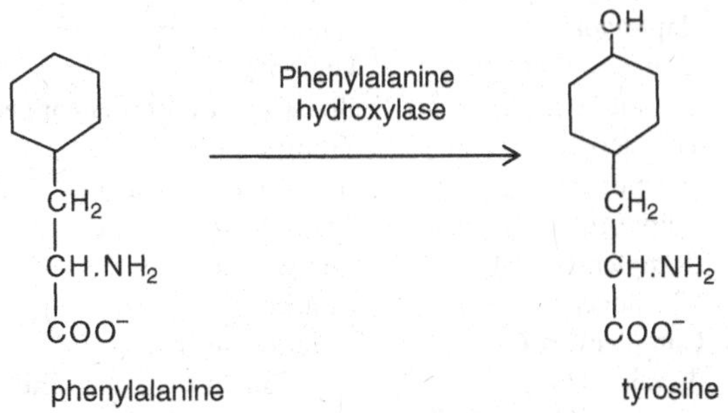

Figure 6.4 Phenylalanine hydroxylase

6.3.1.3 Protein synthesis

With the notable exceptions of the immunoglobulins and various proteins released from necrotic cells, all of the main plasma proteins are synthesized in the liver. The most abundant of the plasma proteins is albumin, accounting for just over half of the plasma total protein concentration of about 75 g/l. All of the remaining plasma proteins are classed as globulins. This very heterogeneous group is subdivided into the α, β and γ globulin fractions, a distinction based on the mobility of the proteins in an electric field (electrophoresis) rather than any useful functional classification. Examples of specific globulins are shown in Table 6.1.

Albumin has a molecular mass of approximately 66 000 and is synthesized at a rate of about 12 g, equal to 3% of total body albumin, per day to replace that which is degraded or lost. Impaired albumin synthesis and therefore a low plasma albumin concentration, is a hallmark of chronic liver disease. Several functions can be ascribed to albumin including osmotic (oncotic) pressure regulation of the plasma and a non-specific transport protein for ligands such as calcium, fatty acids, drugs and bilirubin.

Many of the globulins act as transport proteins. Of particular interest are those proteins which are combined with lipids, themselves synthesized in the liver, to form lipoprotein complexes. High density lipoprotein (HDL), which contains predominantly apoproteins A and C combined with mainly phospholipids (most of the cholesterol found in mature HDL is added later) and very low density lipoprotein

Table 6.1 Some plasma proteins

Protein fraction	Examples	Function
Pre-albumin region	Pre-albumin	Transport of thyroid hormones
	Retinol binding protein	Transport of vitamin A
Albumin	Albumin	Osmotic regulation of plasma
		Non-specific carrier
α_1 globulins	α lipoprotein (apoA)	Carrier of cholesterol and phospholipids (HDL)
	α_1-antitrypsin	Protease inhibitor
	Alpha fetoprotein	?osmotic regulator in fetus
	α_1 acid glycoprotein	Unknown
α_2 globulins	Haptoglobin	Binds free haemoglobin
	α_2-macroglobulin	Unknown
	Caeruloplasmin	Binds and transports copper
β_1 globulins	Transferrin	Transports iron
	Haemopexin	Binds free haem groups
	β-lipoprotein (apoB)	Component of cholesterol-rich lipoproteins
	Complement C4	Immune defence
β_2 globulins	Fibrinogen	Clotting
	Complement C3	Immune defence
	β_2-microglobulin	Component of surface antigen
γ globulins	IgA, IgD, IgE, IgG, IgM	Immunoglobulins

(VLDL) comprising apolipoproteins B and E, combined mainly with triglycerides are secreted from the liver. Apo A is an α globulin and apo B is found in the β fraction, thus HDL and LDL are sometimes referred to as α and β lipoprotein respectively and VLDL because in separation by electrophoresis at pH 8.6, it runs ahead of LDL is called pre-β lipoprotein.

6.3.1.4 Urea synthesis

Urea ($(NH_2)_2CO$), a small and highly water soluble molecule, is an end product of amine and ammonia nitrogen metabolism and as such represents an example of biodetoxification (Section 6.4). The process is discussed in this section because it illustrates a genuine *de novo* biosynthetic pathway rather than detoxification involving chemical modification, via phase I and phase II reactions, of a pre-existing molecule as is the case for haem or steroid hormones.

Urea synthesis occurs in three stages:

1. liberation of ammonia from amino acids;

2. formation of carbamoyl phosphate from NH_3;

3. the reactions of the urea cycle as described by Krebs and Henseleit.

The whole process exemplifies the concept of compartmentalization, occurring partly within mitochondria and partly in the cytosol.

Stage 1. Oxidative deamination.

Glutamate dehydrogenase, the enzyme responsible for the liberation of ammonia from amino acids, occurs in two forms; one (cytosolic) is nicotinamide adenine dinucleotide (NAD^+) dependent whilst the other (mitochondrial) requires $NADP^+$ as coenzyme.

The mitochondrial glutamate dehydrogenase (GLDH) reaction is shown in Figure 6.5.

The mitochondrial glutamate dehydrogenase (GLDH) reaction is:

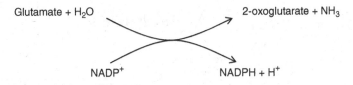

Figure 6.5 Oxidative deamination

This shows that glutamate is the ammonia donor. Glutamate itself may be formed from other amino acids by transamination reactions, thus allowing the amino groups from a range of molecules to contribute to urea synthesis. Before continuing, take a moment to look back to Figure 6.2, which shows the principle of transamination. Glutamate, produced by transamination reactions should be viewed as a metabolic 'focal point' or 'reservoir' of amino groups (Figure 6.6a).

For example, coupling alanine transamination (*via* ALT) with GLDH is shown in Figure 6.6b. A similar scheme can be drawn using, for example, aspartate transaminase in place of alanine transaminase.

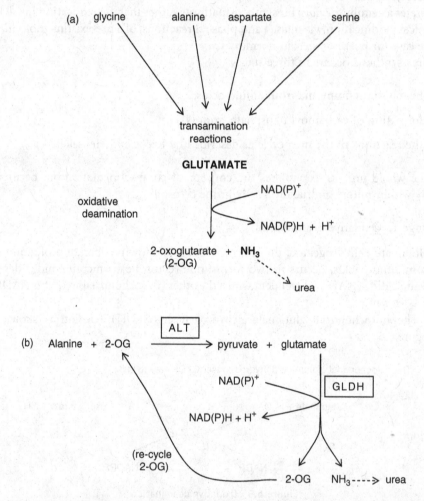

Figure 6.6 (a) Coupling transamination with deamination, the NH_2 groups of many amino acids can be used for urea synthesis. For example, coupling alanine transamination (via ALT) with GLDH; (b). Coupling of alanine transaminase with glutamate dehydrogenase

Stage 2. Carbamoyl phosphate (P-CO·NH$_2$) synthesis

The ammonia liberated by GLDH does not itself enter the urea cycle; it must first be combined with carbon dioxide to form carbamoyl phosphate. This is an energy (ATP) consuming reaction;

$$NH_3 + CO_2 + 2ATP \longrightarrow \text{carbamoyl phosphate} + 2ADP + Pi$$

$$\underset{O}{\overset{\|}{P\text{-}C\text{-}NH_2}}$$

The enzyme carbamoyl phosphate synthase (CPS) is a control point in the process.

Stage 3. The urea cycle (Figure 6.7)

This is a relatively simple cyclical pathway consisting of only four intermediates. However, to complete the structure of urea ((NH$_2$)$_2$CO) a second NH$_2$ group must be introduced and this is achieved by aspartate acting as the donor.

The first enzyme is ornithine transcarbamylase. This enzyme combines ornithine with carbamoyl phosphate to form citrulline. Argininiosuccinate (argininiosuccinic acid, ASA) is formed when argininosuccinate synthetase (ASS) combines citrulline with aspartate. The third step cleaves ASA, releasing succinate and arginine. Finally, urea is produced as arginine is converted to ornithine by arginase and so the cycle is ready to begin again.

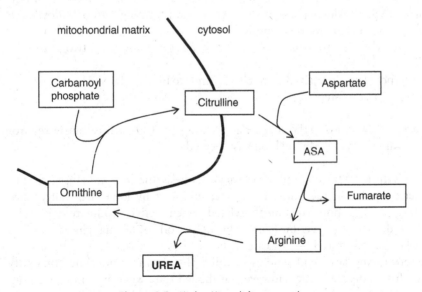

Figure 6.7 Krebs–Henseleit urea cycle

6.3.2 Lipid synthesis in the liver

Lipids are an essential part of our diet and of our biochemical constitution. Cell membranes need phosphoacylglycerides and cholesterol to maintain their structure and function; fat provides thermal and electrical insulation while fatty acids are an important fuel, especially for muscles like the heart. However, have you ever wondered why eating too much carbohydrate causes weight gain? How do jam doughnuts and sweet desserts cause unwanted changes in body shape? The answer, of course, lies in biochemistry.

6.3.2.1 Fatty acid and triglyceride (triacylglycerol) synthesis: acetyl-CoA carboxylase and fatty acid synthase multi-enzyme complex

In common with cholesterol synthesis described in the next section, fatty acids are derived from glucose-derived acetyl-CoA. In the fed state when glucose is plentiful and more than sufficient acetyl-CoA is available to supply the TCA cycle, carbon atoms are transported out of the mitochondrion as citrate (Figure 6.8). Once in the cytosol, citrate lyase forms acetyl-CoA and oxaloacetate (OAA) from the citrate. The OAA cannot re-enter the mitochondrion but is converted into malate by cytosolic malate dehydrogenase (cMDH) and then back into OAA by mitochondrial MDH (mMDH) Acetyl-CoA remains in the cytosol and is available for fatty acid synthesis.

The fatty acid synthesis pathway can be seen to occur in two parts. An initial 'priming' stage in which acetyl-CoA is converted to malonyl-CoA by a carboxylation reaction (Figure 6.9) is followed by a series of reactions which occur on a multi-enzyme complex (MEC), which achieves chain elongation forming C16 palmitoyl-CoA. The whole process occurs in the cytosol.

The acetyl-CoA carboxylase reaction may be seen to occur in two steps:

1.   a carboxyl group derived from HCO_3 binds to biotin attached to the enzyme, ATP is used at this step;

2.   acetyl-CoA binds to the enzyme and accepts the CO_2 as it is transferred from the biotin coenzyme. Malonyl-CoA is released.

This is the key control point of fatty acid synthesis, the enzyme being subject to both allosteric regulation (activated by citrate; inhibited by long chain fatty acid) and covalent modification (dephosphorylated when active, compare with glycogen synthase described in Section 6.3.3). During periods of fasting, glucagon-stimulated phosphorylation of acetyl-CoA carboxylase causes its dissociation into inactive sub-units preventing fatty acid synthesis, whilst at the same time stimulating fatty acid release from adipose tissue. However, in the fed state when insulin concentration is high, acetyl-CoA carboxylase is stimulated and the rise in the cytosolic concentration of citrate becomes a critical 'switch' because in addition to stimulating fatty acid synthesis

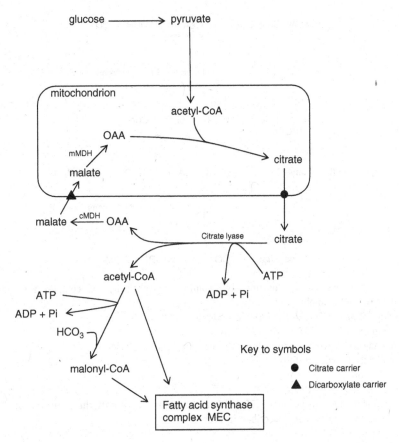

Figure 6.8 Export of acetyl-CoA carbon atoms from mitochondria for fatty acid synthesis

it also inhibits phosphofructokinase-1 (PFK-1). This has the effect of slowing glycolysis and so allowing more glucose-6-phosphate to be diverted through the pentose phosphate pathway and/or glycogen synthesis. To return to the traffic-flow analogy we introduced in Chapter 1, citrate thus acts as a traffic signal at a crossroad, directing the activity of key enzymes and so diverting substrates into preferred pathways. Furthermore, and as we will discover in the next section, fatty acid synthesis requires NADPH at two steps and the major source of this reduced coenzyme is the pentose phosphate pathway.

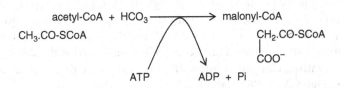

Figure 6.9 Acetyl-CoA carboxylase

(a) acetyl-CoA + malonyl-CoA + 2NADPH + 2H⁺
      (C2)   (C3)

one cycle of FAS complex reactions

butyryl-CoA + CO_2 + 2NADP⁺ + CoA
      (C4)

(b) acetyl-CoA + 7 malonyl-CoA + 14 NADPH + 14 H⁺

seven cycles of FAS complex

palmitate + $7CO_2$ + 14 NADP⁺ + 8CoA

Figure 6.10 (a) FAS reaction outline (b) Palmitate synthesis by FAS complex the individual reactions of the FAS complex are;
1. **B**inding of acetyl-CoA to the FAS; CoA released
2. **B**inding of malonyl-CoA to the FAS; CoA released
3. **C**ombination of acetate and malonate units and decarboxylation;
4. first **R**eduction with NADPH
5. **D**ehydration
6. second **R**eduction with NADPH
    (hint: remember the sequence BB-CRDR and compare this with the sequence of steps in β-oxidation of fatty acids described in Section 7.4).

The second phase of fatty acid synthesis is mediated by fatty acid synthase (FAS) complex, a large aggregate of six functionally different enzymes which work together in sequence, rather like a production line in a manufacturing process, to extend acetate by adding a 2-carbon unit derived from malonyl-CoA (Figure 6.10). Each 2-carbon addition requires 2 molecules of NADPH to provide reducing power. The synthesis terminates with the synthesis of C16 palmitoyl-CoA, thus seven consecutive cycles are required to complete the process but the growing acyl chain remains firmly attached to the fatty acid synthase multi-enzyme complex throughout. The whole process is outlined in Figure 6.11.

After each sequence of reactions the acyl group (C4 after one cycle, C6 after two cycles, C8 after three cycles, etc.) re-enters the process at step 3 and undergoes condensation with the next malonyl-CoA. When the final C16 palmitate molecule has been synthesized it is released but needs to be re-activated with CoA, to palmitoyl-CoA for desaturation, elongation into C18 fatty acid molecule or use for triglyceride synthesis (see Figures 6.12, 6.13 and 6.17).

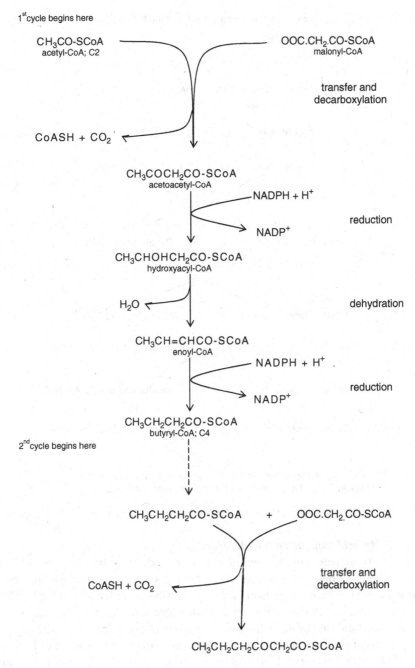

Figure 6.11 Fatty acid synthesis

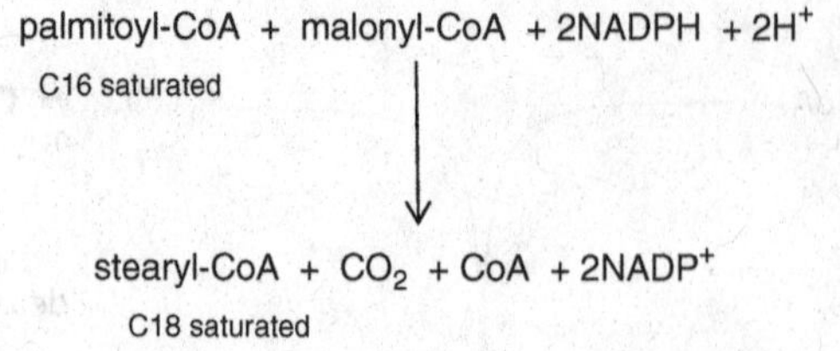

$$\text{palmitoyl-CoA} + \text{malonyl-CoA} + 2\text{NADPH} + 2\text{H}^+$$
C16 saturated

$$\downarrow$$

$$\text{stearyl-CoA} + \text{CO}_2 + \text{CoA} + 2\text{NADP}^+$$
C18 saturated

Figure 6.12　Elongation of C16 fatty acid

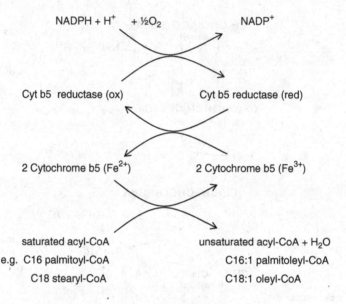

NB: the $\frac{1}{2}\text{O}_2$ shown at the top appears as water at the end of the sequence. This is typical of mixed function oxidase enzymes.

Figure 6.13　Microsomal desaturation of fatty acids

6.3.2.2 Fatty acid elongation and desaturation

The FAS multi-enzyme complex synthesizes saturated C16 fatty acids, but cells and tissues need unsaturated and longer chain fatty acids. The palmitoyl-CoA can be modified by either chain elongation and/or oxidation in order to produce different fatty acid molecules. Both elongation and desaturation occur within the smooth endoplasmic reticulum (SER, microsomal fraction) of the cell.

The process of elongation of C16 to C18 (stearyl-CoA) is independent of FAS but is chemically identical to one cycle of FAS reactions with malonyl-CoA as the C2 donor. Stearyl-CoA may be further elongated if necessary.

The desaturation process is particularly interesting as it provides an example of a microsomal (as opposed to mitochondrial) electron transport system. The enzymes responsible, fatty acyl-CoA desaturases, are examples of mixed function oxidases

(MFO). Similar MFO microsomal redox systems operate for the metabolic detoxification of drugs and some waste products (Section 6.4).

The carbon atoms within fatty acids are identified using two systems, either counting from the carboxyl carbon as No. 1 (the 'delta system') or counting from the methyl end of the molecules. By convention, the methyl carbon of the fatty acid irrespective of chain length is always labelled omega (ω, the last letter of the Greek alphabet). The terms 'omega-3' and 'omega-6' mean that the first double bond is positioned 3 or 6 carbons from the methyl end (the 'omega system') respectively (Figure 6.14). Such

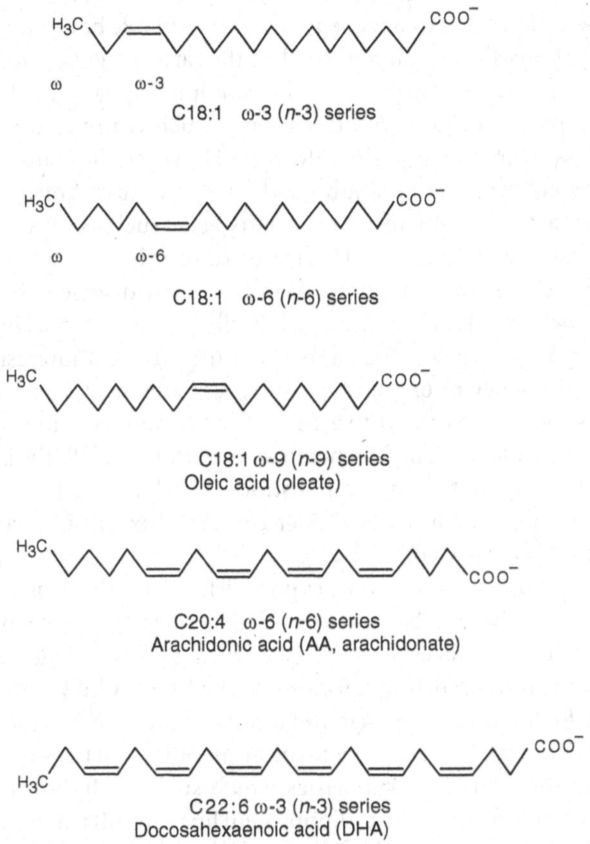

Polyunsaturated fatty acids (PUFA)

α-linolenic acid	(αLA)	C18:3	ω-3,6,9	(= $\Delta^{9,12,15}$)
γ-linolenic acid	(γLA)	C18:3	ω-6,9,12	(= $\Delta^{6,9,12}$)
arachidonic acid	(AA)	C20:4	ω-6,9,12,15	(= $\Delta^{5,8,11,14}$)
docosahexaenoic acid	(DHA)	C22:6	ω-3,6,9,12,15,18	(= $\Delta^{4,7,10,13,16,19}$)
eicosapentanoic acid	(EPA)	C20:5	ω-3,6,9,12,15	(= $\Delta^{5,8,11,14,17}$)

Figure 6.14 Unsaturated fatty acid structures

structures are signified as n-3 or ω-3 and n-6 or ω-6 respectively. Superscript abbreviations such as $\Delta^{9,12}$ (said as 'delta 9,12') indicate position of double bonds, for example between carbons 9 and 10 and 12 and 13.

Arachidonic acid (C20:4 n-6) is the precursor for the synthesis of prostaglandin molecules (Section 4.4.4), which have a wide range of biochemical effects on for example, the perception of pain, inflammation, blood clotting and smooth muscle contraction. Docosahexaenoic acid (DHA, C22:6) and eicosapentaenoic acid (EPA, C20:5) are both n-3 long-chain polyunsaturated fatty acids (PUFA) which have been shown to have significantly beneficial effects on intellectual development and inflammatory conditions such as asthma and cardiovascular disease.

Humans have only a limited capacity to introduce double bonds into long chain fatty acids between carbons 4–5, 5–6 and 9–10 from the carboxyl end, reactions catalysed by Δ^4, Δ^5 and Δ^9 desaturases respectively. The rate limiting step in PUFA synthesis in humans is catalysed by delta-6 (Δ^6) desaturase which is under the control of sterol regulatory element binding protein (SREBP). However, humans do not have the enzymes required to introduce double bonds between the ω carbon and the ω-9 carbon. Thus, the requirement for polyunsaturated fatty acids such as EPA and DHA must be met from the diet, with plant and marine sources being particularly important (Figure 6.15). Indeed, it was the observation that heart disease is comparatively rare in populations such as Eskimos and coastal dwelling populations, who consume large quantities of fish as part of their usual diet that first aroused interest in the effects of PUFA. Dietary deficiency of ω-3 and ω-6 PUFA results in their replacement by ω-9 series molecules (which we can synthesize) leading to abnormalities such as scaly skin and reproductive problems. The biochemical importance of PUFA is well recognized and more details of the nutritional implications of PUFA deficiency can be found by performing an author search for P. Calder or C.H. Ruxton, in a reputable Internet search engine such as PubMed.

Fatty acids are both stored in and exported from the liver as triglycerides. The carbon atoms for the glycerol 'backbone' of triglycerides are also derived from glucose by a diversion of dihydroxyacetone phosphate from glycolysis (Figures 6.16 and 6.17).

Endogenously synthesized triglyceride is packed into VLDL particles in the liver prior to export for long-term storage in adipose tissue (see Section 9.6.1). VLDL is metabolized by lipoprotein lipase, an enzyme immobilised on the surface of endothelial cells as it passes through small capillaries which supply adipose tissue. Progressive removal of triglyceride from VLDL by lipoprotein lipase results in the formation of proatherogenic low-density lipoprotein (LDL), sometimes, and very simplistically, called 'bad cholesterol'. LDL is removed from the circulation by a receptor-mediated process and too much circulating LDL, or a functional defect in the LDL receptor, lead to cholesterol accumulation especially in macrophages within the arterial wall. This is the process which leads to atherosclerosis.

Very low density lipoprotein is not the only lipoprotein to be secreted by the liver. HDL is released into the blood as a nascent (immature) discoid particle. As the HDL circulates within the circulation, it 'matures' by exchanging apoproteins and lipid components with other lipoproteins and cells. Mature spherical HDL is

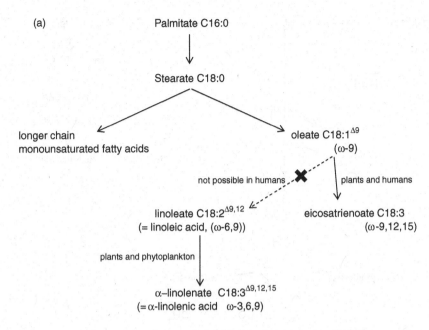

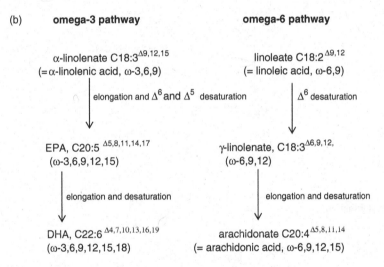

Figure 6.15 (a) Metabolic modification of endogenously synthesized fatty acids. (b) Metabolic modification of essential dietary fatty acids

anti-atherogenic and so sometimes referred to as 'good cholesterol'. Conventional wisdom asserts that HDL's main role is to remove excess cholesterol from tissues and return it to the liver for disposal. Significant research into HDL since the mid-1990s suggests that HDL (probably via its apoA component) has additional roles in

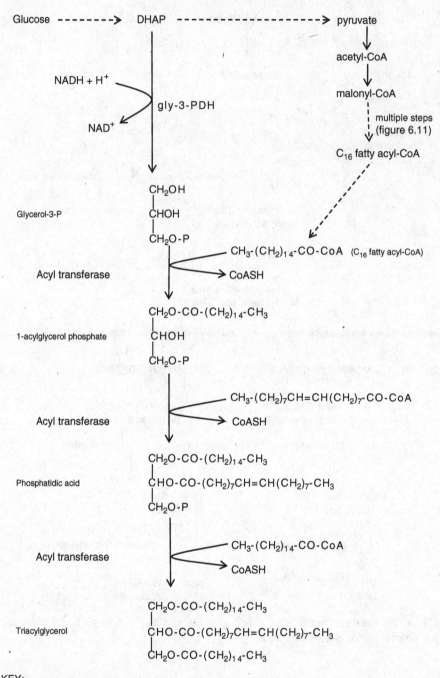

KEY:
Gly-3-PDH   glycerol-3-phosphate dehydrogenase
DHAP dihydroxyacetone phosphate
CoASH Coenzyme A

Figure 6.16 Triglyceride synthesis

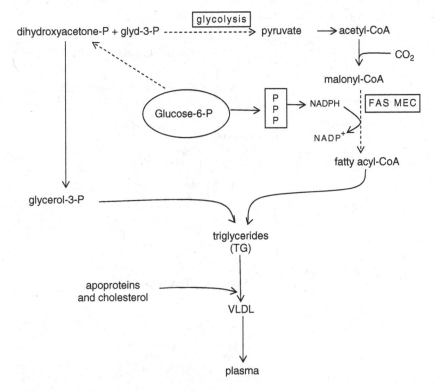

Broken lines indicate several intermediate steps

KEY

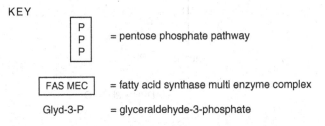

Figure 6.17 Glucose-6-phosphate is at the centre of triglyceride synthesis

protecting against premature cardiovascular disease. The interested reader should consult the research literature for the work of G.W. Cockerill and her colleagues for further details.

6.3.2.3 Cholesterol synthesis and ketogenesis

Several tissues of the body are able to synthesize cholesterol *de novo* (i.e. from its raw materials); the liver is one of these organs. Structurally, cholesterol belongs to the group of compounds called sterols (steroid alcohols) and is derived metabolically from acetate

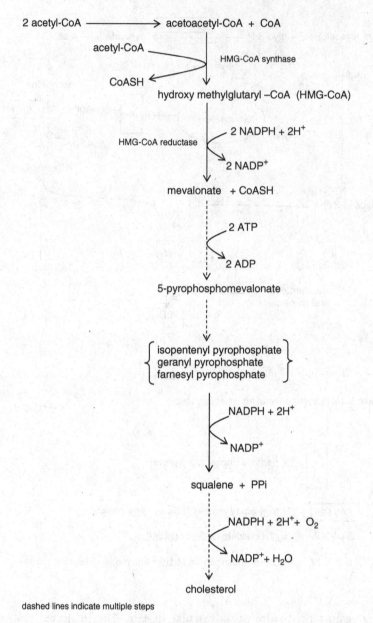

Figure 6.18 Cholesterol biosynthesis: Outline

(acetyl-CoA). The pathway occurs in the cytosol of the cell; compare this with ketogenesis described below which is mitochondrial (Figure 6.18).

A typical adult synthesizes approximately 900 mg of cholesterol each day; this figure compares with only about 400 mg per day coming from the diet. Endogenously synthesized cholesterol is transported from the liver in combination with endogenous triglyceride in VLDL particles, whereas, dietary cholesterol is transported as part of chylomicrons, which carry dietary, also called exogenous, triglycerides.

The key control enzyme in the biosynthetic pathway of cholesterol is hydroxy methylglutaryl CoA reductase (HMGCoA reductase). The enzyme is subject to several control mechanisms including phosphorylation/dephosphorylation initiated by glucagon (inhibits) and insulin (stimulates) and end-product feedback repression. Control of the enzyme is also exerted at the gene level by cholesterol-binding transcription factors (known as SREBPs) that regulate the expression of several genes involved in cholesterol biosynthesis, including genes for HMG CoA synthase and HMGCoA reductase.

It is SREBPs which coordinate the expression of HMG CoA reductase and cell surface receptors for LDL. Cholesterol is an essential component of membranes so if delivery of cholesterol to the cell is limited by low concentrations of LDL-cholesterol, the expression of the genes for both the LDL receptor and HMG CoA reductase are up-regulated allowing the cell to extract as much as possible form the circulation and also to synthesize cholesterol, thus there is an inverse relationship between plasma LDL-cholesterol concentration and HMG CoA reductase activity.

Experiments with fibroblasts *in vitro* have shown that even a very high concentration of LDL in the culture medium does not achieve more than approximately 85% inhibition of HMG-CoA reductase. Only when mevalonate was also added did the activity of the enzyme fall to very low levels. These results indicate that cells require at least two end products of the pathway to bring about repression. This observation should not be surprising, given the proximity to yet another metabolic crossroad (pathway branch point).

Because HMG CoA reductase occurs *before* a branch point in the biosynthetic pathway, complete inhibition of the enzyme by cholesterol would necessarily deprive the cell of many other intermediates, some of which are important in cell growth and division. A group of drugs known as statins are widely used to reduce plasma cholesterol concentration by inhibiting HMG CoA reductase. Interest is now rising in the possible use of statins as anticancer drugs due to their impact on reducing the production of mediators of cell proliferation.

Cholesterol is a non-oxidizable substrate so cannot be utilized for energy generation. The dangers of elevated cholesterol concentrations as a risk factor for cardiovascular disease are widely known and are described in detail in Section 5.5. Plasma cholesterol concentration is reduced by hepatic clearance; in fact, the liver is the only organ which can eliminate cholesterol from the body, a process achieved by excretion through the gut, possibly in the form of bile salts. Cholesterol is the starting material for the synthesis of bile acids from which bile salts are produced. One type of drug known generically as 'resins' for the treatment of hypercholesterolaemia bind to bile salts in the gut and so prevent cholesterol absorption.

Ketogenesis (Figure 6.19) occurs in the liver at most times but is greatly accelerated when acetyl-CoA production from β-oxidation of fatty acids exceeds the capacity of the TCA cycle to form citrate, that is during periods of starvation or in diabetics who have

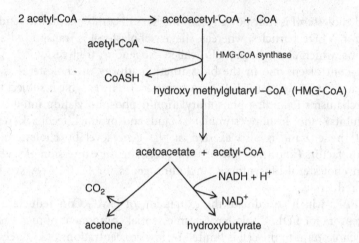

Figure 6.19 Ketogenesis: formation of acetoacetate, hydroxybutyrate and acetone

not taken insulin. In both cases, cells are metabolically glucose-deficient and so triglycerides stored in adipose tissue are hydrolysed and the fatty acids transported to tissues to be used as fuel. The details of triglyceride release from adipose tissue are given in Section 9.6.2 and the process of β-oxidation is described in Section 7.5.

Because of its involvement with many aspects of lipid metabolism so far described, it will be apparent from the discussion so far that acetyl-CoA is an 'axle' around which hepatic lipid metabolism revolves. Indeed, acetyl-CoA links lipid and carbohydrate metabolism. Figure 6.20 summarizes the central role of acetyl-CoA in lipid related pathways in the liver.

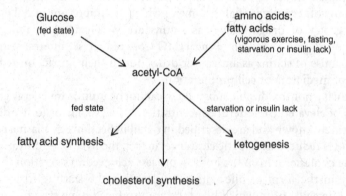

Figure 6.20 Acetyl-CoA as a precursor in lipid-related pathways

6.3.2.4 Glycogen synthesis
Glycogen is a polymer of D-glucose monomers, linked via α 1–4 glycosidic bonds with α1–6 links creating branch points (Figure 6.21).

An overview of the pathway, which is cytosolic, is as follows:

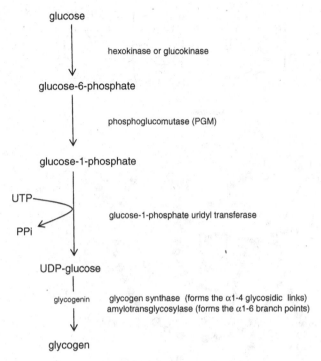

PPi is pyrophosphate, two phosphate groups joined together, P-P.

The introduction of uridyl triphosphate (UTP, a high energy compound similar to ATP) acts to 'energize' the substrate sufficiently to form a new covalent bond.

The first step in *de novo* glycogen synthesis involves the enzyme glycogenin, which autocatalytically attaches to itself six or seven glucose units, all derived from UDP-glucose, to form a short chain of α 1–4 glucosyl links. In this way, glycogenin acts as a 'primer' for the enzyme glycogen synthase which then elongates the primer forming the main α1–4 backbone of mature glycogen.

Glycogen synthase forms α1–4 links thus:

$$\text{UDP-glucose} + \underset{\text{(n glucose units)}}{\text{glycogen}} \rightarrow \text{UDP} + \underset{(n+1 \text{ glucose units})}{\text{glycogen}}$$

The branch points are introduced by amylotransglycosylase which breaks an α 1–4 link approximately 6 residues from the end of the growing chain and re-attaches the removed oligosaccharide as an α 1–6 link. Figure 6.22 illustrates the process.

The key control enzyme in the pathway is glycogen synthase (GS) which occurs in either a 'high activity' state (GS-a) or a 'low activity' state (GS-b); the switch from one to the other is brought about partly by covalent modification of the enzyme in response to stimulation by glucagon and partly by allosteric effects of key metabolites. Glycogen

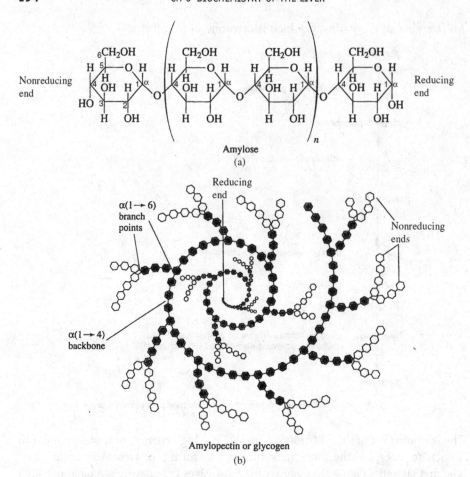

Figure 6.21 Structure of glycogen. Reproduced with permission from Boyer, R, *Concepts in Biochemistry*, John Wiley & Sons, 2006.

metabolism also occurs in skeletal muscle, but there are one or two key differences between the pathways in muscle and liver.

Glycogen synthase (GS) is a dimeric enzyme; each of the two subunits has three 'target' serine residues close to the N- and C- terminals of the protein. These serines are all subject to reversible covalent phosphorylation by the action of a number of protein kinases, notably glycogen synthase kinase-3 (GSK3), an AMP-stimulated kinase, a reaction which is countered by dephosphorylation (and thereby activation of GS) by protein phosphatase 1 (PP1). Phosphorylation of *liver* glycogen synthase (GS-a to GS-b transition) causes inhibition of glycogen synthesis due to a reduction in V_{max} whilst inhibition of *muscle* GS activity is by an increase in K_m for UDP-glucose.

Glycogen synthase kinase-3 is itself subject to control by reversible phosphorylation. Stimulation of the liver cell by insulin, the key hormone of the postprandial state,

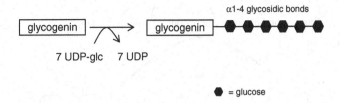

α1-4 glycosidic bonds

7 UDP-glc   7 UDP

● = glucose

(a) glycogenin acts as the base for the assembly of 6-7 glucose residues

(b) glycogen synthase extends the α1-4 linear chain using UDP-glc as glucose donor.

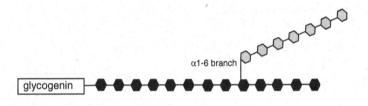

α1-6 branch

(c) amylotransglycosylase removes 6 or 7 glucose residues from the α1-4 chain and re-attaches them as an α1-6 link

Figure 6.22 Glycogen synthase and amylotransglycosylase

activates protein kinases, which phosphorylate and *in*activate GSK, promoting formation of GS-a so glycogen synthesis is favoured in the fed state. Additionally, covalent control of GS is enhanced by allosteric effects mediated on GS by high hepatocellular concentrations of glucose. Glucose-6-phosphate probably promotes glycogen synthesis by allosterically enhancing the action of protein phosphatase-1, which mediates the transition from GS-b to GS-a. Importantly therefore, feeding has two effects: (1) glucose intake and availability are increased, and (2) there is an insulin surge to mediate covalent control and increases in the concentration of the raw materials (glucose and glucose-6-phosphate) for glycogen synthesis.

The outline shown in Figure 6.23 contrasts with glycogen degradation (glycogenolysis) in which the major control step is mediated by glycogen phosphorylase (usually known simply as phosphorylase). It is phosphorylase which breaks the α 1–4 links within glycogen to release glucose-1-phosphate (Figure 6.24 and see Section 6.5 for more details).

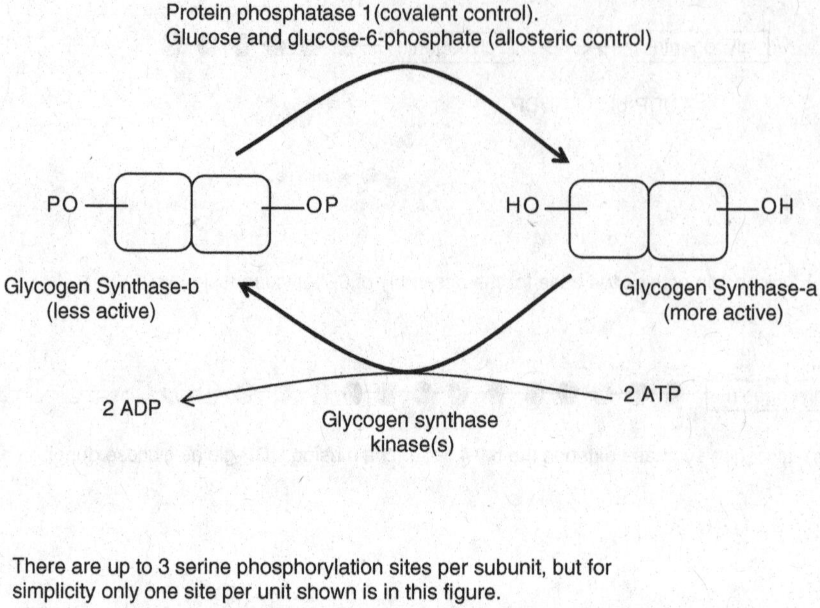

Protein phosphatase 1(covalent control).
Glucose and glucose-6-phosphate (allosteric control)

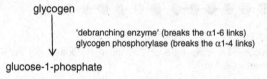

PO—☐☐—OP   HO—☐☐—OH

Glycogen Synthase-b   Glycogen Synthase-a
(less active)   (more active)

2 ADP ←————————————————————— 2 ATP
              Glycogen synthase
                 kinase(s)

There are up to 3 serine phosphorylation sites per subunit, but for
simplicity only one site per unit shown is in this figure.

Figure 6.23   Control of glycogen synthase activity

glycogen

↓   'debranching enzyme' (breaks the α1-6 links)
    glycogen phosphorylase (breaks the α1-4 links)

glucose-1-phosphate

Figure 6.24   First steps in glycogen degradation (compare with Figure 6.39)

Clearly, it would not be advantageous for synthesis and degradation to occur simultaneously, especially as they occur in the same cell compartment, the cytosol. Reciprocal control, that is when one enzyme is '*On*' the other is '*Off*', is achieved via reversible phosphorylation of the key controlling enzyme and allosteric control (see Table 6.2 and Section 3.2.2). However, it is probable that both enzymes retain some residual activity at most times.

Table 6.2   Control steps in glycogenolysis

Enzyme	Activity when phosphorylated	Activity when *de*phosphorylated	Allosteric activators	Allosteric inhibitors
Glycogen synthase	Low	High	Glucose glucose-6-P	AMP Pi (muscle)
Glycogen phosphorylase	High	Low	AMP	ATP glucose glucose-6-P

As stated earlier, glycogen metabolism occurs in skeletal muscle (see Chapter 7) as well as in the liver. Physiologically, the main difference between the two tissues is that liver glycogen is used to 'top-up' blood glucose concentration and thus provide fuel for all tissues of the body whereas muscle glycogen is for sole use within the muscle. In addition to subtle differences in the control of glycogen synthase and glycogen phosphorylase, liver, but not muscle, contains glucose-6-phosphatase. This enzyme is microsomal with its active site facing the lumen of the endoplasmic reticulum of the liver cell, indicating the need for a glucose-6-phosphate transporter to move the substrate from the cytosolic compartment. Here then we see how metabolic differences are reflected in differing physiological roles.

6.3.3 Haem synthesis

All tissues except mature red blood cells are able to manufacture haem for use in the respiratory cytochrome proteins of the electron transport chain. However, the liver is an especially important site of haem synthesis because it (a) is a major organ of erythropoiesis *in utero* and (b) haem-containing cytochrome-P450 (CYP-450) enzymes play significant roles in hepatic detoxification of drugs, toxins and endogenous waste products (Section 6.4).

Chemically, haem is a ferro protoporphyrin whose synthesis begins when succinyl-CoA from the TCA cycle and glycine combine to form aminolevulinic acid (ALA) in a reaction catalysed by ALA synthase within a mitochondrion. The product of this reaction is transported to the cytosol where it is converted into porphobilinogen (PBG), a simple heterocyclic ring structure. Four molecules of PBG are required to form the first recognizable porphyrin ring, uroporphyrinogen. Successive decarboxylations initially in the cytosol but then in the mitochondrion, form coproporphyrinogen and protoporphyrinogen. Haem is formed when the enzyme ferrochelatase inserts an atom of iron into the centre of the protopophyrin ring. Control of the pathway is exerted by the inhibitory action of haem on the gene for ALA synthase. Refer to Figure 5.16 for a diagram of the pathway.

6.4 Detoxification and waste disposal

In addition to some of the synthetic roles played by the liver described in the previous section, the liver also is the prime organ for the metabolism and excretion of unwanted compounds. Such compounds may be exogenous (e.g. drugs and poisons) or of endogenous origin (e.g. steroid or catecholamine hormones and haem groups). The basic biochemical processes, referred to as phase I and phase II reactions, are similar regardless of the starting substrate. The overall effects of phase I and phase II reactions are to (a) increase water solubility of the parent compound, (b) increase the acidity (lower the pKa) of the parent and (c) to decrease the toxicity (increase the LD_{50}).

Table 6.3 Typical data for the metabolism of benzoic acid to hippuric acid

Property	Parent compound (benzoic acid)	Metabolite (hippuric acid)
Solubility (mg/100 ml water)	185	465
Acidity (pKa)	4.2	3.7
LD$_{50}$ (g/kg in mice)^a	2.0	4.2

^{a}LD50 is the dose required to cause death in 50% of a group of experimental animals.
Data based on information given in Gillham, Papachristodoulou and Thomas, 2001, Arnold.

Table 6.3 gives typical data for the metabolism of benzoic acid to its less toxic metabolite hippuric acid.

Phase I reactions (Figure 6.25) increase the polarity of the waste compound by oxidation or hydroxylation in a process which involves an electron transfer and

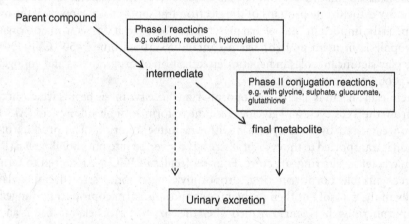

Figure 6.25 Overview of hepatic detoxification reactions

mediated by members of the family of enzymes known as cytochrome P450. Phase II reactions are conjugations: compounds such as glucuronic acid (glucuronate), sulfate, glycine or glutathione are added to the chemically modified waste compound in order to allow excretion via the urine.

6.4.1.1 Phase I reactions

CYP is the generic name for a very large number of chemically related haem-containing enzymes. Many isoenzymes of CYP-450 exist; these are categorized into families (designated by a number), subfamilies (shown as a capital letter) and as particular gene products (a number). For example; CYP1 'family', CYP2E family and sub-family, CYP3A1 a particular gene product.

The CYP enzymes active in phase I reactions are often oxidases or hydroxylases, sometimes called mixed function oxidase (MFO). An oxidase enzyme introduces into the substrate (i.e. the unwanted compound) *both* atoms of an oxygen molecule whilst

$$XH + O_2 \longrightarrow XOOH \qquad \text{oxidase}$$

$$XH + O_2 + 2H^+ \longrightarrow XOH + H_2O \qquad \text{hydroxylase}$$

Figure 6.26   Mixed function oxygenases

an MFO incorporates only one of the two oxygen atoms to form an hydroxyl group. This distinction is shown in Figure 6.26.

In order to function, CYP-450 enzymes require a source of reducing power (shown as $2H^+$ for the hydroxylase reaction above) in the form of NADPH. This coenzyme is a

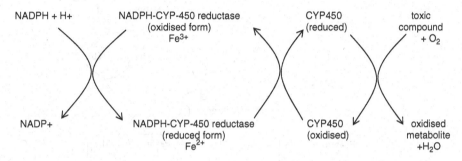

Figure 6.27   Passage of reducing power from NADPH to form water. One atom of oxygen is incorporated into the toxic compound (=oxidation). NADPH-cytochrome P-450 reductase is a flavin-containing protein

product of the pentose phosphate pathway of glucose oxidation (see Figures 5.19 and 6.17), pathway which is also very active in the liver. A carrier molecule, called NADPH-cytochrome P450 reductase is required to transport the electrons from NADPH to CYP450 (Figure 6.27).

Both the CYP and the reductase are firmly embedded within the membrane of the smooth endoplasmic reticulum (SER), which forms microsomes when cells are broken for experimental purposes. Figure 6.28 illustrates the whole process of a generalized phase I hydroxylation.

Many of the specific enzymes involved with phase I reactions are subject to induction (see Section 3.3.3), that is increased synthesis by activation of the particular genes. Two clinical consequences arise from this:

1.   physicians may need to increase the dose of a therapeutic drug to compensate for its more rapid clearance; for example an anticonvulsant such as carbamazepine or phenytoin, given to a patient in order to maintain the desired plasma concentration and therapeutic benefit of that drug, and

2.   co-ingestion of two or more drugs may interfere with the usual metabolism of one or both. The desired therapeutic effect of a drug may be reduced if there is

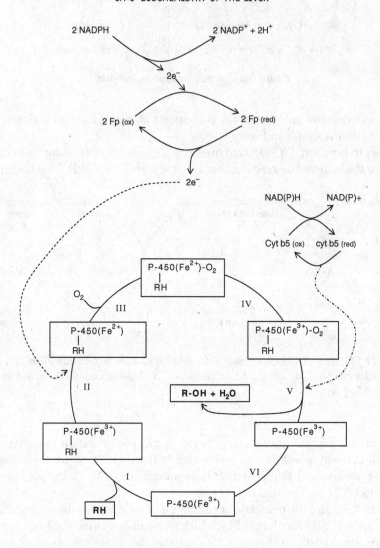

KEY: RH   substrate 'toxin' to be oxidised

            P-450   cytochrome P-450

            Fp   NADPH cytochrome P450 reductase (flavoprotein)

            R-OH   oxidised product

Step I   substrate binds to cytochrome P-450

      II   P-450 accepts one electron from Fp and becomes reduced ($Fe^{3+} \longrightarrow Fe^{2+}$)

      III   molecular oxygen binds

      IV   electron transfer from oxygen to haem group of P-450 ($Fe^{2+} \longrightarrow Fe^{3+}$)

      V   a second electron is accepted from NAD(P)H *via* cytochrome b5 and oxidised

            product released; the other atom of oxygen appears as water

      VI   P-450 ready to accept another substrate and the cycle begins again

Figure 6.28 Cytochrome P450

induction of the CYP responsible for its metabolism, or the drug may reach toxic concentrations if its metabolism is inhibited by another xenobiotic (foreign) compound.

Specific examples of phase I oxidative reactions are shown in Figure 6.29.

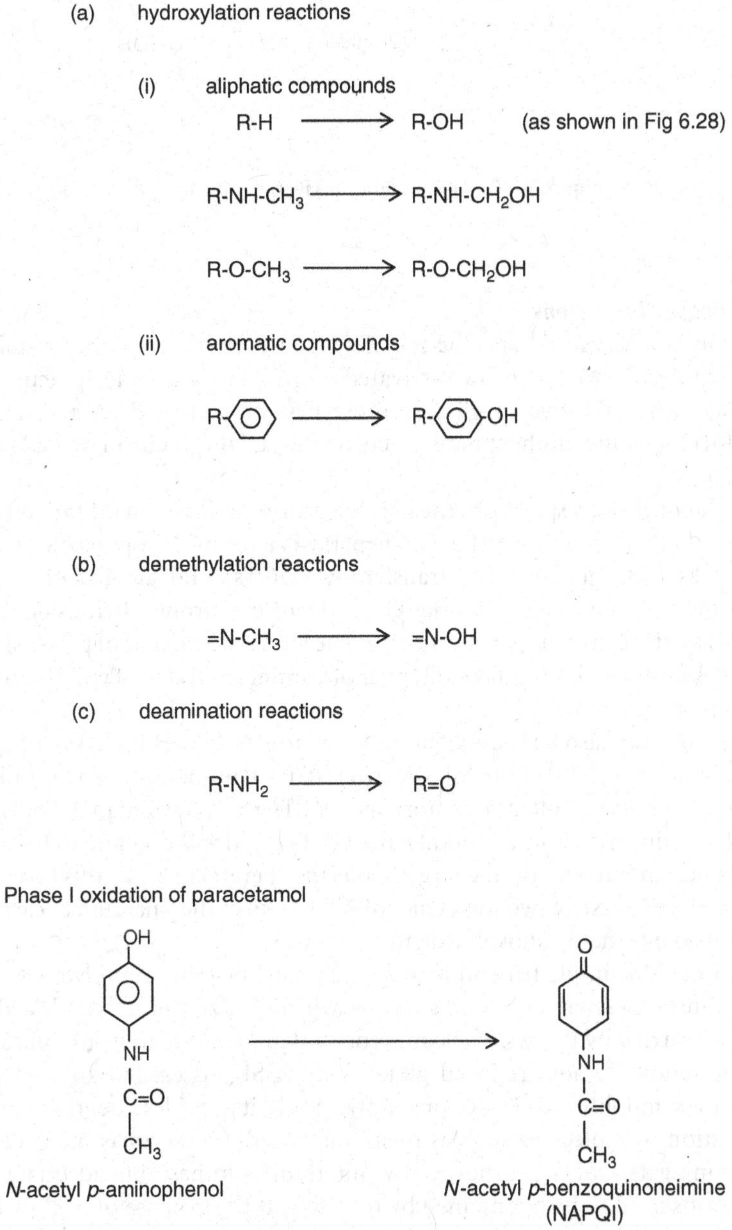

(a) hydroxylation reactions

(i) aliphatic compounds

R-H $\longrightarrow$ R-OH   (as shown in Fig 6.28)

R-NH-CH$_3$ $\longrightarrow$ R-NH-CH$_2$OH

R-O-CH$_3$ $\longrightarrow$ R-O-CH$_2$OH

(ii) aromatic compounds

R-⟨O⟩ $\longrightarrow$ R-⟨O⟩-OH

(b) demethylation reactions

=N-CH$_3$ $\longrightarrow$ =N-OH

(c) deamination reactions

R-NH$_2$ $\longrightarrow$ R=O

Phase I oxidation of paracetamol

N-acetyl *p*-aminophenol   *N*-acetyl *p*-benzoquinoneimine
                                                              (NAPQI)

Figure 6.29 Typical phase I reactions

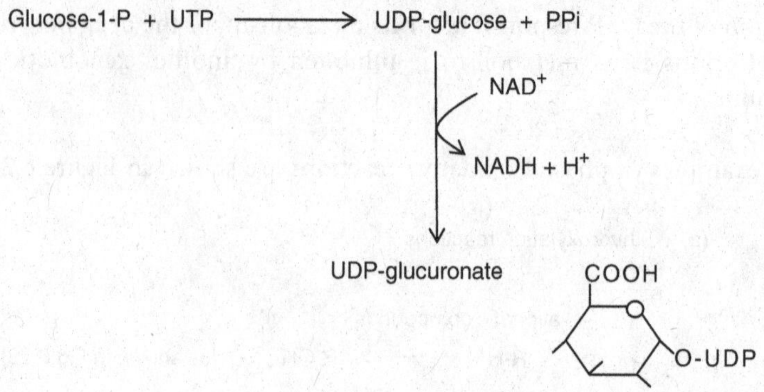

$$\text{Glucose-1-P + UTP} \longrightarrow \text{UDP-glucose + PPi}$$

NAD$^+$

NADH + H$^+$

UDP-glucuronate

COOH

O-UDP

Figure 6.30 Formation of activated glucuronate

6.4.1.2 Phase II reactions

Conjugation is achieved via specific transferases, which may require the substituent group or molecule to be in an 'activated' form. For example, glucuronic acid (glucuronate), a derivative of glucose must first be activated by a reaction with UTP to form uridine diphosphate-glucuronate (UDP-glucuronate) as shown in Figure 6.30.

Although not entirely specific to the liver, glucuronidation is quantitatively the most important phase II reaction. The ten hepatic isoenzymes responsible are known collectively as UDP-glucuronosyl transferases (UGTs) and are found in the SER alongside the CYP enzymes. Like the CYPs, UDP-glucuronosyl transferase activity may be induced by increasing substrate load. The UGTs are capable of processing a wide variety of substrates including alcohols, phenols, amines and acids Typical reactions are illustrated in Figure 6.31.

Conjugations can also be brought about by sulfotransferases (SULTs) and glutathione-S-transferases (GSTs), both of which exist in a number of isoenzymic forms. Amines and alcohols are sulfate acceptors and SULTs are important in steroid hormone and catecholamine metabolism and like the UGTs require the sulfate to be 'activated' prior to its incorporation into the target molecule (Figure 6.32). In this case, sulfate is activated at the expense of two molecules of ATP to form the final sulfate carrier PAPS (3'-phosphoadenosine-5'-phosphosulfate).

Glutathione is a simple tripeptide of γ-glutamate, cysteine and glycine and glutathione-S-transferases) are cytosolic enzymes which utilize the sulfydryl (–SH) functional group carried by the cysteine component of glutathione to form thioesters with toxic compounds. Because reduced glutathione (GSH) is easily oxidized (oxidized glutathione is symbolised as GS-SG, or simply GSSG), it provides a degree of protection from oxidation of molecules such as membrane lipids so serves as more than just a potential conjugate. GSH is synthesized by glutathione synthase, but adequate cytosolic concentrations need to be maintained by recycling at the expense of NADPH derived from the pentose phosphate pathway (Figure 6.33).

Formation of UDP-glucose

$$\text{glucose-1-phosphate} + \text{UTP} \xrightarrow{\text{2 step reaction}} \text{UDP-glucuronate} + \text{PPi} + \text{Pi}$$

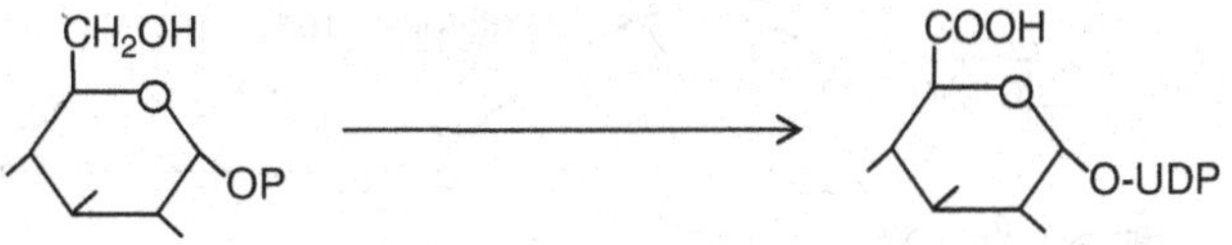

(a)   generic equation

UDP-glucuronate + R-OH   glucuronide product

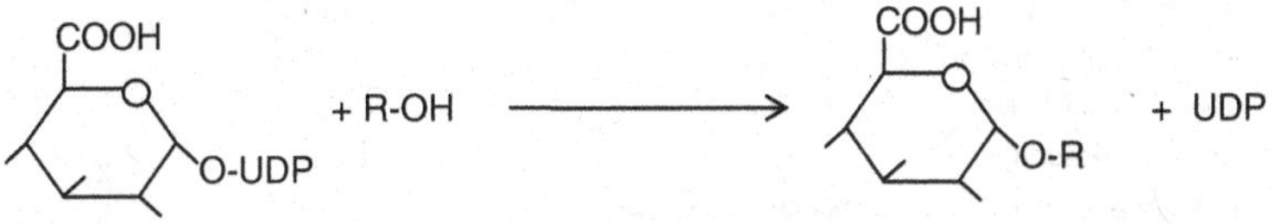

(b)   phenols

phenolic   +   UDP-glucuronate ⟶ phenolic glucuronide

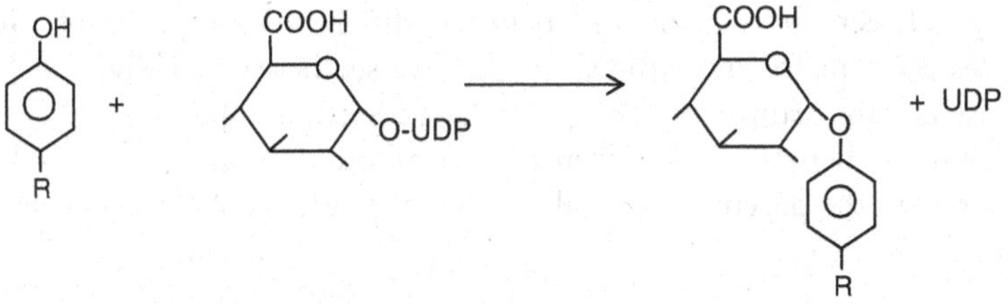

Figure 6.31 Glucuronidation reactions (catalysed by glucuronosyl transferases)

activation steps

(i)   sulphate + ATP ⟶ adenosine phosphate

(ii)   adenosine phosphate + ATP ⟶ PAPS

conjugation

(iii)   PAPS + R-OH ⟶ R-OSO$_4$ + phosphoadenosine phosphate

Figure 6.32 Sulphation activation steps

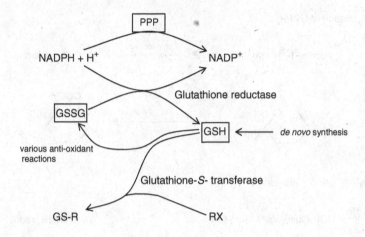

GSSG = oxidised glutathione
GSH   = reduced glutathione
RX   = toxic compound
GS-R = glutathione conjugate (e.g. thioester derivative)

Figure 6.33 Glutathione conjugation

The existence of various isoforms of many of the enzymes involved with phase I and phase II reactions has been noted above. Key enzymes are also subject to variation by genetic polymorphisms so there may be considerable difference in the metabolic efficiency between individuals and between different ethnic groups. Such genetic differences account for most the variability we see between individuals' capacity to metabolise certain drugs. For example, refer to Section 6.4.3.

Occasionally, hepatic metabolism of a compound creates a metabolite which is *more* toxic than the parent; this is called a lethal synthesis. An example is given in the following section.

6.4.2   Examples of phase I and phase II reactions for the metabolism of exogenous and endogenous compounds

Two examples will be used to illustrate phase I and phase II reactions; paracetamol and the catabolism of haem groups.

6.4.2.1 Paracetamol metabolism
When taken in therapeutic doses (approximately 4 g; ~60 mg/kg body weight) paracetamol (*N*-acetyl-*para*-aminophenol, acetaminophen) is a safe and effective analgesic, but overdosage, possibly with the intent of self-harm, is a major cause of drug-induced hepatic toxicity. The cellular damage which may not be evident for up to

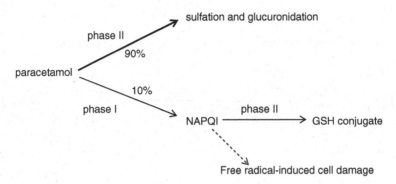

Figure 6.34 Paracetamol metabolism

60 h post ingestion is due to the failure to eliminate a toxic metabolic intermediate of the drug known as *N*-acetyl-*p*-benzoquinoneimine (NAPQI).

More than 90% of a paracetamol dose is metabolized by phase II reactions only; 5–10% of the drug is metabolized by CYP2E1 to produce NAPQI which is further metabolized by conjugation with glutathione. When concentrations of NAPQI exceed the availability of glutathione, cellular damage leading to cell death occurs (Figures 6.34 and 6.35).

Treatment of paracetamol overdose is based on replenishment of antioxidant thiols to supplement the role of glutathione; the most commonly used antidote is *N*-acetyl cysteine, but is only effective if given within a particular time window after ingestion.

6.4.2.2 Haem catabolism. The formation of bilirubin

Approximately 80–85% of the haem which undergoes catabolism each day derives from red blood cells. The remainder is from haem-containing enzymes such as cytochromes, peroxidases and catalase.

Free haem groups are ferroporphyrins (cyclic tetrapyrroles). The first reaction of haem catabolism is the release of iron; this is followed by the opening of the ring to produce a linear tetrapyrrole called biliverdin. A molecule of carbon monoxide is released as the ring opens. Biliverdin is converted to bilirubin by reduction. These initial reactions may occur in the liver or in other tissues of the reticuloendothelial system, notably the spleen.

Because of its very low water solubility, bilirubin formed outside the liver must be transported through the circulation bound to albumin. This is called *unconjugated* bilirubin, indicating that it has not yet passed through the phase II reactions in the liver.

Unconjugated bilirubin is taken into the hepatocytes by binding to membrane transport proteins and transported through the liver cells to the SER by proteins called ligandins. The SER is the location of a specific bilirubin-UDP-glucuronosyl transferase

(a)   direct glucuronidation of parent compound

paracetamol + UDP-glucuronate ⟶ glucuronide +UDP

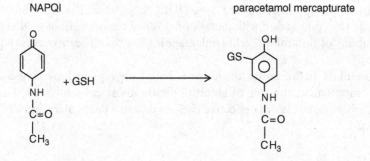

(b)   conjugation of phase I product with glutathione

Figure 6.35 Biotransformation of paracetamol (*N*-acetyl *p*-aminophenol)

whose function is to attach two molecules of glucuronate to the bilirubin (now called *conjugated* bilirubin), which is water soluble (Figure 6.36). Most of the conjugated bilirubin enters the bile canaliculi (very small vessels of the biliary system which drain the liver lobules) and passes to the gut via the gall bladder and the common bile duct. A small amount of the conjugated bilirubin leaks into the plasma. In health, only about 10% of the total plasma bilirubin is conjugated and sometimes, for a reason related to its quantitative estimation in the clinical laboratory, may be referred to as 'direct reacting' bilirubin.

Bacterial action in the large gut converts the conjugated bilirubin into bilinogens and then to yellow-brown coloured bilins which finally leave the body through in the faeces (as stercobilin) and urine (as urobilin). A small proportion of the bilin produced in the gut is passively reabsorbed into the portal system and re-excreted as the blood flows through the liver.

The over-production of bilirubin to the point at which the liver's capacity to metabolize is exceeded or if there is dysfunction of the liver itself due to damage or metabolic immaturity, can lead to a yellow discolouration of tissues called jaundice. The accumulation of unconjugated bilirubin in neonates, often as a result of antibody-mediated destruction of the baby's red cells is dangerous as serious and irreversible brain damage can occur. Acute or chronic damage to the adult liver (hepatitis) may cause jaundice but not brain damage.

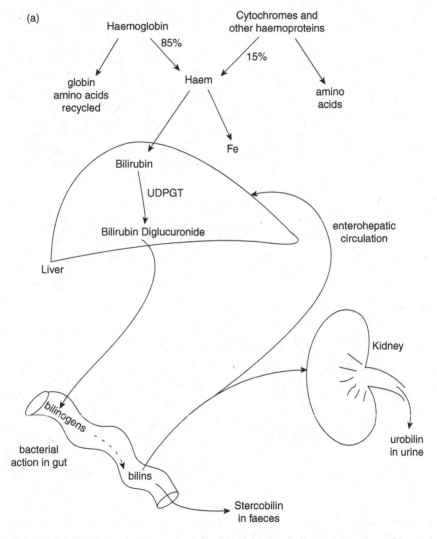

Figure 6.36 (a) Haem catabolism; overview. (b) Haem catabolism; conversion of haem into bilirubin

(b)

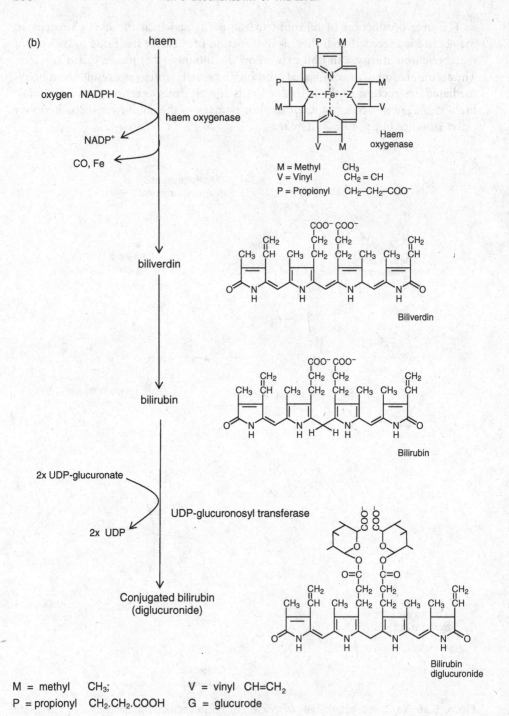

M = methyl   CH₃;   V = vinyl CH=CH₂
P = propionyl CH₂.CH₂.COOH   G = glucurode

Figure 6.36 (*Continued*)

6.4.3 Alcohol metabolism

Alcohol (ethanol) is a commonly used drug many human societies. Although the consumption of alcohol in moderation has some beneficial physiological effects, alcohol abuse and over indulgence lead to serious consequences to the individual, to the fetus during pregnancy and to society as a whole. A study of the metabolism of alcohol is instructive as the processes act as useful models of enzymology and genetics.

Blood alcohol concentration (BAC), typically in the range 4–20 mmol/l (equivalent to approximately 20–100 mg/100 ml) following 'social' drinking, is determined by the balance between the quantity consumed and the rate of absorption on the one hand and, on the other, its dilution in body fluids and the rate of elimination via metabolism (which accounts for about 98%) of the ingested load. Metabolism occurs mainly in the liver but many tissues including the brain and muscle have some catabolic capacity and a little is excreted unchanged (only about 2% of the load; in exhaled breath, through the skin and in the urine). Absorption from the stomach and upper small intestine is very rapid, reaching peak BAC after 30–45 min. This is much more rapid than the rate of ethanol elimination so BAC rises quickly following intake and may remain elevated for several hours thereafter.

Alcohol metabolism (Figure 6.37) occurs mainly through oxidative pathways involving the enzymes alcohol dehydrogenase (ADH), acetaldehyde dehydrogenase (ALDH),

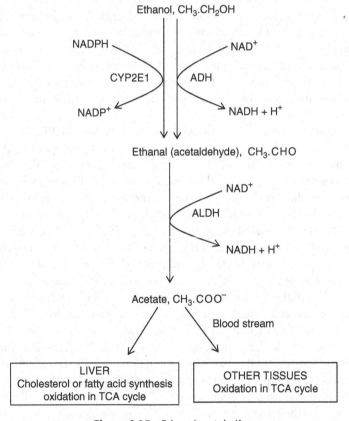

Figure 6.37 Ethanol metabolism

CYP-450 isoenzymes, principally CYP2E1, but also to a lesser extent, the CYP1A2 and CYP3A4, and catalase. The genes encoding ADH, ALDH and CYP2E1 all exhibit polymorphism resulting in enzyme proteins with varying metabolic efficiency as described later. Non-oxidative metabolism of alcohol is minimal but the formation of fatty acid ethyl esters or phosphatidyl ethanol may have significant pathological effects such as interference with cell signalling and proliferation and development of foetal brain cells. A small amount of ethanol escapes phase I reactions and is excreted as ethyl or glucuronyl conjugates.

ADH is a non-specific enzyme, able to use a wide range of simple alcohols, hydroxysteroids and retinol as substrates. Clearly therefore, we can conclude that the evolutionary role of ADH was not simply to metabolize ethanol in alcoholic beverages. The enzyme requires zinc as a cofactor and is found located in the cytosolic compartment of cells of several tissue types. At least seven genes encoding ADH have been described and two of these (*ADH2* and *ADH3*) also exhibit allelic polymorphisms resulting in 10 different proteins, each with a molecular weight of approximately 40 000 Da. Note here that genes are indicated in italic typeface and enzyme/protein in roman type.

The liver expresses *ADH1*, *ADH2*, *ADH3*, *ADH4* and *ADH7*; most tissues express *ADH5*, whilst *ADH3* and *ADH7* are found in the gut. *ADH3* in the gut is thought to play an important role in initial metabolism of ethanol and the lower activity of γADH isoforms in females may contribute to gender differences in alcohol metabolism.

Gene frequencies and allelic mutations for ADH vary considerably and significant interpersonal and ethnic variations have been described. Furthermore, functional ADH is a dimer, usually of identical subunits but heterodimers are also possible and to date approximately 20 ADH isoenzymes are known. There are notable kinetic differences between the various functional ADH isoenzymes; for example, K_m values differ almost 1000-fold and V_{max} by approximately 40-fold (see Table 6.4). These differences account, to some extent, for variation in alcohol tolerance between individuals.

The microsomal ethanol-oxidizing system (MEOS) utilizes CYP2E1 and accounts for approximately 10% of the metabolism when small quantities of ethanol are ingested. However, although CYP2E1 has a higher K_m for ethanol (approximately 12 mmol/l), compared with most ADH isoenzymes, CYP2E1, like many cytochrome P450 enzymes, is inducible and long-term consumption of alcohol will increase enzyme production and decrease its degradation. This has potentially significant consequences on the CYP-mediated metabolism of other drugs, whose clearance may be affected. CYP2E1 but not ADH is expressed in the central nervous system. Excessive maternal alcohol intake in pregnancy leads to induction of CYP2E1 in the fetal brain at a critical time of organogenesis (around 55 days gestation). This has been associated with a range of physical and mental abnormalities known as fetal alcohol syndrome (FAS) or fetal alcohol effects (FAE). Individuals exposed to alcohol *in utero* but without obvious signs of FAS may be prone to mental health and behavioural problems later in life.

Ethanal (acetaldehyde), the product of ADH and CYP action on ethanol is the substrate for isoforms of ALDH. Like ADH, ALDH isoenzymes are not entirely substrate specific and will act on aliphatic and aromatic aldehydes to generate the corresponding carboxylic acid. There are nine genes encoding ALDH isoforms and all are subject to polymorphism and the enzyme products of *ALDH* gene expression are found widely in

Table 6.4 ADH isoenzymes

Gene	Protein subunit	Sequence variation	Enzyme structure	Approx K_m for ethanol (mmol/l)	Approx. K_m For NAD$^+$ (mmol/l)	V_{max} (U/mg protein)	Notes
ADH1	α		αα	4	13	0.5	
ADH2/1	β1	Arg^{47} Arg^{369} (= wild type)	β1β1	0.05	7.5	0.25	Common in Caucasians and ethnic Africans
ADH2/2	β2	His^{47} Arg^{369}	β2β2	1	180	10	Common in Oriental races and Jews
ADH2/3	β3	Arg^{47} Cys^{369}	β3β3	37	700	8	Common in ethnic Africans
ADH3/1	γ1		γ1γ1	non MM[a]	8	2.5	Common in Asians
ADH3/2	γ2	Gln^{271} Val^{349}	γ2γ2	non MM[a]	8.5	0.9	
ADH4	π		ππ	34	14	0.5	Liver only
ADH5	χ		χχ	??	25	11	Most tissues

[a]Non MM = non Michaelis–Menten kinetics. Note also that the kinetic differences between γ1 and γ2 containing ADH are functionally negligible. Alterations in the β and γ subunits are seen primarily to affect binding of NAD$^+$. ?? = not certain.

many human tissues. In particular, two main isoforms (ALDH-1 and ALDH-2) enzymes are located in the cytosol and mitochondria of hepatocytes respectively.

ALDH-2 is of prime importance as its K_m for acetaldehyde is very low (about 2 μmol/l), well below the expected cellular concentration during the metabolism of even moderate quantities of ethanol. Two allelic variants of the *ALDH-2* gene exist; wild type *ALHD2/1* and *ALDH2/2*, which carries a point mutation (guanine to adenine transition) resulting in substitution of lysine for glutamate at position 487. Here we have an excellent example of how crucial primary structure is to enzyme action. The substitution is close to the presumed active site and individuals homozygous for the *ALDH-2/2* variant are completely lacking acetalde-hyde dehydrogenase activity whereas heterozygotes have typically 30–50% of normal enzyme activity. Any individuals who are homozygous for *ALDH-2/2* experience discomforting side effects of drinking even quite small volumes of alcohol and so are deterred from over indulgence. Heterozygotes however seem to have a higher risk of developing alcoholism.

An estimated 50% of oriental people have the *ALDH-2/2* variant and thus the inactive isoenzyme, which probably accounts for the far lower incidence of heavy drinkers found amongst Chinese and Japanese than, say, Caucasians. Disulfiram is a drug used to treat alcoholics; its action is to inhibit ALDH thus creating in heavy or compulsive drinkers an unpleasant and hopefully deterrent reaction.

As implied above, acetaldehyde is toxic; the accumulation of this metabolite leads to many of the deleterious consequences of alcohol intoxication because it is able to alter the function of many proteins by binding to certain amino acids such lysine, cysteine and those with aromatic side chains. Apolipoproteins, haemoglobin, collagen, cytochromes and albumin are all subject to chemical modification by acetaldehyde; so too are proteins found associated with the cell membrane and cytoskeleton. A characteristic laboratory finding in alcoholics is macrocytosis, that is a high number of enlarged red cells in the circulation, a result of damage to the erythrocyte membrane. In experimental rats, similar acetaldehyde-induced damage to proteins in hepatocyte cell membranes led to an auto-immune antibody response; evidence of comparable events has been found in humans with alcoholic liver disease (ALD).

Oxidative metabolism of ethanol in the cytosol, mitochondria, microsomes and peroxisomes carries with it an increased risk for the generation of reactive oxygen species (ROS) as by-products. These short-lived but very reactive chemicals such as hydrogen peroxide (H_2O_2), and oxygen-containing free radicals (superoxide, hydroxyethyl and hydroxyl radicals) damage DNA, proteins and lipids resulting in cell death and generate compounds derived from lipid peroxidation which like acetaldehyde form protein adducts that initiate an inflammatory response. Liver pathology associated with excessive alcohol intake is due in part to the production of ROS and is often exacerbated by the poor diet (especially the low intake of antioxidant vitamins and minerals) of alcoholics.

6.5 Maintenance of blood glucose concentration

Although many tissues can utilize fat as a source of fuel, an adequate supply of glucose is essential for normal metabolism in some tissues. The total amount of glucose in body fluids and that which is stored as glycogen in the liver equates to about 180 g, that is only a little more than 1 day's supply in the fasting state, as each day, approximately 160 g of glucose are oxidized by the typical 70 kg adult male. Of that quantity, approximately 75% is consumed by the brain alone. Not surprisingly therefore, hypoglycaemia (usually defined as a blood glucose concentration of less than 2.2 mmol/l compared with a typical minimum fasting value of approximately 3.5 mmol/l) has serious effects on the brain in particular. Rapid and dramatic falls in blood glucose concentration lead to acute neuroglycopaenia (= low neurocellular glucose concentration).

Because glucose is the preferred fuel for the brain, an individual who experiences a rapid fall in glucose concentration leading to acute neuroglycopenia will initially feel confusion and may progress to coma and even death. In the event that the person survives 3–4 days, the brain can adapt its metabolism to utilize ketone bodies, metabolically derived from acetyl-CoA (see Figure 6.17), as a source of energy.

The CNS is not the only vulnerable tissue as red cells also rely upon a constant supply of glucose to maintain structure and function. Because they lack mitochondria, and therefore the mechanism to produce ATP via oxidative phosphorylation, RBCs are entirely dependent upon anaerobic glucose metabolism to synthesize ATP through substrate level phosphorylation.

The liver not only extracts glucose from the blood in the postprandial state and stores it as glycogen, but is also able to synthesize glucose from non-carbohydrate sources via gluconeogenesis, therefore the liver is crucial in regulating glucose homeostasis.

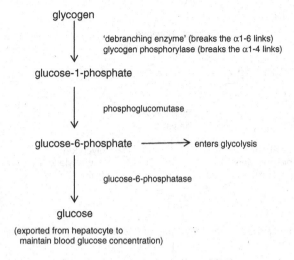

Figure 6.38 Glycogenolysis (compare with Figure 6.24)

6.5.1 Glycogenolysis

The catabolism of glycogen is not simply the reversal of its synthesis. The process is outlined in Figure 6.38.

As indicated in Section 6.3.3 and Table 6.2 the key control step is mediated by glycogen phosphorylase, a homodimeric enzyme which requires vitamin B_6 (pyridoxal phosphate) for maximum activity, and like glycogen synthase (Section 6.2) is subject to both allosteric modulation and covalent modification.

Glycogen phosphorylase isoenzymes have been isolated from liver, brain and skeletal muscle. All forms are subject to covalent control with conversion of the inactive forms (GP-b) to the active forms (GP-a) by phosphorylation on specific serine residues. This phosphorylation step, mediated by the enzyme phosphorylase kinase, is initiated by glucagon stimulation of the hepatocyte. Indeed, the same cAMP cascade which inhibits glycogen synthesis simultaneously stimulates glycogenolysis, giving us an excellent example of reciprocal control.

Phosphorylase kinase is a very large multi-subunit protein and is tightly regulated by covalent and non-covalent mechanisms. The conversion of inactive phosphorylase kinase-b (PK-b) in to active PK-a is brought about by cAMP dependent protein kinase A (PKA) and requires calcium for full activity. Phosphorylase kinase is rendered inactive by phosphatases including protein PP1; this is the same protein phosphatase which converts active glycogen phosphorylase-a to inactive glycogen phosphorylase-b and glycogen synthase-b (low activity form) to glycogen synthase-a (highly active form). Refer to Figures 6.39 and 6.40.

Reversible phosphorylation is the main control mechanism of *liver* phosphorylase; allosteric effects being much less pronounced. This is in contrast with *muscle* phosphorylase, which is also controlled by phosphorylation, stimulated by an adrenaline-

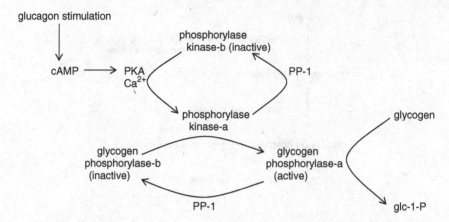

Figure 6.39 Interconversion of liver glycogen phosphorylase

induced cAMP cascade, and by allosteric effects mediated by the key metabolic intermediates, AMP and glucose-6-phosphate. In muscle, a rise in the cellular concentration of AMP as would occur for example during exercise, indicates a fall in ATP concentration, so glycogen phosphorylase activity is stimulated by AMP to provide more fuel. Given that adrenaline is involved with the 'fight or flight reaction', increased supply of fuel in the form of glucose-6-phosphate to active muscles allows the flight (running away) from a potential danger. The regulatory effects of AMP and glucose-6-phospahte on *liver* phosphorylase are much less significant. In short, the muscle isoenzyme meets local intracellular metabolic needs, whereas liver phosphory-lase is adapted to respond to 'whole body' changes in fuel metabolism, largely determined by the feeding/fasting pattern of the individual. Physiologically, feeding (the postprandial state) is associated with insulin which promotes glycogen synthesis, whilst the body's fasting signal, glucagon, promotes glycogenolysis.

6.5.2   Gluconeogenesis (GNG)

Literally, gluconeogenesis (GNG) means the 'synthesis of new glucose'. A more exact metabolic description would be that GNG synthesizes glucose from carbon atoms derived from non-carbohydrate precursors, that is certain so-called glucogenic amino acids, oxaloacetate (a substrate of the TCA cycle), from glycerol and also lactate. The liver and the kidney are the only tissues which undertake GNG and the contribution of the liver is approximately 10 times greater than that of the kidney. Gluconeogenesis (Figure 6.41) is accelerated when carbohydrate stores are diminished and fat is utilized as fuel, during fasting or vigorous exercise, and glucose generated by GNG can be used to supply the central nervous system. Fat is far more calorific than carbohydrate and its use as the primary fuel generates large amounts of substrates such as acetyl-CoA, NADH and ATP, which stimulate GNG and suppress glycolysis. Acetyl-CoA, NADH and ATP signal that the cell has sufficient 'energy', or the materials from which to generate energy, so catabolism of glucose is not required. This is a glucose-sparing effect. During fasting and starvation, muscle provides significant quantities of amino acids for GNG.

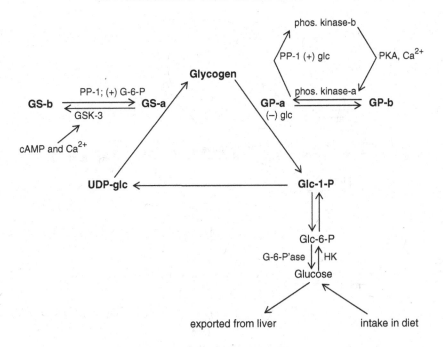

KEY: GS-a, GS-b   glycogen synthase-a; glycogen synthase-b

GP-a, GP-b   glycogen phosphorylase-a; glycogen phosphorylase- b

GSK-3   glycogen synthase kinase-3

phos. kinase phosphorylase kinase (-*a* & -*b* forms)

cAMP   cyclic AMP

HK   hexokinase

G-6-P'ase   glucose-6-phosphatase

glc   glucose

(+)   activates

(−)   inhibits

Figure 6.40 Reciprocal control of glycogen phosphorylase and glycogen synthase

GNG exploits the fact that *most* of the reactions of glycolysis are reversible so the enzymes are shared between the two pathways. There are three kinase reactions (glucokinase/hexokinase, PFK and pyruvate kinase), which are not physiologically reversible are therefore the problem steps in the synthesis of glucose; these three steps are overcome using alternative enzymes (Table 6.5, see also Section 1.7.1).

As can be seen in Figure 6.42, pyruvate is very much a focal point in GNG. Normally, once pyruvate has entered a mitochondrion, it is converted into acetyl-CoA by pyruvate dehydrogenase complex, but for GNG the pyruvate is 'diverted' in to oxaloacetate (OAA) by pyruvate carboxylase (see Figure 6.43).

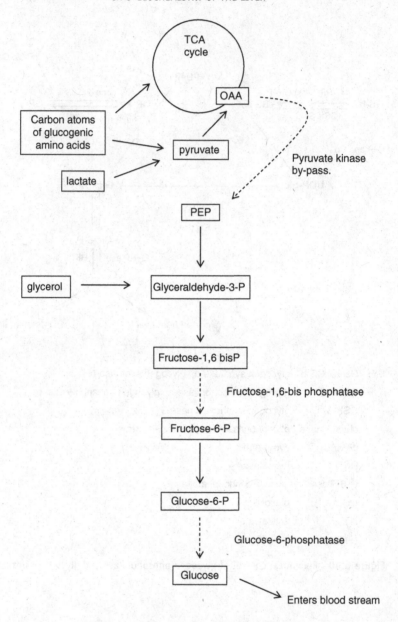

dashed lines indicate reactions which are unique to gluconeogenesis.

Figure 6.41 Gluconeogenesis: outline

Pyruvate carboxylase is a mitochondrial enzyme and like other carboxylase or decarboxylase enzymes requires biotin as coenzyme. The biotin is firmly attached to the enzyme protein (i.e. a prosthetic group) via a lysine residue. The role of biotin is to 'hold' the CO_2 in the correct orientation to allow its incorporation into the pyruvate.

Table 6.5 Irreversible reactions of glycolysis and the enzymes needed in gluconeogenesis to overcome them

	Enzymes	
Reaction	Glycolysis	GNG
PEP to pyruvate	Pyruvate kinase (PK)	Pyruvate carboxylase (PC) and Phosphoenolpyruvate carboxykinase (PEP CK) ['bypass' route]
Fructose-6-phosphate to fructose-1,6 bisphosphate	phosphofructokinase (PFK)	Fructose-1,6 bisphosphatase (FBP'ase)
Glucose to glucose-6-phosphate	hexokinase (HK) and glucokinase (GK)	Glucose-6-phosphatase (Glc-6-Pase)

Pyruvate dehydrogenase is a MEC consisting of three separate catalytic proteins; (i) a component with combined pyruvate decarboxylase/dehydrogenase activity; (ii) a dihydrolipoyl transacetylase (also called acetyl transferase) unit and (iii) linked dihydrolipoyl dehydrogenase. This is clearly a very big protein; the pyruvate decarboxylase/

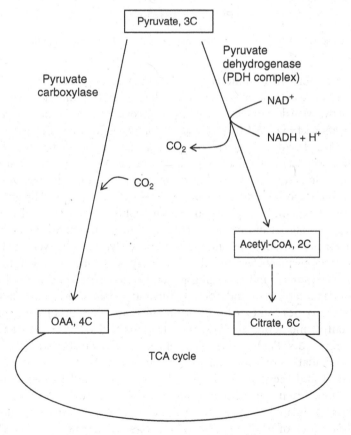

Figure 6.42 PDH complex and pyruvate carboxylase both act on pyruvate

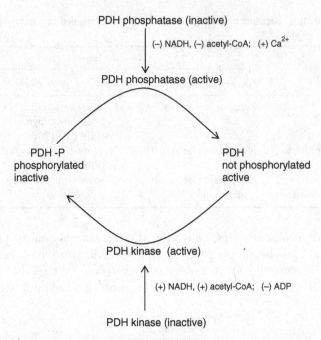

PDH phosphatase (inactive)

$(-)$ NADH, $(-)$ acetyl-CoA; $(+)$ Ca^{2+}

PDH phosphatase (active)

PDH -P
phosphorylated
inactive

PDH
not phosphorylated
active

PDH kinase (active)

$(+)$ NADH, $(+)$ acetyl-CoA; $(-)$ ADP

PDH kinase (inactive)

Figure 6.43 Regulation of PDH complex activity by covalent modification

dehydrogenase component alone is a tetrameric protein of 2α (M_r 35 000) and 2β (M_r 40 000) protomers. The enzyme requires thiamine, FAD and biotin as coenzymes.

Control of pyruvate dehydrogenase activity is via covalent modification; a specific kinase causes *in*activation of the PDH by phosphorylation of three serine residues located in the pyruvate decarboxylase/dehydrogenase component whilst a phosphatase activates PDH by removing the phosphates. The kinase and phosphatase enzymes are non-covalently associated with the transacetylase unit of the complex. Here again we have an example of simultaneous but opposite control of enzyme activity, that is, reciprocal regulation.

There are many examples of phosphorylation/dephosphorylation control of enzymes found in carbohydrate, fat and amino acid metabolism and most are ultimately under the control of a hormone induced second messenger usually, cytosolic cyclic AMP (cAMP). PDH is one of the relatively few mitochondrial enzymes to show covalent modification control, but PDH kinase and PDH phosphatase are controlled primarily by allosteric effects of NADH, acetyl-CoA and calcium ions rather than cAMP (see Table 6.6).

Where two enzymes compete for the same substrate, we expect to see some form of metabolic control and in this case the concentrations of NADH and acetyl-CoA are the key controlling factors (Figure 6.44). When glucose is *not* available as a fuel, metabolism switches to β-oxidation of fatty acids, which generates more than sufficient quantities of both NADH and acetyl-CoA to drive the TCA cycle and to maintain oxidative phosphorylation. Pyruvate dehydrogenase activity is suppressed and pyruvate carboxylase is stimulated by ATP, NADH and acetyl-CoA (strictly speaking by low mitochondrial ratios of ADP/ATP, NAD^+/NADH and coenzyme A/acetyl-CoA), so

Table 6.6 Control of PDH kinase and PDH phosphatase

	Activated by	Inhibited by
PDH kinase (inhibits PDH activity)	NADH acetyl-CoA ATP	NAD^+ CoA ADP Pyruvate
PDH phosphatase (activates PDH)	NAD^+ calcium ions	NADH

Note the complementary effects of NAD and NADH

channelling carbon atoms as pyruvate, but derived from lactate or some amino acids, into OAA and then to glucose synthesis (Figure 6.45).

Pyruvate carboxylase plays another important role in regulating metabolism by ensuring an adequate supply of OAA for the TCA cycle. Take a moment to stop and

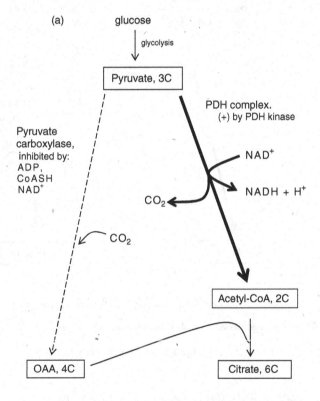

Pyruvate derived from glucose is converted to acetyl-CoA by PDH;
PC activity is low but helps to ensure adequate OAA to combine with acetyl-CoA for entry into the TCA cycle.

Figure 6.44 Relative activities of PDH and PC during the fed and fasting states

(b) no recent glucose ingestion: glucagon concentration HIGH, insulin LOW

Fat used as main energy source.
Carbon atoms from amino acids are usedin GNG during fasting and starvation, but
during vigorous exercise, skeletal muscle provides substantial amounts of lactate

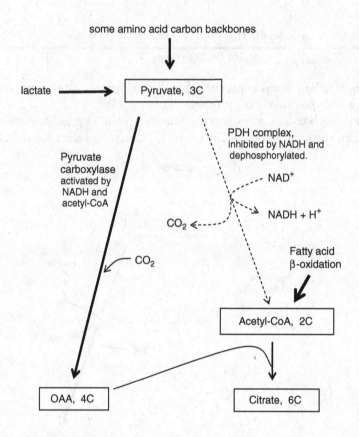

Acetyl-CoA is derived from the oxidation of fatty acids and pyruvate carbons are diverted to OAA.

Dotted lines indicate low activity; heavy lines represent accelerated activity.

Figure 6.44   (*Continued*)

think about the dynamics of this process. Acetyl-CoA can only enter the TCA cycle if
there is sufficient oxaloacetate to combine with it to form citrate (catalysed by citrate
synthase). An increase in pyruvate carboxylase activity when acetyl–CoA concentration
is high will ensure that OAA availability keeps pace with acetyl-CoA production; this is
an example of a 'topping-up' or anaplerotic reaction.

Following the conversion of pyruvate into oxaloacetate, phosphoenolpyruvate
carboxykinase (PEP CK) catalyses a decarboxylation to form PEP (Figure 6.42). This

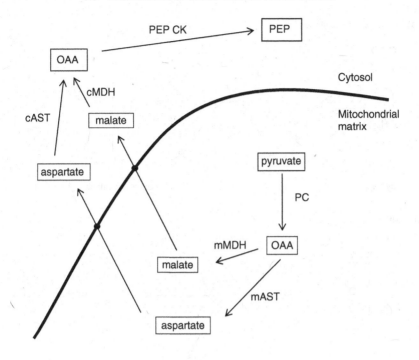

mMSD = mitochondrial malate dehydrogenase
mAST = mitochondrial aspartate transaminase

cMDH = cytsolic malate dehydrogenase
cAST = cytosolic aspartate transaminase

Translocase

Figure 6.45 Malate-OAA-aspartate loop

by-pass reaction is complicated by the fact that OAA, the product of the PC reaction, is mitochondrial but PEP and PEP-CK are cytosolic so there has to be a translocation between compartments. Oxaloacetate is converted into either malate or aspartate to allow transport of the carbon atoms across the inner mitochondrial membrane (Figure 6.46). Before PEP CK can complete the sequence either malate or aspartate must be reconverted into OAA. This by-pass route is 'expensive' as it requires the hydrolysis of both ATP and GTP. Figure 6.47 shows the PC and PEP-CK reactions.

The GNG pathway from PEP to fructose-1,6-bisphosphate is a reversal of the glycolytic enzyme reactions (Figure 6.42, compare with Figure 1.20). The final two

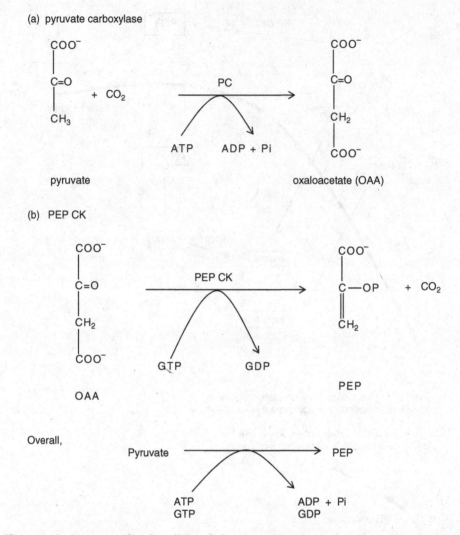

Figure 6.46 Pyruvate carboxylase (PC) and phosphoenolpyruvate carboxykinase (PEP CK) bypass reactions

kinase reactions of glycolysis, PFK and HK/GK are reversed by specific phosphatases: fructose-1,6 bisphosphatase (FBPase) and glucose-6-phosphatase. Note that the phosphatases release inorganic phosphate (Pi) and do not generate ATP when the substrates are dephosphorylated. This combination of an ATP consuming kinase coupled with a phosphatase has been called a futile cycle.

The first phosphatase step is very important: FBPase converts fructose,1-6-bisphosphate into fructose-6-phosphate under allosteric control of several factors but during fasting, glucagon-induced regulation is crucial. One effect of glucagon stimulation of liver cells is to reduce the concentration of fructose-2,6-bisphosphate, an isomer that activates PFK-1 and is itself synthesized by PFK-2 when fructose-6-phosphate concentration rises

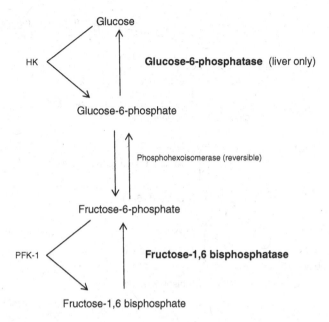

Key: HK   = hexokinase
     PFK-1 = phosphofructokinase-1

Figure 6.47 Two phosphatase reactions to overcome kinase reactions

(see Figure 3.4). Therefore, if the concentration of fructose-2,6-bisphosphate falls, its stimulatory effect on PFK-1 also falls and not only is less fructose-6-phosphate converted into fructose-1,6-bisphosphate, but the activity of fructose bisphosphatase increases and so fructose-6-phosphate concentration rises. Glucagon therefore suppresses glycolysis and permits gluconeogenesis to proceed.

In addition to the hormonal control of FBP, allosteric activation of the enzyme is mediated by ATP and citrate, two of the modulators which along with long-chain fatty acids inhibit PFK. When concentrations of citrate and ATP rise glycolysis is slowed and GNG is simultaneously accelerated. A high concentration of ATP is clearly a signal that the cell has sufficient energy currency; long chain fatty acids are a fuel and an important potential source of ATP. The cytosolic concentration of citrate will rise when fatty acids are being metabolized because a product of β-oxidation is acetyl-CoA, which along with oxaloacetate forms citrate in the mitochondria; recall that ready availability of fat as a fuel effectively accelerates gluconeogenesis. Recent dietary intake of fatty acids will slow the metabolism of glucose through glycolysis ('glucose sparing effect', Figure 6.48) and furthermore, by regulating glycolysis at a point after the formation of glucose-6-phosphate, a larger amount of glucose may be diverted to glycogen synthesis (Section 6.2.3). If, on the other hand, fat metabolism has been initiated by glucose deficiency, that is hormone-stimulated mobilization of stored triglyceride in fasting or exercise, glucose is not only synthesized from lactate, glycerol or amino acids (described

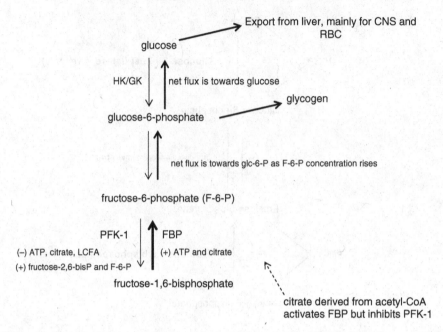

Figure 6.48 Glucose sparing effect and net glucose synthesis when fatty acids are primary fuel

below) but whatever glucose may be available is routed to the CNS and red blood cells whose metabolism relies heavily on this sugar.

6.5.2.1   GNG from the carbon skeletons of glucogenic amino acids and lactate

Following their deamination, several amino acids can be converted into oxaloacetate via pyruvate or other components of the TCA cycle these are termed glucogenic amino acids because their carbon atoms may eventually appear in glucose (see Table 6.7). The carbon skeletons of other amino acids may be converted into acetyl-CoA or acetoacetyl-CoA and these are ketogenic amino acids. A few amino acids be metabolized through both glucogenic and ketogenic pathways.

As we have seen, normally pyruvate would be the substrate for pyruvate dehydrogenase complex to form acetyl-CoA, but during fasting in the absence of glucose, acetyl-CoA for the TCA cycle is derived from fatty acid β-oxidation (see Section 7.5.2) so pyruvate is 'diverted' into oxaloacetate by the enzyme pyruvate carboxylase. Thus any amino acids whose carbon skeletons can be converted into pyruvate, OAA or another substrate of the TCA cycle, can be used for glucose synthesis.

The principal source of amino acids for gluconeogenesis is skeletal muscle and it is significant that two important enzymes ALT (see Figure 6.2) and AST occur in both the liver and the skeletal muscle as it is these enzymes that are responsible for re-cycling carbon atoms into glucose. In the liver, alanine may be converted into pyruvate and then, as shown in Figure 6.2, into oxaloacetate. Refer also to Figure 7.11, which shows the glucose–alanine cycle.

Table 6.7 Glucogenic and ketogenic amino acids

Glucogenic amino acids	Ketogenic amino acids	Amino acids that are glucogenic *and* ketogenic
Alanine, arginine	Isoleucine	Phenylalanine
Aspartate, cysteine	Leucine	Threonine
Glutamate, glutamine	Lysine	Tryptophan
Glycine, histidine		Tyrosine
Methionine, proline		
Serine, valine		

Additionally, several amino acids may undergo transamination to produce glutamate which in the liver is oxidatively deaminated to form 2-oxoglutarate (2-OG, see Figure 6.6), a substrate of the TCA cycle. Alternatively, glutamate may be converted into glutamine, an important but often overlooked fuel substrate.

The Cori cycle. The production of pyruvate in the liver can also arise from lactate released from exercising muscle. The anaerobic conditions created in muscles during strenuous exercise cause much of the pyruvate generated from glucose metabolism in glycolysis to be converted into lactate rather than acetyl-CoA. This metabolic adaptation has two effects; firstly, generation of NADH from the TCA cycle is slowed to balance the reduced rate of re-oxidation of coenzyme by the electron transport chain arising from the relative lack of oxygen, and secondly, NAD^+ is recycled in the cytosol to maintain the activity of glyceraldehyde-3-dehydrogenase in glycolysis, because if the action of this enzyme becomes jeopardized through lack of NAD^+, glycolysis in the muscle would be severely compromised. Lactate is exported from the muscle and transported to the liver where it is acted upon by lactate dehydrogenase (LD);

$$\text{lactate} + NAD^+ \xrightarrow{\text{LD}} \text{pyruvate} + NADH + H^+$$

LD occurs in various tissues of the body as five isoenzymes; details are given in Table 6.7. Hepatic LD has a relatively high K_m for lactate so only becomes significantly active when lactate concentration rises above a typical cellular level.

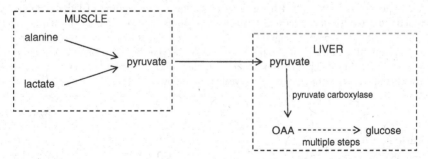

Glucose produced by GNG from alanine or lactate may then be recycled to the tissues, including the muscle which provided the initial alanine and lactate.

Chapter summary

The liver is metabolically the most complex and diverse organ in the body. Anabolic, catabolic and interconversion reactions occur extensively within the liver to achieve provision of important molecules to all tissues of the body. After feeding and prompted by insulin, glucose is stored as glycogen and converted into fatty acid for export and long term storage as triglyceride in adipose tissue. During short periods of fasting, blood glucose concentration is maintained by glycogenolysis, but as the period of fasting progresses, gluconeogenesis and ketogenesis become more prominent as means of supplying fuels to tissues, especially the central nervous system. In effect, the liver is the body's 'fuel valve', helping to regulate the supply of raw materials to other tissues. Partly as a result of the metabolic complexity we find in the liver, hepatic pathways illustrate the full range of control mechanisms; substrate inhibition, allosteric effects, covalent modification and induction.

The liver is the body's major site of waste processing prior to its disposal via the gut or kidneys. A wide variety of unwanted molecules, both endogenous in origin and xeno-biotics, are rendered less harmful by enzyme systems most of which are associated with the smooth endoplasmic reticulum of the liver cells. The majority of the proteins found in the plasma are synthesized in the liver, including those known collectively as acute phase proteins and clotting factors, all of which are involved with the body's response to trauma and infection, carrier proteins, albumin which regulates fluid balance, and those many proteins whose functions are not yet known. Aside from glycogen, the storage functions of the liver have not been discussed in detail in this chapter but many important micro-nutrients such as vitamins and trace elements are stored in the liver.

Overall, it should not be surprising that damage to the liver (e.g. viral or toxic hepatitis) or genetic defects affecting pathways within the tissue often lead to pathology.

Case notes

1. A 26-year-old male consulted his general practitioner complaining of lethargy and loss of appetite following a short period of 'flu-like symptoms. The young man showed no obvious clinical signs but on questioning, the he informed the GP that he had recently returned from a prolonged visit to India where, for the last 8 weeks of his stay, he had been doing voluntary work in a rural community. The GP took a blood sample and asked the patient to return in a week to discuss the results. Initial blood tests revealed the following results;

		reference range:
Bilirubin	22 µmol/l	5–17 µmol/l
Aspartate transaminase (AST)	680 IU/l	10–50 IU/l
Alanine transaminase (ALT)	565 IU/l	10–40 IU/l
Albumin	42 g/l	35–45 g/l
Alkaline phosphatase	120 U/l	<250 U/l

This pattern of results is typical of hepatocellular damage (necrosis). The high activity of AST and ALT are due to leakage from damaged cells; the normal albumin value indicates that this is an acute (recent) condition. The modest rise in bilirubin concentration in plasma is not itself diagnostic at this stage.

At the second consultation, the patient said that he felt better but the GP noticed he was now clinically jaundiced (yellow discolouration of the skin and the whites of the eyes) which supported the suspicion of viral hepatitis. To confirm the diagnosis, the GP took another blood sample for analysis. The results were;

		reference range:
Bilirubin	68 µmol/l	5–17 µmol/l
Aspartate transaminase (AST)	350 IU/l	10–50 IU/l
Alanine transaminase (ALT)	215 IU/l	10–40 IU/l
Albumin	44 g/l	35–45 g/l
Alkaline phosphatase	145 U/l	<250 U/l
Viral serology	Positive for HBsAg (hepatitis B surface antigen), and anti-HBc (hepatitis B core antigen) IgM antibody	

The activity of AST and ALT had by this time passed their peak but the bilirubin concentration was elevated, often good signs that the acute condition is beginning to reside. Viral serology provided definitive evidence of hepatitis B infection; this is not uncommon in conditions where sanitation is rudimentary. The patient made a full recovery about 4 months after diagnosis, but was advised to abstain from alcohol for several more months.

2. A mother became increasing concerned about the health of her 6-month-old son. Within a few weeks of birth he developed an enlarged abdomen. The child often appeared pale and had several episodes of excessive sweating and weakness. The symptoms were relieved by eating. A fasting blood sample gave, among others, the following results;

		reference range
Glucose	1.9 mmol/l	3.2–5.5 mmol/l
Free fatty acids	1.7 mmol/l	<1.0 mmol/l

The paediatrician suspected an inborn error of metabolism and further studies showed the activity of glucose-6-phosphatase to be greatly reduced, confirming a diagnosis of type Ia glycogen storage disease (von Gierkes disease). Hepatomegaly (enlarged liver) the cause of the protuberant abdomen, is characteristic of this autosomal recessive condition.

Fasting hypoglycaemia occurs in a number of genetically determined disorders of carbohydrate metabolism. In this case, the cause is due to the inability to top-up blood glucose concentrations. The low glucose concentration initiated adrenalin secretion (the fight or flight reaction or more accurately in this case in response to hypoglycaemic stress) which mobilized fatty acids from adipose tissue to provide fuel for metabolism. The only remedy other than a liver transplant is frequent feeding to maintain blood glucose concentration.

3. S.A. was a 21-year-old who was surprised to discover that she was approximately 7 weeks pregnant. Although this pregnancy had occurred somewhat earlier in their marriage than she and

her husband had anticipated, both were happy with the news even though they realized it meant that their busy social lives would come to an end.

At her first consultation, S. stated that for the previous 3 years she had been an occasional smoker, often drank 'fairly heavily' at weekends with friends but denied having a 'drink problem'. She was informed of the potential dangers of smoking and heavy drinking during pregnancy and strongly advised to reconsider both habits.

The pregnancy went to term, although the obstetricians grew a little concerned in the last few weeks S as did not gain weight normally suggesting some fetal growth retardation. At birth, the baby was indeed underweight for gestational age, had a slightly smaller head than usual and showed mild facial abnormalities around the mouth and eyes. A diagnosis of fetal alcohol spectrum disorder (FASD) was made,

FAS is just one, but an extreme, example of FASD in which those affected suffered poor growth and development during pregnancy and during childhood, congenital abnormalities of the head and face, behavioural disturbances such as short attention span and hyperactivity often with intellectual retardation. Some of these early problems were apparent in SA.'s baby, but only to a mild degree. The incidence of full-blown FAS is quite low, estimates vary from less than 1/1000 in the United Kingdom and Canada to nearly 10/1000 in the United States. Damage to the developing fetus may be caused by 'binge drinking' at around the time of conception until about 10 weeks into the term (was apparently the case with S.A.) or in chronic alcoholics who drink heavily throughout pregnancy. Confounding factors such as poor nutrition, smoking and use of other drugs during pregnancy are believed to be important as not all, even very heavy, drinkers deliver an affected child. The mechanisms involved with damage are not entirely resolved but toxic effects of acetaldehyde, poor oxygen delivery due to abnormalities with erythrocytes and disturbance with prostaglandin production have all been suggested as contributing factors.

Alcohol consumption is very difficult to assess. There is widespread belief that individuals under-report their intake and there are no reliable laboratory tests available for definitive diagnosis of alcohol abuse. A combination of abnormalities in the plasma activity of gamma-glutamyl transferase (GGT or γGT), AST and reduction in erythrocyte mean cell volume (MCV) may be useful and all are 'routine' lab. tests. A potential marker of interest is carbohydrate-deficient transferrin (CDT) which is an abnormal isoform of serum transferrin arising due to defects in the attachment of carbohydrate chains to the protein core. Unfortunately, CDT is a somewhat specialized test, not performed by most laboratories. Other markers which have attracted some research interest are ethyl sulphate and ethyl glucuronide. Excretion in the urine of these metabolites occurs for up to 50 hours after 'binge' drinking so they offer a useful index of recent heavy alcohol intake.

For current guidelines and recommendations of alcohol consumption during pregnancy, refer to the web site of the Royal College of Obstetricians and Gynaecologists.

7

Biochemistry of muscle

Overview of the chapter

This chapter considers the structure of muscles and biochemical processes that operate to generate energy for contraction and the contribution muscle makes to the supply of substrates such as amino acids and lactate (which are used by the liver to generate glucose) and ketone bodies (which act as fuels for other tissues of the body). Histologically, there are three types of muscle; striated skeletal, striated cardiac and smooth muscle. The mechanism of contraction of all three types is powered by ATP and is explained by the sliding filament hypothesis of actin and myosin interaction; the physiological control of contraction varies between striated and smooth muscle. Metabolically, different types of muscle cells are distinguished by their reliance on mainly oxidative or non-oxidative pathways of ATP generation. Muscle is a significant consumer of the body's fuel, using mainly fatty acids during basal conditions supplemented by carbohydrate during exercise. Muscle is quantitatively the body's most important insulin-sensitive tissue.

Key pathways

Glycogenolysis and glycogen synthesis; β-oxidation of fatty acids; transamination and deamination of amino acids; Cori cycle and glucose–alanine cycle, which recycles substrates between muscle and liver.

7.1 Introduction

Contributing up to 40% of the total body mass, muscle is the most abundant tissue in the human body. Furthermore, it is, along with adipose tissue, one of the few tissues whose mass may change dramatically during adulthood. Good practices (e.g. regular exercise) and also bad practices (anabolic steroid abuse) can increase muscle mass, physique and stamina. On the other hand, chronic illness, especially if an individual is bed-ridden, and the general aging processes reduce muscle bulk. Muscles require a continual supply of energy for contraction but inherent differences in the metabolic characteristics between myocytes of skeletal and smooth fibres, plus the ability of muscle cells to adapt to changing circumstances mean that this tissue, like the liver, has been a valuable one for the study of fuel metabolism.

7.2 Physiology of muscles

The anatomical unit of muscle is an elongated cell called a fibre. Each individual fibre cell consists of myofibrils which are bundles of contractile protein filaments composed of actin and myosin (Figure 7.1). Differences in structure indicate that muscles have evolved to perform particular functions. Although the structure of fibres, myofibrils and filaments of actin and myosin, is similar in all muscle types, their arrangement, action and control allow identification of three tissue types:

- striated muscle (skeletal or voluntary),

- smooth muscle (involuntary or visceral),

- cardiac muscle.

Skeletal muscle, also known as striated muscle because of the microscopic appearance, is responsible for locomotion and those fine, voluntary movements of the body which are under conscious control. Smooth muscle exerts automatic, involuntary control on physiological systems such as controlling tone of the blood vascular system and peristalsis in the gut. Cardiac muscle, although striated like skeletal muscle, maintains the automatic regular rhythm of the heart beat and so functions without conscious or deliberate control rather like smooth muscle. Some important characteristics of these three tissues are summarized in Table 7.1.

Functional differences are reflected by metabolic characteristics of the skeletal musculature in particular where specialized cell types have evolved to permit activity of different intensity and duration to accommodate the wide range of actions for which skeletal muscle is responsible.

The contractile filaments of smooth muscle run length-wise through the cell but because they are neither grouped together into fibrils nor of the same length, there is no regularly arranged pattern and thus no striations as seen in skeletal muscle. Smooth muscle contracts relatively slowly but is able to maintain an even tension for long periods of time, a feature which explains its importance in regulating the tone of blood vessels, peristalsis in the gut, regular contraction/relaxation of the respiratory muscles and allowing the filling and subsequent emptying of the urinary bladder. Pain of labour and childbirth is due to regular and strong contraction of uterine smooth muscle followed by an expulsive force at parturition.

In contrast, skeletal muscle contraction is more rapid than that of smooth muscle but skeletal muscle cannot maintain the same tone for long periods of time. As indicated in Table 7.1, we can distinguish sub-types of muscle fibre within

(a) actin

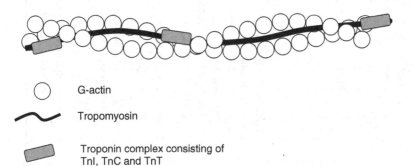

      ◯   G-actin

      〜   Tropomyosin

      ▭   Troponin complex consisting of
                TnI, TnC and TnT

(b) single myosin molecule consisting of two intertwined chains.

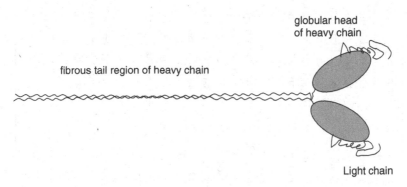

globular head
of heavy chain

fibrous tail region of heavy chain

Light chain

(c)   A thick myosin filament is an assembly of many myosin
      molecules arranged in a staggered side-by-side and
      end-to-end fashion

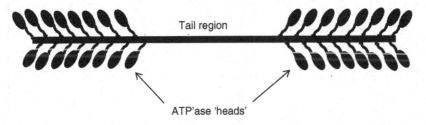

Tail region

ATP'ase 'heads'

Figure 7.1 Actin and myosin filaments

Table 7.1 Features of different types of muscle

Feature	Skeletal muscle	Cardiac muscle	Smooth muscle
Organization, morphology and appearance	Highly organized; striations arise due to the alignment of the fibrils within the sarcomeres of the fibres (cells) Long fibres (up to 30 cm in length) Type I (slow) and type II (fast) fibres; red muscle contains myoglobin, white muscle does not contain myoglobin	Highly organized; striations arise due to the alignment of the fibrils within the sarcomeres of the fibres (cells) Fibres are branched and interdigitate High oxidative capacity relying on aerobic metabolism	No striations as there are no sarcomeres Type I fibres
Excitation–contraction coupling	Induced by a rise in intracellular calcium ions Most of the Ca^{2+} is released from intracellular stores in the sarcoplasmic reticulum Very fast repolarization (1 ms)	Induced by a rise in intracellular calcium ions Ca^{2+} released from the sarcoplasmic reticulum and by an influx from the outside the cell via specific voltage-activated channels Fast repolarization (200 ms)	Induced by a rise in intracellular calcium ions Most of the Ca^{2+} enters the cell via voltage- activated or ligand-activated membrane channels Continuous waves of depolarization with slow repolarization
Neural innervation	Conscious control via spinal motor nerves. Direct contact between a motor nerve and a few fibres where fine control is needed; in larger muscles one motor neuron may innervate several hundred fibres	Intrinsic 'pacemaker' activity which may be modulated by sympathetic neural activity	Intrinsic activity with diffuse neural modulation. No conscious voluntary control is possible but neural activity affects force or timing of contraction

skeletal muscles. Type I fibres take approximately 100 ms to contract fully and are known as 'slow-twitch' fibres. Because type I fibres utilize mainly metabolically efficient oxidative metabolism, that is the TCA cycle and oxidative phosphorylation, to obtain energy they are also slow to fatigue but cannot produce an energy surge if suddenly called upon for high power output.

There are two subgroups of type II fibres both of which are known as 'fast-twitch' requiring only about 40 ms to reach maximum tension. Type IIa fibres are like type I in that they are fatigue-resistant but derive their energy from oxidative *and* glycolytic metabolism to provide moderate power output. Type IIb fibres rely on glycolytic metabolism which enables high power generation but over a short period of time before tiring. The glycolytic activity of type I fibres is low in comparison with type II fibres. It should not be assumed that a particular whole muscle, say, the biceps of the upper arm or the gastrocnemius in the lower leg will be composed of only one type of fibre. All skeletal muscles contain all three types of fibre. Such an organization of fibre types allows a muscle to exhibit a graded response to physiological demands, and to optimize and adapt substrate utilization accordingly. The proportion of fast and slow fibres any individual possess is genetically determined and dictates whether the person is suited to sporting events which require endurance (marathon, slow type I fibres) or explosive power (e.g. sprinting, fast type IIb fibres). Athletic training programmes can optimize the metabolic potential of muscles (see Section 7.7).

7.2.1 Muscle contraction

Contractile proteins which form the myofibrils are of two types: myosin ('thick' filaments each approximately 12 nm in diameter and 1.5 μm long) and actin ('thin' filaments 6 nm diameter and 1 μm in length). These two proteins are found not only in muscle cells but widely throughout tissues being part of the cytoskeleton of all cell types. Filamentous actin (F-actin) is a polymer composed of two entwined chains each composed of globular actin (G-actin) monomers. Skeletal muscle F-actin has associated with it two accessory proteins, tropomyosin and troponin complex which are not found in smooth muscle, and which act to regulate the contraction cycle (Figure 7.1).

Tropomyosin is a fibrous molecule which twists around the F-actin strands. The troponin (Tn) complex is composed of three proteins; TnI (I = inhibitory) which prevents myosin binding to actin in the resting muscle, TnT which binds tropomyosin and TnC (C for calcium-binding). Cardiac muscle troponins are different from those of skeletal muscle and are designated cTnI, cTnT and cTnC.

Myosin's quaternary structure consists of two heavy chains (M_r 200 kDa), each with head and a tail, and two pairs of light chains (20 kDa). The heavy chain tails entwine with each other whilst the globular 'heads' of the myosin associate with the light chains and have an F-actin binding domain and inherent ATPase activity. A key difference between the subtypes (I, IIa and IIb) of fibre is the particular isoforms of myosin ATPase they posses. Bundles of myosin molecules aggregate together in a staggered fashion so that the heads form a helical arrangement.

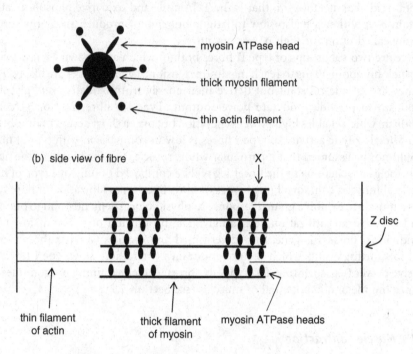

(a) cross-section through fibre at X on figure b) below

myosin ATPase head

thick myosin filament

thin actin filament

(b) side view of fibre

X

Z disc

thin filament
of actin

thick filament
of myosin

myosin ATPase heads

Figure 7.2 Arrangement of actin and myosin filaments

Our understanding of muscle cell contraction is based on the sliding filament model proposed in the mid-1950s. In this model, actin and myosin filaments are arranged in parallel with each thick myosin filament being surrounded by six thin actin filaments as shown in Figure 7.2. The pattern of striations seen by light microscopy is due to the regular arrangement of actin and myosin filaments. Contraction requires the two types of filament to become cross-linked by 'bridges' formed by the myosin heads where the ATPase activity is located. In relaxed skeletal and cardiac muscle, actin-bound tropomyosin prevents the interaction between the actin and the myosin.

Muscle fibre shortening occurs when the actin filaments slide between the myosin filaments. The staggered helical arrangement of the myosin heads ensures multiple cross bridge interactions between the myosin and F-actin at different positions along the length of the actin filament. To try to visualize the process, imagine that the thick myosin containing filaments are fixed and immobile and the myosin head attaches to the thin filament and pulls it along the length of the thick filament in an analogous motion to collapsing an extendable ladder. The overall length of the fibril shortens but the length of each filament remains the same as they slide over each other.

Hydrolysis of ATP provides the energy required for each stroke of myosin heads to pull on the actin filament so shortening the muscle fibre (Figure 7.3). In the resting

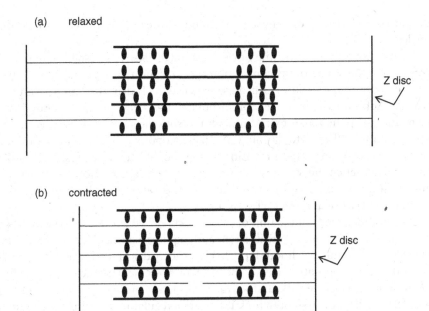

Thin actin filaments are pulled inwards by myosin heads so
shortening the fibre. The Z disc is an 'anchorage' point for actin

Figure 7.3 Skeletal muscle contraction

state, ATP binds to the head and the ATPase in the myosin head forms ADP and
inorganic phosphate (Pi). In this 'high energy' conformation, the head is held in a
'cocked' position but tropomyosin prevents contact between the myosin head and actin
filament. However, when the muscle fibre receives a neural stimulus mediated by
acetylcholine released at the neuromuscular junction, the muscle membrane is
electrically depolarized as sodium ions enter the cell. Calcium ions are released from
the sarcoplasmic reticulum; the calcium binds to troponin-C, causing the tropomyosin
to reposition itself on the actin, so relieving the inhibition to myosin binding and
allowing the formation of a cross-bridge.

   The power stroke actually occurs when first Pi and then ADP are released causing a
conformational change in the myosin as the cocked head moves to a lower energy
conformation; this movement exerts a pulling action on the actin filament and the
fibre shortens. The binding of another ATP to the myosin weakens the attachment
between the myosin head and the actin filament allowing the repositioning of the
head at another part of the actin; another bridge is made between the myosin head and
actin, providing of course that cytosolic Ca^{2+} concentration remains high enough to
prevent tropomyosin blocking the interaction, and the stroke cycle repeats. Each cycle
takes typically about 30 μs to complete. It is worth just taking a minute to stop
and consider the high degree of physiological coordination required to ensure that
the process described occurs simultaneously at multiple sites in multiple fibres to

achieve the smooth flexing of a muscle controlling, for example, the movement of an arm or leg.

ATP is used not only to power muscle contraction, but also to re-establish the resting state of the cell. At the end of the contraction cycle, calcium must be transported back into the sarcoplasmic reticulum, a process which is ATP driven by an active pump mechanism. Additionally, an active sodium-potassium ATPase pump is required to re-set the membrane potential by extruding sodium from the sarcoplasm after each wave of depolarization. When cytoplasmic Ca^{2+} falls, tropomyosin takes up its original position on the actin and prevents myosin binding and the muscle relaxes. Once back in the sarcoplasmic reticulum, calcium binds with a protein called calsequestrin, where it remains until the muscle is again stimulated by a neural impulse leading to calcium release into the cytosol and the cycle repeats.

The calcium mediated contraction of smooth muscle, which unlike striated muscle does not contain troponin, is quite different and requires a particular calcium-binding protein called calmodulin. Calmodulin (CM) is a widely distributed regulatory protein able to bind, with high affinity, four Ca^{2+} per protein molecule. The calcium–calmodulin (CaCM) complex associates with, and activates, regulatory proteins, usually enzymes, in many different cell types; in smooth muscle the target regulatory proteins are caldesmon (CDM) and the enzyme myosin light chain kinase (MLCK). As described below, CaCM impacts on both actin and myosin filaments.

In smooth muscle, caldesmon plays an analogous role to that of troponin in striated muscle in that it 'blocks' the myosin binding sites. The CaCM complex removes caldesmon from its binding on the thin actin filaments allowing tropomyosin to reposition in the helical grooves of F-actin leading to myosin ATP'ase activation.

In addition to the displacement of caldesmon, smooth muscle cell contraction requires kinase-induced phosphorylation of myosin. Smooth muscle has a unique type of myosin filament called p-light chains which are the target (substrate) for MLCK, but MLCK is only active when complexed with CaCM. Myosin light chain phosphatase reverses the PKA-mediated process and when cytosolic calcium ion concentration falls, CDM is released from CaCM and re-associates with the actin. The central role of calcium-calmodulin in smooth muscle contraction is shown in Figure 7.4.

Furthermore, as well as CaCM-induced phosphorylation, MLCK is also subject to control via a cAMP-dependent protein kinase, PKA. Phosphorylated MLCK binds CaCM only weakly, thus contraction is impaired. This explains the relaxation of smooth muscle when challenged with adrenaline (epinephrine), a hormone whose receptor is functionally linked with adenylyl cyclase (AC), the enzyme that generates cAMP from ATP.

7.3 Fuel metabolism within muscles

Because of its very active nature, skeletal muscle has long been a favourite tissue for biochemists to study metabolism and its regulation. Although skeletal muscle uses

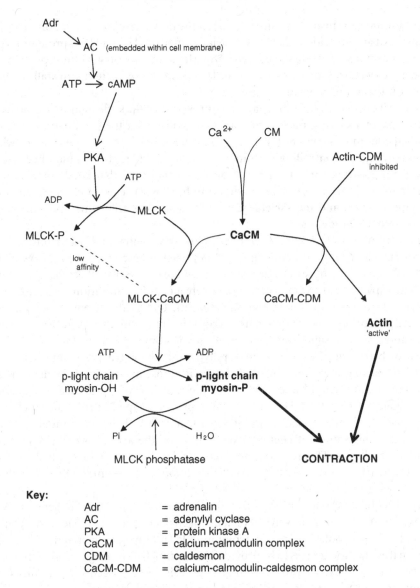

Figure 7.4 Control of smooth muscle contraction

pathways found in tissues such as liver and adipose in particular, the structure and kinetic properties (i.e. K_m and V_{max}) of enzymes within common pathways are often different reflecting the particular circumstances found in muscles.

All muscle types require ATP to achieve contraction. Glucose, fatty acids and amino acids may all be used as oxidizable substrates to produce ATP and all three energy sources may be obtained from stored intracellular sources (glycogen, triglyceride and protein) or imported from the blood stream. In quantitative terms, skeletal muscle is the most important insulin-sensitive tissue in the body. Insulin stimulation promotes the uptake of amino acids and glucose, the latter mediated by GLUT-4 proteins (type 4 glucose transporters) in the cell surface, for the synthesis of structural protein and glycogen respectively, providing an excellent example of insulin's overall anabolic actions on tissue metabolism.

As you sit quietly reading this chapter, your heart and the skeletal muscles you use to turn the pages are using mostly fatty acid catabolized via the β-oxidation pathway, described later in this chapter (Section 7.5) to meet their energy requirements, whilst the smooth muscle controlling your internal organs such as gut and blood vessels are using glucose as their primary energy source. In the event that you should need to generate energy to allow a burst of activity, your heart will use glucose to supplement its energy production and the skeletal muscles in your legs will draw upon their own internal reserves of glycogen.

Recall from Section 7.2 that type I fibres are slow-twitch and use mainly aerobic oxidative mechanisms to generate the ATP they need to function. Type II fibres are fast twitch. The type IIa fibres are also aerobic whereas type IIb fibres have a greater glycolytic capacity and so can tolerate relatively anaerobic conditions. Because type I and type IIa fibres utilize oxidative metabolism, that is the Krebs TCA cycle and oxidative phosphorylation, to generate energy for contraction they require a rich blood supply and contain a significant amounts of myoglobin, a protein which consists of a haem prosthetic group bound to a single polypeptide chain to act as an internal oxygen depot. The haem groups in myoglobin along with those in respiratory cytochromes in the numerous mitochondria contribute to the red–brown appearance of the tissue. Because glucose is readily taken up via GLUT proteins, type I fibres require relatively little glycogen storage capacity, but they do, however, have substantial stores of triglyceride. This source of energy is augmented by the action of endothelial lipase in muscle capillaries. This is an enzyme that enables fibres to extract fatty acids from blood-borne triglyceride-rich very low density lipoproteins from the liver and chylomicrons synthesized in the gut following a meal.

Type IIb fibres exploit non-oxidative, that is anaerobic glycolysis, to generate ATP. Because glycolysis is a relatively inefficient mechanism for ATP generation (only 2 moles of ATP per mole of glucose), white muscle (no myoglobin) requires a store of glycogen if adequate energy is to be provided for contraction during periods of maximal activity. Conceptually, the processes of glycogenesis and glycogenolysis in muscle are the same as in the liver (see Figure 7.5 and Chapter 6) but there are important metabolic differences between the two tissues.

Muscle glycogen phosphorylase is one of the most well studied enzymes and was also one of the first enzymes discovered to be controlled by reversible phosphorylation (by E.G. Krebs and E. Fischer in 1956). Phosphorylase is also controlled allosterically by ATP, AMP, glucose and glucose-6-phosphate. Structurally, muscle glycogen phosphorylase is similar to its hepatic isoenzyme counterpart composed of identical subunits each with a molecular mass of approximately 110 kDa. To achieve full activity, the enzyme requires the binding of one molecule of pyridoxal phosphate, the active form of vitamin B_6, to each subunit.

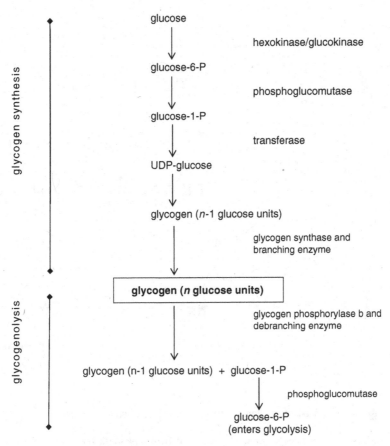

glucose

hexokinase/glucokinase

glucose-6-P

phosphoglucomutase

glucose-1-P

transferase

UDP-glucose

glycogen (*n*-1 glucose units)

glycogen synthase and branching enzyme

glycogen (*n* glucose units)

glycogen phosphorylase b and debranching enzyme

glycogen (n-1 glucose units) + glucose-1-P

phosphoglucomutase

glucose-6-P (enters glycolysis)

glycogen synthesis

glycogenolysis

Figure 7.5 Glycogen metabolism in muscle (compare with Figure 6.22)

The role, and therefore the control of activity of muscle glycogen phosphorylase (GP) is different to that of the liver isoenzyme described in Section 6.5. Both of the tissue specific isoenzymes are subject to hormone-induced phosphorylation in response to systemic physiological conditions. Muscle glycogen phosphorylase is activated via a signalling cascade mechanism initiated by adrenaline stimulation ('fight or flight' response) but the liver is stimulated by glucagon, a hormone signal indicating that the body is a fasting state. In liver and muscle, phosphorylation converts the 'b' form to the 'a' form (GPb and GPa respectively), both of which are in the T allosteric state and so essentially inactive (Figure 7.6 and refer to Section 3.2.2 also).

The interconversion of GPa and GPb from the T allosteric state (inactive) to the R allosteric state (active) is meditated by AMP (*high* cellular AMP concentration can be equated with a *low* concentration of ATP). In contrast, glucose, glucose-6-phosphate and ATP (signs of an 'energy-rich' cell), all suppress glycogen phosphorylase activity by causing an R to T transition. In short, muscle glycogen phosphorylase is very

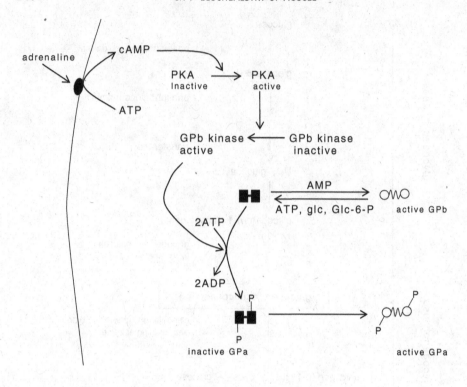

Key:

■	T allosteric state of glycogen phosphorylase (inactive)
○	R allosteric state of glycogen phosphorylase (active)
◖	Adenylyl cyclase bound to the myocyte membrane
glc	glucose
Glc-6-P	glucose-6-phosphate

Figure 7.6 Activation of muscle glycogen phosphorylase

sensitive to local intracellular metabolite concentrations, reflecting changes in physical activity. It should be noted that glycogen stores in muscle are limited and will support contraction for only a few minutes, after which time glucose supplied by gluconeogenesis from lactate generated during exercise and fatty acids become more important fuel sources.

Furthermore, the liver, but *not* muscle, contains the enzyme glucose-6-phosphatase which allows glycogen catabolism to continue to glucose, which is exported from the liver to sustain metabolism in other tissues, especially the central nervous system and the red blood cells. However, glycogen stored in muscle is intended exclusively for use by muscle during exercise when glucose-6-phosphate is fed directly into glycolysis to power contraction, hepatic glycogen is a depot store of glucose potentially for use by all tissues.

A means of co-ordinating muscle contraction with glycogenolysis is required. A dramatic 100-fold increase in cytosolic Ca^{2+} concentration from 10^{-7} to 10^{-5} molar initiates *both* glycogenolysis *and* muscle contraction. This increase in cytosolic calcium concentration is mainly due to release of calcium from the sarcoplasmic reticulum in response to acetylcholine stimulation of the muscle fibre (Figure 7.7).

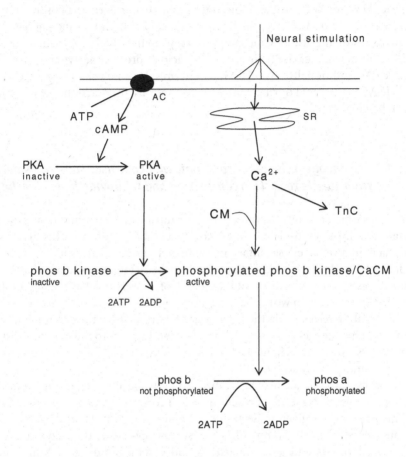

Calcium ions released from the SR are required to initiate contraction and activate glycogen phosphorylase kinase

Key:

AC	adenylyl cyclase
PKA	protein kinase A
phos	glycogen phosphorylase
CM	calmodulin
TnC	troponin C
SR	sarcoplasmic reticulum

Figure 7.7 Coordination of glycogenolysis and muscle contraction

7.4　Maintenance of ATP availability in active muscle

ATP is the only suitable source of energy for the contracting muscle cell; however the ATP concentration in muscle cells is very limited (e.g. approximately 25 mmol/kg of dry skeletal muscle tissue). This amount is sufficient to keep muscles contracting at their maximal capacity for only a few seconds, thus active muscle cells must be able to recycle ADP to ATP to maintain contraction. If the need for contraction extends beyond a few seconds, fibres which require glucose as their fuel and energy source will begin to rely on hepatic gluconeogenesis to top-up their fuel supply and oxidative-type muscle fibres will increase their use of fatty acid to produce acetyl-coenzyme A for the TCA cycle. Additionally, myocytes have enzyme-driven mechanisms which efficiently recycle ADP generated during contraction. These are the reactions we will consider in the following section.

7.4.1　Gluconeogenesis: an hepatic pathway that uses substrates derived from muscle to sustain contraction and to provide fuels for other tissues

Fast fibres which rely on anaerobic glycolysis during sudden bursts of activity generate lactate and tire easily are said to have a low 'anaerobic threshold'. This situation occurs especially in a relatively sedentary individual, and one of the important adaptations induced by athletic training is the reduction in lactate accumulation after exercise. Slow fibres do not provide a sudden burst of power and have a higher anaerobic threshold so are able to work effectively for longer periods of time. Some of the lactate can be usefully recycled via the Cori cycle (Figure 7.8) for the production of glucose as part of gluconeogenesis or carried by the blood stream to the liver or heart where it may be oxidized and channelled into the Krebs cycle following conversion into pyruvate and then to acetyl-CoA.

In the liver, the lactate may be oxidized to pyruvate and the carbon atoms routed to glucose synthesis in a series of reactions which constitute gluconeogenesis. The details of this pathway are given in Section 6.5.2 and outlined in Figure 7.9, but in essence because most of the reactions of glycolysis are reversible, the enzymes are 'shared' between glycolysis and gluconeogenesis and carbon atoms of certain amino acids (notably alanine), glycerol or muscle-derived lactate can be used to generate glucose, which in turn is released from the liver and may be used by actively contracting anaerobic muscles to maintain a supply of ATP via substrate level phosphorylation.

A key enzyme in the process of recycling lactate from muscle to liver is lactate dehydrogenase (LD) which catalyses the reversible interconversion of lactate and pyruvate. This important reaction is shown in Figure 7.10.

LD occurs in five isoenzymic forms and is widespread in cells around the body. The five isoenzymes arise due to the quaternary arrangement of the four subunits which comprise the enzyme. The subunits are of two types, H (heart) and M (muscle) and arranged as shown in Table 7.2.

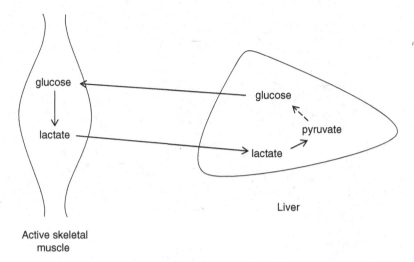

Figure 7.8 The Cori cycle

As might be expected, the kinetic properties of the five LD isoenzymes vary widely. LD1 has a low K_m for lactate thus ensuring that lactate does not accumulate in heart muscle and so cause fatigue. Expressing this in another way, we can say that cardiac muscle is a net consumer of lactate generated by erythrocytes and skeletal muscle. LD5 has a low K_m for pyruvate to allow rapid reduction to lactate and simultaneous generation of reduced nicotinamide adenine dinucleotide (NADH), which is required for full activity of glyceraldehyde-3-phosphate dehydrogenase, a key enzyme in glycolysis.

If food is unavailable for more than approximately 24 h, glycogen reserves in the liver will become depleted and the individual would enter a state of biochemical starvation. Progressive loss of muscle protein (wasting) would occur in order to generate sufficient glucose to maintain the metabolic activity of, in particular, the central nervous system.

Muscle protein catabolism generates amino acids some of which may be oxidized within the muscle. Alanine released from muscle protein or which has been synthesized from pyruvate via transamination, passes into the blood stream and is delivered to the liver. Transamination in the liver converts alanine back into pyruvate which is in turn used to synthesise glucose; the glucose is exported to tissues via the blood. This is the glucose-alanine cycle (Figure 7.11). In effect, muscle protein is sacrificed in order to maintain blood adequate glucose concentrations to sustain metabolism of red cells and the central nervous system.

7.4.2 Recycling of ADP

Muscle contraction produces ADP; if this cannot be recycled to ATP contraction will cease. Rephosphorylation of ADP by mitochondrial oxidative phosphorylation is an

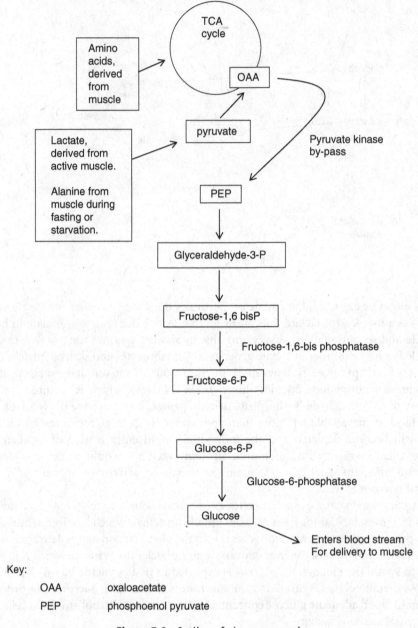

Figure 7.9 Outline of gluconeogenesis

obvious option for regenerating ATP, but this applies mainly to *oxidative* type I fibres. The predominantly non-oxidative type IIa fibres may use glycolysis to generate ATP from ADP via substrate level phosphorylation but this is quantitatively a relatively inefficient process. To prevent premature fatigue arising from a lack of energy,

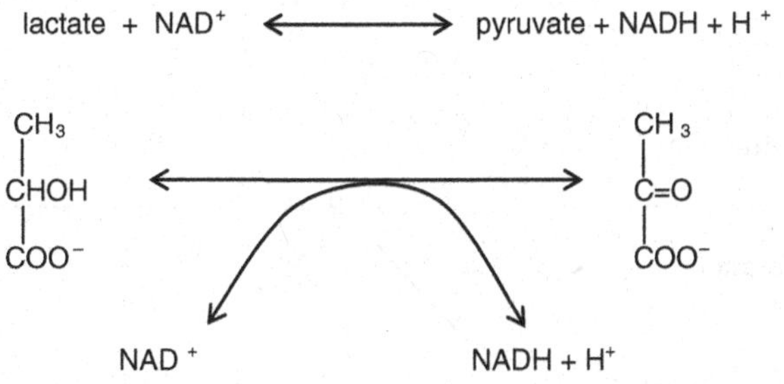

$$\text{lactate} + NAD^+ \longleftrightarrow \text{pyruvate} + NADH + H^+$$

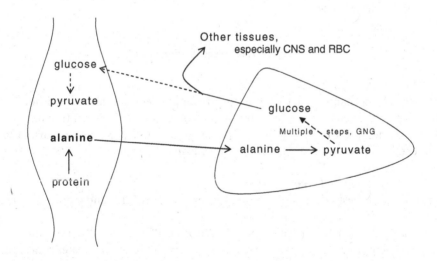

Figure 7.10   Lactate dehydrogenase (LD)

Table 7.2   Isoenzymic forms of lactate dehydrogenase

Designation	Subunit arrangement	Approximate % of tissue total LD activity (values rounded to nearest 5%)			
		Cardiac muscle	Erythrocyte	Skeletal muscle	Hepatocyte
LD1	H_4	60	40	<5	<5
LD2	H_3M	30	45	<5	<5
LD3	H_2M_2	<10	10	20	15
LD4	HM_3	<5	<5	15	15
LD5	M_4	<5	<5	55	65

Key: CNS   central nervous system
       RBC   red blood cells
       GNG   gluconeogenesis, see figure 7.9

Figure 7.11   Glucose–alanine cycle

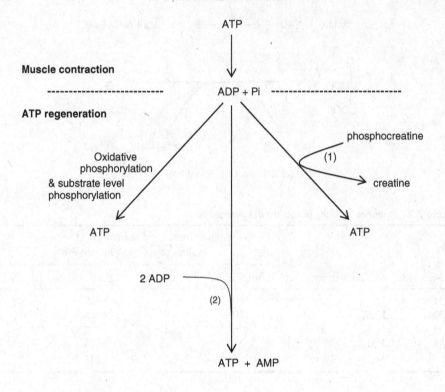

(1)   creatine kinase

(2)   adenylate kinase

Figure 7.12 Regeneration of ATP

specialized mechanisms have evolved to supplement oxidative phosphorylation and glycolysis to ensure an adequate resupply of ATP to actively contracting muscles. Two enzymes, creatine kinase (CK) and adenylate kinase (AK), are important in this context: These reactions are summarized in Figure 7.12.

7.4.2.1 Creatine and creatine kinase

Creatine is synthesized from glycine and arginine (Figure 7.13) and requires S-adenosyl methionine (SAM) as a methyl group donor.

Creatine phosphate (also called phosphocreatine, PCr) is a small compound which is a more 'energy rich' (has a more negative $\Delta G^{\circ\prime}$) and is present in greater concentration ($\sim$75 mmol/kg) than ATP and so is able to rephosphorylate ADP so acts as an 'energy buffer':

$$ADP + Pi + H^+ \rightarrow ATP \qquad \Delta G^{\circ\prime} \sim +30.5\,\text{kJ/mol}$$
$$\text{Creatine phosphate} \rightarrow \text{creatine} + Pi \qquad \Delta G^{\circ\prime} \sim -45\,\text{kJ/mol}$$

Adding:

$$ADP + H^+ + \text{creatine phosphate} \rightarrow ATP + \text{creatine} \qquad \Delta G^{o'} \sim -14.5 \, kJ/mol$$

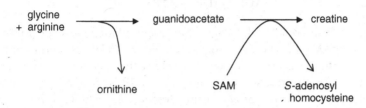

Figure 7.13 Creatine synthesis

The enzyme responsible for this 'topping-up' ATP in active muscle is CK. CK is found in high concentration in muscle cells, both free within the sarcoplasm and also associated with membranes of mitochondria and the sarcoplasmic reticulum. Structurally, creatine kinase is a dimeric enzyme of B and/or M subunits, each of about 40 kDa. Three quaternary structure isoenzyme forms arise: CK-MM, CK-BB and CK-MB. The predominant form in all muscles is CK-MM, but cardiac muscle also contains a significant amount of CK-MB and this isoenzyme can be used as a specific marker of myocardial damage (see Case Notes at the end of this chapter).

During periods of recovery following exercise, creatine phosphate is regenerated at the expense of ATP synthesized from mitochondrial oxidative phosphorylation; energy 'currency' is paid into a reserve account, or reservoir, for the next period of sustained exercise.

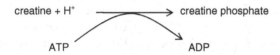

Under standard conditions, this reaction would be unfavourable but physiological conditions during recovery phase after exercise are such as to allow creatine phosphate formation to occur.

The rate of creatine phosphate resynthesis differs between type I and type II fibres and according to the type and duration of exercise undertaken. A key controlling factor of the enzyme is pH. Severe exercise reduces cellular pH, due largely to excessive production of lactate. Only when the pH has been restored to approximately 7.0 can CK begin to resynthesize PCr. Evidence exists to suggest that dietary supplementation with creatine leads to an increase in cellular concentrations, which may bring about improvement in sporting performance, probably by reducing recovery time after exercise. Creatine is excreted from the body as creatinine, one of the nitrogenous waste products found in urine.

7.4.2.2 Adenylate kinase (AK)

Whereas CK rephosphorylates ADP using PCr as the phosphate donor, AK (also called myokinase or AMP kinase) uses another molecule of ADP as the phosphate donor.

$$ADP + ADP \rightarrow ATP + AMP$$

The AK reaction could lead to significant accumulation of AMP. Ideally rephosphorylation of AMP to ATP would occur to maintain ATP concentration but this is not really an option because the bioenergetics of such a reaction are so unfavourable, therefore an alternative mechanism of recycling AMP is required. An enzyme called AMP deaminase (AD) is important in this process. The activity of AMP deaminase is really one of AMP removal rather than recycling of AMP to ATP.

$$AMP + H_2O \xrightarrow{AD} IMP + NH_3$$

The enzyme AMP deaminase is inactive in the sarcoplasm in a resting muscle but is activated by (i) low pH and (ii) a reduction in the [ATP]-to-[AMP] ratio; both of these changes are consistent with actively working skeletal muscle.

Combining the AK and AMP deaminase reactions:

$$2\,ADP \rightarrow ATP + AMP \quad \text{(AK reaction)}$$
$$AMP + H_2O \rightarrow IMP + NH_3 \quad \text{(AMP deaminase reaction)}$$
$$\text{Overall, } 2\,ADP + H_2O \rightarrow ATP + IMP + NH_3$$

Clearly therefore, some adenosine nucleotides are effectively 'lost' from the cell during normal ATP turnover by conversion into inosine or inosine monophosphate (IMP) (see Figure 7.14), so although ATP has been regenerated by the AK and AD reactions, the overall reaction appears to be a wasteful process because a purine nucleotide (IMP) which has little or no value as a high energy compound has been formed. However, the production of ammonia which occurs during the deaminase reaction may be physiologically significant because (a) it is a base and given that cellular pH is likely to drop during periods of anaerobic metabolism, the NH_3 *may* help to buffer the sarcoplasm and (b) *in vitro* at least NH_3 is an activator of phosphofructokinase (PFK), a key regulatory enzyme in glycolysis.

The loss of AMP especially in active muscles is partly ameliorated by recycling of IMP via adenylosuccinate. Furthermore, because AMP is an important allosteric activator of PFK, regeneration of AMP ensures that glycolysis is fully active and able to provide pyruvate for the TCA cycle and some ATP via substrate level phosphorylation.

7.5　Fatty acid as a fuel in muscle

Fatty acid oxidation occurs in mitochondria and peroxisomes in most tissues but quantitatively muscle is a major consumer of fat. Although carbohydrates and fatty acids may both be used as fuels for muscle contraction, fatty acids are more calorific

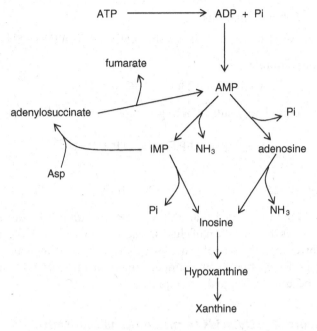

Key:
    IMP   inosine monophosphate
    Asp   aspartate

Figure 7.14 Adenosine nucleotide loss and recycling of IMP to AMP

(energy producing) yielding approximately 38 kJ/mol compared with approximately 16 kJ/mol for glucose. The reasons for this are:

1. fatty acids are larger molecules, that is have more carbon atoms, and

2. they are chemically more reduced, that is contain fewer oxygen atoms and are thus more oxidizable.

These points can be better appreciated by comparing the complete oxidation of glucose with that of a typical saturated fatty acid, palmitate.
    Oxidation of glucose

$$C_6H_{12}O_6 + 6O_2 \rightarrow 6CO_2 + 6H_2O$$

the overall change in free energy being sufficient to produce 32 moles[1] of ATP per mole of glucose oxidized

$$32\,ADP + 32Pi \rightarrow 32\,ATP + 32\,H_2O$$

---

[1] Some calculations put this figure as high as 38 moles of ATP depending on assumptions made about the use of NADH generated in the cytosol to contribute to ATP production in the mitochondria.

So, adding the above equations

$$C_6H_{12}O_6 + 6O_2 + 32ADP + 32Pi \rightarrow 6CO_2 + 32\,ATP + 38H_2O$$

Oxidation of palmitate

$$C_{18}H_{32}O_2 + 23O_2 \rightarrow 16CO_2 + 16H_2O$$

With a free energy change sufficient to generate 106 moles of ATP per mole of fatty acid

$$106ADP + 106Pi \rightarrow 106ATP + 106H_2O$$

adding

$$C_{18}H_{32}O_2 + 23O_2 + 106ADP + 106Pi \rightarrow 16CO_2 + 106ATP + 122H_2O$$

However, it should be noted that although fat oxidation provides quantitatively more ATP than does the oxidation of glucose, the *rate* at which ATP is generated via β-oxidation is slower than the rate of generation by glycolysis; a fact which explains why glucose is the preferred fuel during sudden bursts of muscular activity when ATP concentration need to be 'topped-up' rapidly.

7.5.1 Transport of fatty acids to the mitochondrial matrix

Fatty acid utilized by muscle may arise from storage triglycerides from either adipose tissue depot or from lipid stores within the muscle itself. Lipolysis of adipose triglyceride in response to hormonal stimulation liberates free fatty acids (see Section 9.6.2) which are transported through the bloodstream to the muscle bound to albumin. Because the enzymes of fatty acid oxidation are located within subcellular organelles (peroxisomes and mitochondria), there is also need for transport of the fatty acid within the muscle cell; this is achieved by fatty acid binding proteins (FABPs). Finally, the fatty acid molecules must be translocated across the mitochondrial membranes into the matrix where their catabolism occurs. To achieve this transfer, the fatty acids must first be activated by formation of a coenzyme A derivative, fatty acyl CoA, in a reaction catalysed by acyl CoA synthetase.

$$\text{Fatty acid} + \text{CoASH} + \text{ATP} \xrightarrow{\text{Acyl CoA synthetase}} \text{fatty acyl CoA} + \text{AMP} + \text{PPi}$$

For example palmitate (C16)

$$CH_3(CH_2)_{14}COO^- + CoASH + ATP \rightarrow CH_3(CH_2)_{14}CO\text{-}CoA + AMP + PPi$$
$$\text{Palmitoyl CoA}$$

CoASH = coenzyme A

PPi = pyrophosphate (= 'diphosphate').

Note that ATP is consumed in this activation process so the net gain of ATP is reduced marginally as a result.

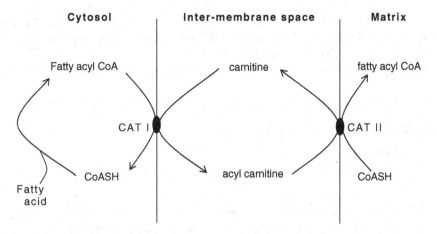

Figure 7.15 Translocation of fatty acid to the mitochondrial matrix

At this point, the acyl-CoA is still in the cytosol of the muscle cell. Entry of the acyl-CoA into the mitochondrial matrix requires two translocase enzymes, carnitine acyl transferase I and carnitine acyl transferase II (CAT I and CAT II), and a carrier molecule called carnitine; the carnitine shuttles between the two membranes. The process of transporting fatty acyl-CoA into mitochondria is shown in Figure 7.15.

Once in the mitochondrial matrix, acyl-CoA (e.g. palmitoyl-CoA), is degraded by β-oxidation generating acetyl-CoA for the TCA cycle and reduced coenzymes which supply hydrogen atoms and electrons for oxidative phosphorylation.

7.5.2 Catabolism of saturated fatty acyl CoA: β-oxidation

β-Oxidation is a cyclical pathway which removes a C2 (acetyl) unit form the fatty acyl-CoA on each cycle. The designation β derives from the traditional system of labelling atoms within fatty acid molecules where the carbon attached to the carboxyl group is α and the methyl carbon is always ω (omega):

$$\underset{\omega}{C\,H_3} \cdot CH_2 \; K \; \underset{\gamma}{C\,H_2} \cdot \underset{\beta}{CH_2} \cdot \underset{\alpha}{C\,H_2} \cdot COO^-$$

The process of β-oxidation can be summarized fairly simply as follows:

$$Palmitoyl\;CoA + 7CoASH + 7\,FAD + 7\,NAD^+ + 7H_2O$$

$$\downarrow$$

$$8\;acetyl\;CoA + 7\,FADH_2 + 7\,NADH + 7\,H^+$$

The flavin adenine dinucleotide (FAD) and NADH deliver electrons and hydrogen directly to the respiratory cytochromes whilst acetyl-CoA enters the TCA cycle.

Recall from Figure 3.14 in Section 3.3.2, that each turn of the TCA generates three NADH and an FAD so the total amount of reduced coenzyme made available for oxidative phosphorylation is considerable.

This overall equation represents four individual steps per cycle (see Figure 7.16):

- First **O**xidation (using FAD),

- **H**ydration,

- Second **O**xidation (using NAD^+),

- **C**leavage.

You may wish to memorize this sequence as OHOC. Compare with the sequence of reactions which constitute fatty acid synthesis described in Figure 6.11.

Most of our fat intake will consist of fatty acids with an even number of carbon atoms, but not all dietary fatty acids nor all those synthesized in the liver are saturated. A variable, but probably not inconsiderable, proportion of dietary fatty acids are unsaturated, partly perhaps because a high intake of unsaturated fat is recommended to help reduce the risk for diseases of the heart and vascular system. Unsaturated and odd-numbered fatty acids pose particular chemical problems to the β-oxidation pathway and additional enzymes are required for their metabolism.

First, most naturally occurring unsaturated fatty acids have double bonds in the *cis* isomeric configuration . . .

```
                              H
                              |
   -C = C-                -C = C-
    |   |                  |
    H   H                  H
```

cis isomer   *trans* isomer

. . . but enoyl hydratase (step 2 in Figure 7.16) can only operate with *trans* isomers, so an isomerase is required to 'flip' the molecule into the *trans* form.

Second, the position of double bonds within the acyl chain is a potential problem. Note in Figure 7.16 that the *trans* double bond is introduced between the α and β carbons. If a fatty acyl derivative is unsaturated in the equivalent to the β–γ position, a reductase enzyme is required to reposition the double bond appropriately (Figure 7.17).

The situation is simpler for odd numbered fatty acyl derivatives as β-oxidation proceeds normally until a 5-carbon unit remains, rather than the usual 4-carbon unit. The C5 moiety is cleaved to yield acetyl-CoA (C2) and propionyl-CoA (C3). Propionyl CoA can be converted to succinyl CoA and enter the TCA cycle so the entire molecule is utilized but with a slight reduction in ATP yield as the opportunity to generate two molecules of NADH by isocitrate dehydrogenase and 2-oxoglutarate dehydrogenase is lost because succinyl-CoA occurs after these steps in the Krebs cycle (Figure 7.18).

During periods of fasting, muscles may also derive energy from the metabolism of ketone bodies (3-hydroxybutyrate and acetoacetate). These intermediates are

• indicates the β carbon atom of the original molecule

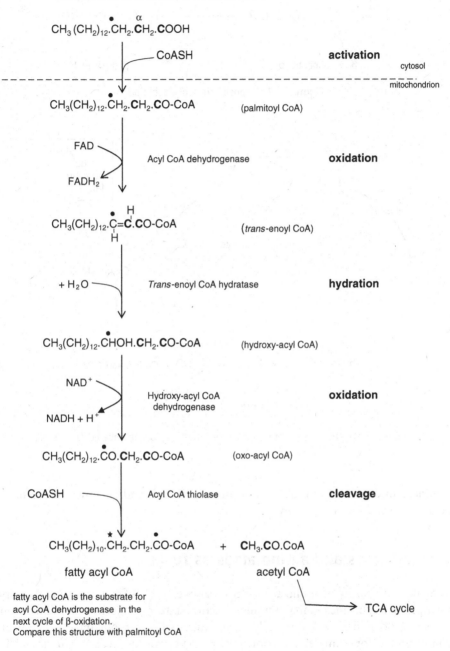

fatty acyl CoA is the substrate for
acyl CoA dehydrogenase in the
next cycle of β-oxidation.
Compare this structure with palmitoyl CoA

★ 'new' β carbon atom, the original (•) having been oxidised to a carbonyl
   (C=O) and which is now attached to coenzyme A.

Figure 7.16 Reaction steps of one cycle of β-oxidation

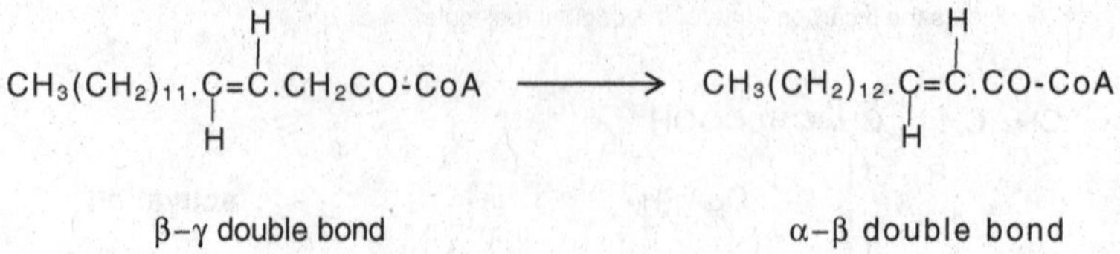

$$CH_3(CH_2)_{11}.\overset{\overset{\textstyle H}{|}}{C}=\overset{\overset{\textstyle }{|}}{C}.CH_2CO\text{-}CoA \qquad \longrightarrow \qquad CH_3(CH_2)_{12}.\overset{\overset{\textstyle H}{|}}{C}=\overset{\overset{\textstyle }{|}}{C}.CO\text{-}CoA$$

$$\beta-\gamma \text{ double bond} \qquad\qquad\qquad \alpha-\beta \text{ double bond}$$

Figure 7.17 Repositioning of double bond

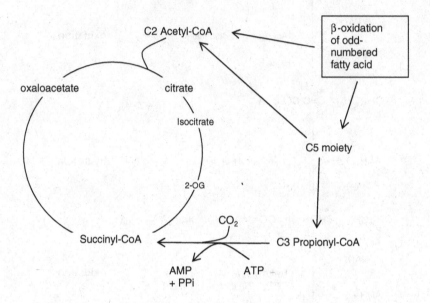

Figure 7.18 TCA entry of terminal carbon atoms of odd-numbered fatty acids

produced in the liver from the catabolism of fatty acids and as shown in Figure 7.19 below, converted to acetyl-CoA.

7.6   Proteins and amino acids as fuels

The calorific capacity of amino acids is comparable to that of carbohydrates so despite their prime importance in maintaining structural integrity of cells as proteins, amino acids may be used as fuels especially during times when carbohydrate metabolism is compromised, for example, starvation or prolonged vigorous exercise. Muscle and liver are particularly important in the metabolism of amino acids as both have transaminase enzymes (see Figures 6.2 and 6.3 and Section 6.4.2) which convert the carbon skeletons of several different amino acids into intermediates of glycolysis (e.g. pyruvate) or the TCA cycle (e.g. oxaloacetate). Not all amino acids are catabolized to the same extent

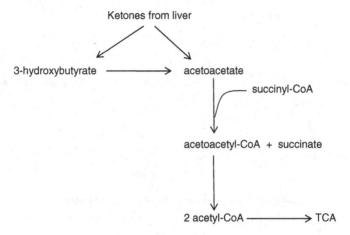

Figure 7.19   Metabolism of ketones as fuel

and in muscle the branched chain molecules (leucine, isoleucine and valine) are preferentially oxidized along with alanine, aspartate and glutamate.

The generic transamination reaction is;

$$\text{amino acid} + \text{2-oxoglutarate} \leftrightarrow \text{oxoacid} + \text{glutamate}$$

$$\text{(amino donor) (amino acceptor)}$$

Transaminase enzymes (also called aminotransferases) specifically use 2-oxoglutarate as the amino group acceptor to generate glutamate but some have a wide specificity with respect to the amino donor. For example, the three branched-chain amino acids leucine, isoleucine and valine, all serve as substrates for the same enzyme, branched-chain amino acid transaminase, BCAAT;

$$\text{Leucine} + \text{2-oxoglutarate} \rightarrow \text{2-oxo-4-methylvalerate} + \text{glutamate}$$

$$\text{Isoleucine} + \text{2-oxoglutarate} \rightarrow \text{2-oxo-3-methylvalerate} + \text{glutamate}$$

$$\text{Valine} + \text{2-oxoglutarate} \rightarrow \text{2-oxo-3-methylbutyrate} + \text{glutamate}$$

In a muscle at rest, most of the 2-oxo acids produced from transamination of branched chain amino acids are transported to the liver and become subject to oxidation in reactions catalysed by branched-chain 2-oxo acid dehydrogenase complex. During periods of exercise, however, the skeletal muscle itself is able to utilize the oxo-acids by conversion into either acetyl-CoA (leucine and isoleucine) or succinyl-CoA (valine and isoleucine).

Transamination of alanine yields pyruvate catalysed by alanine transaminase (ALT) whilst aspartate produces oxaloacetate catalysed by aspartate transaminase (AST). All transaminase enzymes operate close to a true equilibrium ($K'_{eq} \sim 1$, see Chapter 2) and

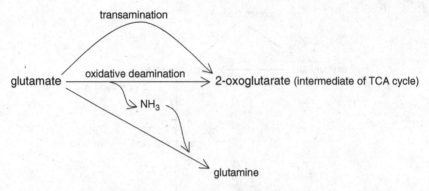

Figure 7.20 Glutamate metabolism in muscle

are freely reversible. Glutamate in muscle may be converted to 2-oxoglutarate by either a reversal of the transamination or by oxidative deamination, catalysed by glutamate dehydrogenase (GlD):

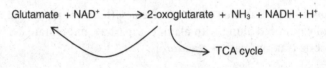

Compare with figure 6.8.

The 2-oxoglutarate produced is recycled for transamination or may enter the TCA cycle. The ammonia liberated by oxidative deamination is used to form glutamine (from glutamate, catalysed by glutamine synthase) prior to export from the muscle cell;

$$\text{Glutamate} + NH_3 + ATP \rightarrow \text{glutamine} + ADP + Pi$$

Glutamine is exported from the muscle and extracted from blood mainly by the kidneys or the gut; hepatic uptake of glutamine is relatively low in comparison. In the renal tubular cells, glutamine is deaminated in the processes of urinary acidification (see Figure 8.11) or used by the intestinal cells as a fuel.

Summarizing, muscle protein provides a number of compounds which can generate fuel for other tissues, a fact which explains the severe wasting seen in people who are literally starving, as muscle protein is sacrificed to maintain function of the central nervous system in particular. We have also seen how glutamate plays a central role in amino acid metabolism in muscle (Figure 7.20).

7.7 Fuel utilization by muscle: adaptation to exercise and training

Numerous factors are known affect the ability of muscles to adapt to and cope with exercise. These include general state of nutrition, age, genetics, the availability of fuel substrates and oxygen and the local cellular environment, for example pH. As indicated previously, fitness training can bring about changes to the physiology and biochemistry of skeletal muscles leading to improved sporting performance. Studies have shown significant differences in muscle fibre composition following several weeks of intensive endurance training. The trend was for an increase in the number oxidative fibres (type I and IIa) and a reduction in type IIb (glycolytic) fibres. This change is mirrored by increases in:

1.   blood supply to muscles as shown by a greater capillary density;

2.   myoglobin content of red fibres;

3.   the size and number of mitochondria and the activity of enzymes of the TCA cycle;

4.   upregulation of GLUT-4 glucose transporters allowing more efficient entry of glucose into the muscle cell, and upregulation of MCT (monocarboxylate transporter) which exports lactate from active muscle into the bloodstream for delivery to the liver and the heart for oxidation.

The transition from rest to mild activity and then to strenuous activity provides us with a good example of metabolism adapting to changes in the physiological situation. Exercise-related biochemistry is a major subject of research and a detailed description is beyond the scope of this text and the interested reader is referred to a specialized source. A brief overview is given below.

At rest, the human body as a whole utilizes mainly fat as fuel. Given that muscle constitutes 40% of body mass, it is reasonable to assume that resting muscle also relies mainly on oxidative metabolism of fatty acids. Predictably, the type and intensity of exercise determines cellular metabolism. The energy demands of a short (less than 30 s) sprint would be met almost exclusively by anaerobic glycolysis and phosphocreatine hydrolysis (i.e. drawing upon intracellular stores) whereas more prolonged but less strenuous exercise would depend mainly on aerobic metabolism. Moderate exercise lasting less than about 60 min would utilize fat and carbohydrate approximately equally, but as glycogen stores become depleted during longer periods of activity, β-oxidation of fat liberated from adipose tissue depots as a result of adrenaline or glucagon stimulation becomes more significant.

A measure of the intensity of exercise intensity is given by the rate of oxygen uptake per minute, symbolized by $\dot{V}O_2$. A typical value for $\dot{V}O_2$ for an 'untrained' person is 40 ml oxygen/kg/minute; this value may double for a person who has undergone a period of endurance training. As activity increases from rest to mild, then moderate and finally strenuous exercise (so $\dot{V}O_2 \cong \dot{V}O_2max$), there is a progressive recruitment of type I (low power output but high fatigue threshold), IIa and then IIb fibres (high power but short time, typically 3 min, to exhaustion), reflecting the increasing reliance on glycolytic ATP generation.

Transition from rest to even moderately intense exercise will cause the rate of glycolysis in skeletal muscle to rise significantly and rapidly (reaching a peak response of more than a 10-fold acceleration in less than 5 s). This metabolic adaptation is mediated by allosteric control of, mainly, PFK. Allosteric activators of PFK include ADP and AMP, ammonium ion and Pi. A rise in cytosolic ADP and Pi concentrations is due to increased hydrolysis of ATP; AMP is derived from ADP via the adenylate kinase reaction (Figure 7.14) and ammonium ion arises from the adenosine deaminase reaction (Section 7.4.2). The increased glycolytic flux causes an increase in lactate production to re-oxidize NADH to match NAD^+ utilization (by glyceraldehyde-3-phosphate dehydrogenase); the lactate is exported from the cell and may be oxidized by the heart or taken up by the liver and re-cycled as glucose (Cori cycle, Figure 7.8). Note here also that reduction in the ATP-to-ADP and NAD^+-to-NADH ratios promotes the activity of pyruvate dehydrogenase and therefore the production of acetyl-CoA for entry into the TCA cycle.

Energy generation from amino acid metabolism during exercise is small but not insignificant. As described in Section 7.5, skeletal muscle efficiently imports from the blood stream, branched-chain amino acids, released from the liver in response to prolonged exercise, which undergo transamination forming glutamate and branched chain oxo acids (BCOA). Oxidation by 2-oxoacid dehydrogenase of the BCOA allows the generation of succinyl-CoA or acetyl-CoA whilst glutamate may be used either to generate alanine (also a transamination reaction) or is converted into glutamine. Thus, the net effect is that exercising skeletal muscle greatly increases its secretion of alanine and glutamine whilst increasing its uptake of branched chain amino acids.

Chapter summary

Muscle is a metabolically highly active and adaptable tissue making up a large part of our body mass. The three types of muscle contain myocytes, which are long cylindrical cells with different metabolic and functional capabilities; type I slow-twitch fibres use mitochondrial oxidative metabolism whereas type II fast-twitch fibres rely on glycolysis for ATP production. Fuels used to drive contraction may be stored as glycogen or triglyceride within the myocytes or extracted from the blood stream in the form of glucose and fatty acids. During periods of fasting and starvation, muscle protein may be sacrificed in order to liberate amino acids which along with lactate can be used in gluconeogenesis to generate glucose to ensure near normal functioning of other tissues notably the central nervous system. Relatively few of the metabolic pathways associated with muscle are unique to the tissue as many also found in other organs but the control of the regulatory enzymes is often quite different allowing muscle to meet changing physiological demands in particular ways.

Case notes

1. **Muscular dystrophy.**

   Non-identical twins, Duncan and Elspeth McCracken, were born at term after an uneventful pregnancy. The development of the twins appeared normal at first; a few months later, Mrs McCracken noticed that Duncan seemed to experience difficulty lifting his head and remaining upright. By the age of 18 months he was not walking properly, doing so with a waddling gait and often falling to the floor. Strangely, or so it seemed to the parents given the apparent physical weakness, he had enlarged calf muscles in his legs. At 2 years of age he was generally less developed and having more communication difficulties than his sister and boys of a similar age. The family doctor arranged for a blood test which showed that the activity of CK was more than 10 times above the normal value for a child of Duncan's age. This evidence, plus the physical signs suggested a diagnosis of muscular dystrophy (MD). A blood test on Mrs McCracken showed that she too had an elevated CK activity, approximately five times above the normal.

   The term muscular dystrophy is a generic one which encompasses several different but related disorders. Overall, the incidence of MD is approximately $1 : 35\,000$ with the two commonest forms of MD being Duchenne (DMD) and Beckers (BMD). The cause of MD can be traced to a defect in the gene carried on the X chromosome coding for an important protein called dystrophin. Most mutations leading to MD can be traced through the family pedigree but a significant number arise spontaneously. Because the dystrophin gene is X-linked, the disorder only presents clinically in boys, but girls (like Duncan's mother and possibly his sister) are carriers.

   Functionally, dystrophin is associated with a glycoprotein complex embedded within the sarcolemma, where it acts to help maintain the shape and integrity of each myocyte and is also involved with cell signalling. Boys with DMD typically have less than 5% of the normal amount of functionally active dystrophin whereas in the less severe BMD there may be more than 20% of the protein present. All muscles are affected so not only movement but also breathing becomes impaired.

   Muscle damage and subsequent myocyte death are progressive and not yet preventable. Various lines of research are investigating the potential of gene and oligonucleotide therapy, cell therapy, the use of inhibitors of enzymes involved with cell necrosis, and the upregulation of utrophin which is structurally similar to dystrophin and may be able to substitute functionally for the defective protein. Young men are usually wheelchair-bound by the age of 20 and few survive long into their fourth decade of life.

2. **Acute myocardial infarction (AMI, heart attack)**

   Mr Patel came to the UK from his native India in 1969 when he was 22 years old. For almost 30 years he has run, very successfully with the help of his family, a small business which now employs 30 people in the suburbs of Leicester. At the age of 38 years, Mr Patel was diagnosed with diabetes mellitus (Type 1 diabetes) and now has trouble controlling his weight and is on medication to regulate his plasma triglyceride and cholesterol concentrations.

   Just as he and his wife were closing their premises at the end of the day, Mr Patel collapsed to the floor experiencing severe 'crushing' chest pain. Mrs Patel called an ambulance and her husband was admitted with a provisional diagnosis of '? AMI'. One of the vessels supplying blood to the heart muscle (myocardium) had become blocked, probably by atheroma (see Chapter 5) so preventing blood flow to an area of tissue, which consequently suffered necrotic damage due to hypoxia (diminished oxygen availability).

An ECG tracing was made at the time of admission, but no obvious abnormality was apparent. On admission and at an estimated 12 h after the onset of symptoms, blood samples were taken. The concentrations of myoglobin and troponin T were elevated in the first sample and the activity of CK second blood sample was beginning to show increase above normal. Despite the apparently normal ECG, a not uncommon finding in AMI, the pattern of changes in myocardial marker proteins confirmed the provisional diagnosis. A third blood sample taken approximately 24 h after admission showed significant elevation in CK. Mr Patel remained in hospital where his condition was monitored until his discharge.

Mr Patel returned to work but, reluctantly, agreed to allow his eldest son to take more responsibility for the day to day management of the business.

8

Biochemistry of the kidneys

Overview of the chapter

The excretion of water soluble waste via the kidneys requires filtration followed by selective reabsorption from and secretion into the renal tubules. Regulation of normal blood pH within very strict limits due to proton secretion and bicarbonate reabsorption is a major role of the kidney.

Structure and physiology of the kidney: glomerular filtration; tubular activity; selective reabsorption and secretion, often using specific carrier mechanisms; carbonic anhydrase and acid-base balance. The kidney also produces, and is sensitive to, hormones; actions of the hormones ADH, aldosterone and PTH; the kidney as a secretory organ; erythropoietin, the renin–angiotensin system; vitamin D_3.

Key pathways

Gluconeogenesis; glutamine and glutamate metabolism. Part synthesis of vitamin D.

8.1 Introduction

The gross appearance (macrostructure) of a kidney is recognizable to most people, even those who have no detailed knowledge of mammalian physiology. Furthermore, many people would be able to state that the role of the kidneys is to excrete waste materials in the urine. What is less likely to be so widely appreciated is the importance and complexity of action of the kidney in regulating the chemical composition and volume of the body fluids, a key aspect of homeostasis. Receiving approximately 25% of cardiac output per minute, the kidneys are adapted to monitoring blood pressure.

Essential Physiological Biochemistry: An organ-based approach   Stephen Reed
© 2009 John Wiley & Sons, Ltd

8.2 Renal physiology

The kidneys are paired encapsulated organs, each weighing approximately 150 g, and typically $11 \times 6 \times 3$ cm with a smooth outer surface. A longitudinal cut reveals two distinct layers; the dark reddish coloured outer cortex which makes up about 70% of the tissue mass and the paler coloured inner medulla.

Some basic 'facts and figures' about the kidneys reveal their dynamic nature:

- Each kidney consists of approximately one million functional units called nephrons;

- The total blood flow to the kidneys is approximately 1200 ml/minute (equivalent to 750 ml of plasma) but this can increase to 1500 ml/minute if the renal blood vessels are fully dilated. This means that the entire blood volume passes through the kidneys in les than 5 min. Only the liver receives a greater total volume of blood per minute (1300 ml) but the larger mass (1.5 kg) of the liver means that the volume of blood reaching the kidneys per gram of tissue per minute is approximately five times greater than the liver;

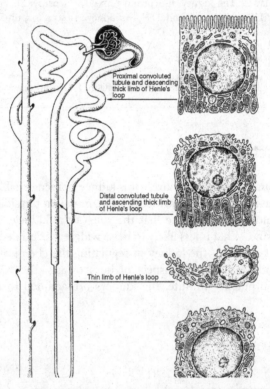

Figure 8.1 Microstructure of the kidney (Reproduced from Basic Histology by Junqueira, LC., Carneiro, J. and Kelley RO 1995 with permission of McGraw-Hill)

- An adult will filter approximately 150 l of blood plasma in 24 h but eliminate only about 1.5 l[1] of urine in the same period;

- Each kidney has a large functional reserve such that each organ can, if necessary, do the work of two and individuals with only one kidney can live normally. The diagnosis of renal disease is often delayed because a significant amount of tissue deterioration usually occurs before there are clinical or biochemical signs of dysfunction.

Renal microstructure is based on the nephron: the glomeruli (singular = glomerulus) are knots of blood capillaries enclosed with a cup-like structure composed of epithelial cells and called the Bowman's capsule. They are located in the cortex along with the proximal and distal convoluted tubules. The medulla contains the loops of Henle and the collecting ducts (Figure 8.1). The cytology of the tubules varies along the length of the nephron according to the nature of the biochemical activity occurring. For example, cells lining the proximal tubule and part of the distal tubule are relatively thick containing many mitochondria, a sure sign that these are metabolically active cells.

Expressed in the simplest terms, the glomeruli are filters and the tubules execute active and passive transport between the tubular fluid (glomerular filtrate) and the blood. The combined and coordinated function of the glomeruli and tubules constitutes the renal waste disposal and nutrient recycling system.

To the physiologist, removal of waste is known as 'clearance', a process which is also carried out by the liver (via the gut), the skin and the lungs. The aqueous nature of urine means that only hydrophilic compounds, notably the endogenous nitrogenous molecules urea (see Section 6.2), creatinine (see Section 7.4.2), some amino acids and urate (or uric acid) are excreted via this route, along with exogenous waste (e.g. drug metabolites) and ions. Indeed, the only reason we excrete water, a valuable biocommodity, is to keep waste solutes in solution to ensure their removal from the body. A minimum urine volume of approximately 500 ml per day is required to excrete a typical load of solute waste without risk of precipitation within the urinary system. Clearance from the blood stream is brought about by the concerted actions of both the glomeruli (selective filtration) and the tubules (selective reabsorption and secretion).

8.2.1 Glomerular function

Blood is supplied to the kidneys via the renal vein, a branch of the descending vena cava, at relatively high pressure to ensure rapid filtration of plasma across the membranes of the blood vessels in the glomeruli and the epithelial cells of the Bowman's capsule. The net filtration pressure of about 5–6 kPa, is the difference between the blood pressure forcing plasma water across the filtration barrier and the opposing osmotic and

---

[1] The 1.5 l of urine quoted is very variable depending mainly on fluid intake over the 24 h time interval. Figures relate to a typical 70 kg adult male.

hydrostatic pressures within the Bowman's capsule. The osmotic effect of the higher protein concentration in the plasma relative to that in the glomerular filtrate tends to pull water back into the blood vessel whilst the hydrostatic pressure is maintained due to a 'traffic jam' effect created by the anatomy of the microvessels associated with the Bowman's capsule. The internal diameter of the efferent ('leaving') capillary is narrower than that of the afferent ('incoming') capillary thus establishing a back-pressure within the glomerular knot of capillaries.

Assuming the capsular pressures opposing the movement of water out of the blood and into the top of the nephron are constant, the net filtration pressure is due largely to the blood pressure. Any fall in blood pressure can have a dramatic effect on the efficiency of filtration and therefore clearance of waste materials. So important is the pressure within the renal vasculature that the kidney is critical in regulating systemic blood pressure via the renin–angiotensin–aldosterone (RAA) axis, a physiological process which relies on transport mechanisms within the renal tubules.

The fluid which passes into the proximal tubule is often referred to a 'protein-free filtrate of plasma'. This is not strictly true as even completely healthy kidneys will allow about 2–5 g of small molecular weight protein, and even the occasional red blood cell, to escape into the filtrate. To view the selectivity of the glomerular barrier simply in terms of a sieve with (very) small holes is not tenable. Rather, selectivity of the filtration process is regulated largely by the biochemical nature of the filtration barrier. The pores which create the physical gaps in the barrier are highly negatively charged due to extensive sulfation (SO_4^- groups) of glycoproteins, so proteins which are themselves negatively charged at blood pH will be repelled by the glycoproteins and retained in the circulation.

The glomerular filtration rate (GFR) defines how much plasma water passes from the blood into the top of the nephron per minute. In health, the true GFR for a 70 kg adult is typically 100–120 ml/minute. Expressed another way, we can say that, in health, every minute each of the approximately 2 million glomeruli present in both adult kidneys filters between 0.05 and 0.06 μl of plasma water. The GFR is a good overall measure of renal function and the clinical laboratory has many ways of estimating its value.

Except for its lower protein concentration, glomerular filtrate at the top of the nephron is chemically identical to the plasma. The chemical composition of the urine is however quantitatively very different to that of plasma, the difference is due to the actions of the tubules. Cells of the proximal convoluted tubule (PCT) are responsible for bulk transfer and reclamation of most of the filtered water, sodium, amino acids and glucose (for example) whereas the distal convoluted tubule (DCT) and the collecting duct are concerned more with 'fine tuning' the composition to suit the needs of the body.

8.2.2 Tubular function

8.2.2.1 Tubular transport mechanisms: an overview
Transport in the nephron is mechanistically similar to that in the gut and indeed the two tissue types have essentially identical properties. Substances moving between the

glomerular filtrate and the blood must negotiate two barriers, namely the membranes of the tubular cells. Tubular transport is usually viewed in terms of reclamation of systemically important compounds or excretion of unwanted solutes, the implication being that such substances move into the tubular cell at one membrane and out through another. However, it is worth remembering that renal cells themselves need to sustain their own metabolism requiring an inward transport of nutrients obtained from the general circulation.

The luminal membrane is in contact with the tubular fluid and the basolateral membrane with extracellular (tissue) fluid and indirectly therefore, the blood. Transport is achieved predominantly via carrier-mediated (active or passive) processes with fewer examples of simple diffusion. Passive diffusion implies that substances pass relatively freely through the lipid-bilayer down a concentration gradient (an exergonic process) and without the need for a membrane-bound binding protein. Gases, notably carbon dioxide, are able to diffuse through membranes and water too crosses the barrier without carrier binding but through pores called aquaporins. The 'leakiness' of these pores may be regulated by pituitary-derived antidiuretic hormone (ADH) to allow more or less water to cross, depending upon the physiological conditions.

Carrier proteins transporting single compounds (uniports) or two compounds (cotransporters; symports or antiports) are found associated with both the luminal and basolateral membranes of cells in the PCT and DCT. Carrier-mediated mechanisms may be active (endergonic, operating against an electrical or chemical gradient and thus requiring an energy source) or passive, also called facilitated diffusion, operating down a gradient. Active pump mechanisms such as the H^+-ATPase and the Na^+/K^+-ATPase consume a significant amount of the ATP produced by tubular cells. Removal of Na^+ from the tubular cells by a transporter on the basolateral membrane is important not only to facilitate Na^+ reabsorption in to the bloodstream, but also to create a sodium gradient across the luminal surface of the cell. This Na^+ gradient is used to transport glucose and amino acids into the cell via symport proteins (Figure 8.2).

The quantity of any given solute being presented to the reabsorptive mechanisms is determined by the product of the GFR and the solute concentration in plasma. One of the features of any carrier-mediated process is its limited capacity. Binding of a substance to its transport protein follows the same principles as substrate binding to an enzyme or hormone binding to its receptor so we may appropriately liken the dynamics to Michaelis–Menten kinetics.

A plot of rate of transport against solute concentration in the tubule (Figure 8.3) shows t_m, the tubular transport maximum to be analogous with V_{max} for an enzyme, which is a maximum rate of solute transport across tubular cells. Assuming a fixed GFR, the point at which the plotted line begins to deviate from linearity, indicates that the substance exceeds a critical threshold concentration and begins to be excreted in the urine. When the plotted line reaches a plateau indicating that saturation point, that is t_m has been reached, the rate of excretion is linear with increase in plasma concentration. The concept of t_m as described here for tubular reabsorption applies equally well to carrier-mediated secretory processes. If the t_m value for a particular is exceeded for any reason, there will be excretion of that solute in the urine.

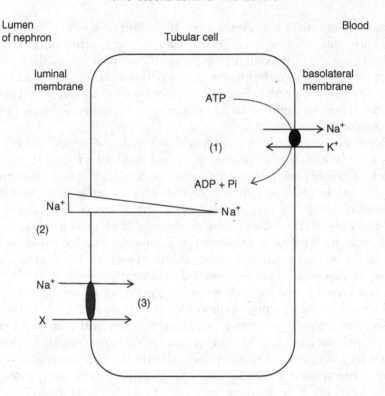

(1)   Na⁺/K⁺-ATP'ase extrudes sodium at the basolateral membrane, reducing intracellular [Na⁺]

(2)   Na⁺ gradient is established; higher [Na⁺] in lumen than inside cell

(3)   Na⁺ co-transporter (symport) allows uptake of X (e.g. amino acid or glucose)

Figure 8.2 Sodium gradient

Proximal tubule Cells of the PCT are responsible for bulk transport of solutes, with approximately 70–80% of the filtered load of sodium chloride (active processes) and water (passive, down the osmotic gradient established by sodium reabsorption) and essentially all of the amino acids, bicarbonate, glucose and potassium being reabsorbed in this region.

Carbonic anhydrase (CA, also called carbonate dehydratase) is an enzyme found in most human tissues. As well as its renal role in regulating pH homeostasis (described below) CA is required in other tissues to generate bicarbonate needed as a co-substrate for carboxylase enzymes, for example pyruvate carboxylase and acetyl-CoA carboxylase, and some synthase enzymes such as carbamoyl phosphate synthases I and II. At least 12 isoenzymes of CA (CA I–XII) have been identified with molecular masses varying between 29 000 and 58 000; some isoenzymes are found free in the cytosol, others are membrane-bound and two are mitochondrial.

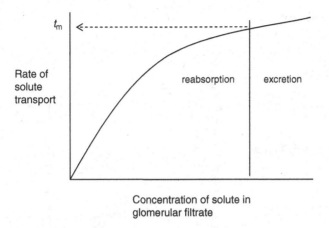

Excretion of a solute occurs if its rate of delivery to the tubules exceeds the t_m.

Figure 8.3 t_m: tubular transport kinetics

The reaction catalysed by CA is the hydration of carbon dioxide thus:

$$CO_2 + H_2O \xleftarrow{\text{carbonic anhydrase}} H_2CO_3$$

The carbonic acid so produced may spontaneously, but weakly, dissociate:

$$H_2CO_3 \rightarrow H^+ + HCO_3^-$$

Carbonic anhydrase is a metalloprotein with a co-ordinate bonded zinc atom immobilized at three histidine residues (His 94, His 96 and His119) close to the active site of the enzyme. The catalytic activity of the different isoenzymes varies but cytosolic CA II is notable for its very high turnover number (K_{cat}) of approximately 1.5 million reactions per second.

Cytosolic CA II is widespread through tissues, the kidney possesses CA IV which is anchored to the cell membrane of the luminal PCT brush border by linkage with a membrane phospholipid, glycosylphosphatidylinositol. Such luminal positioning allows the enzyme to act upon filtered bicarbonate ions as they enter the tubule.

CA IV splits H_2CO_3 formed by protonation of filtered HCO_3^- into CO_2 and H_2O. The carbon dioxide diffuses into the PCT cell and is rehydrated by the action of CA II. The protons generated by the dissociation of the H_2CO_3 are secreted via a Na^+/H^+ exchanger in the apical (luminal) membrane and used to protonate more filtered HCO_3^-, whilst the HCO_3^- is passed via a basolateral Na^+/HCO_3^- cotransporter into the blood. Figure 8.4 shows the location and roles of CA II and CA IV.

Overall, for each bicarbonate ion filtered, one has been returned to the blood; that is sodium bicarbonate has been reabsorbed and the glomerular filtrate leaving the PCT has a greatly reduced bicarbonate concentration. The relatively small amount of proton secreted by the PCT cell and not used to protonate filtered bicarbonate

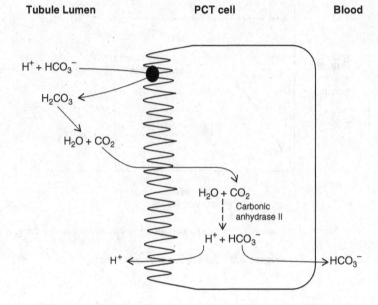

● = Carbonic anhydrase IV embedded in brush border of the PCT cell

Figure 8.4 Role of carbonic anhydrase in proton secretion and bicarbonate reabsorption

is buffered with HPO_4^{2-} which also derives from the plasma. The pH of the glomerular filtrate at this point is only slightly lower than that of blood.

Carbonic anhydrase is clearly responsible for generating H^+ for secretion and HCO_3^- ions for reabsorption, it is not the only enzyme which is important in renal acid-base regulation. Glutaminase, an enzyme which is intimately involved with ammoniagenesis is located in the proximal tubule. Glutamine, the substrate for the enzyme, enters the PCT cell mainly from the glomerular filtrate, with only a little coming from the bloodstream, before passing into the mitochondria where glutaminase is situated. The reaction catalysed by glutaminase is a deamidation as shown below (Figure 8.5).

The glutamate which results form the reaction remains in the mitochondria where it is oxidatively deaminated by glutamate dehydrogenase to form 2-oxoglutarate.

The reaction shown in Figure 8.6 is also important in the liver where glutamate dehydrogenase is involved in the catabolism of amino acids and the entry of nitrogen into the urea cycle, as explained in Chapter 6.

Combining reactions 8.5 and 8.6:

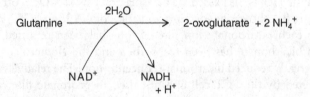

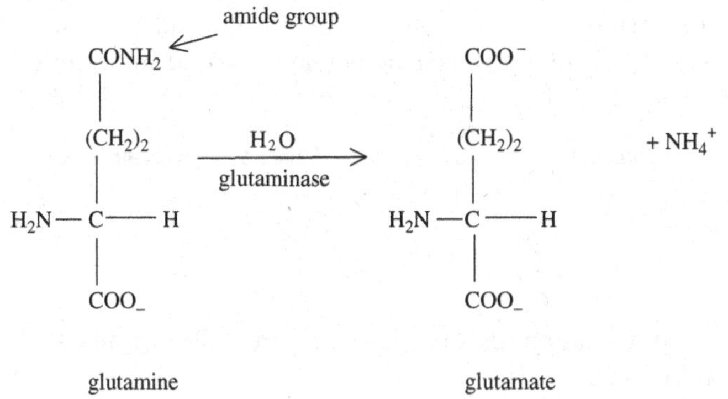

amide group

CONH$_2$ COO$^-$
|
(CH$_2$)$_2$ H$_2$O   (CH$_2$)$_2$ + NH$_4^+$
| glutaminase
H$_2$N—C——H   H$_2$N—C——H
|
COO$_-$ COO$_-$

glutamine   glutamate

Figure 8.5 Glutaminase

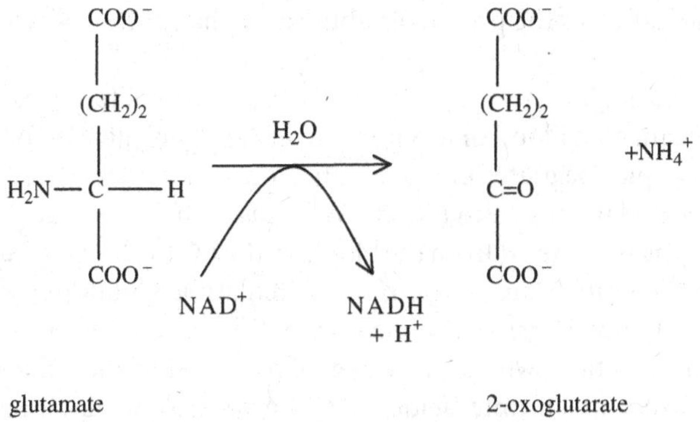

COO$^-$ COO$^-$
|
(CH$_2$)$_2$ H$_2$O   (CH$_2$)$_2$
| +NH$_4^+$
H$_2$N—C——H   C=O
|
COO$^-$ COO$^-$
NAD$^+$ NADH + H$^+$

glutamate   2-oxoglutarate

Figure 8.6 Glutamate dehydrogenase

The two ammonium ions produced from glutamine as illustrated in Figures 8.4 to 8.6 are secreted into the PCT lumen the by a Na$^+$/H$^+$ antiport (the NH$_4^+$ substitutes for H$^+$). Subsequent metabolism of 2-oxoglutarate has the potential to generate two bicarbonate ions from the hydration of carbon dioxide by carbonic anhydrase:

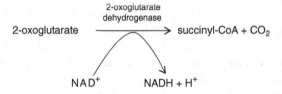

2-oxoglutarate $\xrightarrow[\text{dehydrogenase}]{\text{2-oxoglutarate}}$ succinyl-CoA + CO$_2$

NAD$^+$ NADH + H$^+$

The oxidative decarboxylation reaction above is part of the TCA cycle and leads to the formation of oxaloacetate, which may be used to synthesize citrate (with acetyl-CoA) or may be used as a substrate by phosphoenol pyruvate carboxykinase, PEPCK. It should be noted that the phosphoenolpyruvate generated by PEPCK reaction shown above is available for conversion to glucose so this explains the contribution of the kidney to gluconeogenesis. Details of gluconeogenesis may be found in Section 6.5.2.

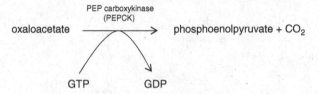

Thus, two decarboxylation reactions have occurred; the CO_2 liberated is a substrate for carbonic anhydrase:

$$2CO_2 + 2H_2O \xrightarrow{\text{CA II}} 2HCO_3^- + 2H^+$$

Bicarbonate ions secreted into the blood stream help maintain the normal plasma bicarbonate concentration of approximately 25 mmol/l, whilst the two protons are secreted into the lumen of the proximal tubule in exchange for sodium via a Na^+/H^+ antiport.

The importance of glutamine metabolism in the proximal tubule is partly due to its adaptability to changes to blood pH. Within 3 h of a sudden fall in blood pH (i.e. during an acute acidosis), plasma glutamine concentration rises approximately twofold as the amino acid is released mainly from muscle. Whereas in the normal acid-base state, only very little glutamine is removed from the blood by the PCT cells, in acute acidosis, up to 30% of the now elevated concentration of circulating glutamine is extracted and ammoniagenesis is accelerated. Coupled with this is a slight increase in plasma bicarbonate concentration, which of course helps to buffer the acidosis.

During situations of chronic acidosis, the activities of glutaminase, glutamate dehydrogenase and PEPCK are all increased and the PCT transporters for protons and bicarbonate ions are all up-regulated. These changes require between 2 and 3 days to become fully effective, meaning that the acidosis is not compensated during that period of time. However, once fully operational, the excretion of ammonium, a major urinary buffer, may rise 5–10 times above normal, reflecting increased conversion of glutamine to glutamate. Figure 8.7 summarizes the roles of carbonic anhydrase and glutaminase in renal acid–base homeostasis.

Amino acid reabsorption in the renal tubules Amino acids are small, easily filtered molecules. Efficient reabsorption mechanisms are vital to conserve amino acids which are metabolically valuable resources. Transport of individual amino acids and small peptides is symport carrier mediated mechanisms in which sodium is co-transported. The process is indirectly ATP dependent because Na is returned to the lumen of the nephron by the 'sodium pump', Na^+/K^+ dependent ATPase.

Several symport proteins have been identified in the luminal and basolateral surfaces of the proximal tubule cells, each with a specific transport function. For example, mechanisms exist for transport of: (i) neutral amino acids, except glycine, (ii) glycine alone, (iii) acidic amino acids (glutamate and aspartate), (iv) basic amino acids

| Tubule Lumen | PCT cell | Blood |

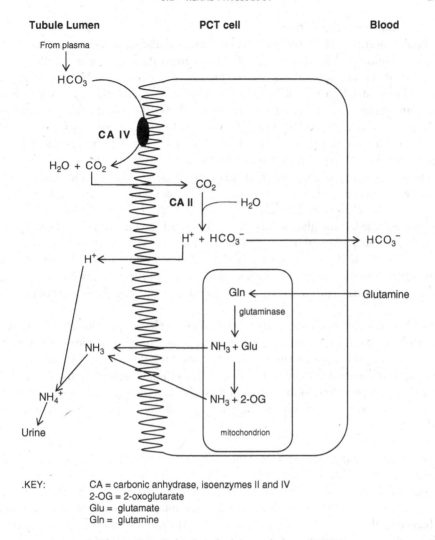

Figure 8.7 Carbonic anhydrase and glutaminase

(arginine, lysine) and (v) cysteine and cysteine. Several genetic defects of amino acid reclamation are known. Some, for example cystinuria, affect specifically amino transport, whilst others, for example Fanconi syndrome are more generalized disorders of renal tubular reabsorption.

Additionally, amino acids may be reclaimed as dipeptides. The transport mechanisms for dipeptides are less specific than those for individual amino acids but require the dipeptide to carry a net positive charge so there is cotransport of protons, rather than of Na^+ as for free amino acids. A potential advantage of dipeptide transport process is the favourable cell–lumen concentration gradient, which exists for peptides compared with free amino acids.

8.2.2.2 Glucose reabsorption

Glucose concentration in the glomerular filtrate is the same as that in the plasma at about 5 mmol/ll. The movement of glucose from the lumen into the PCT cell is, like that of the amino acids, an active process which uses the Na electrochemical gradient generated by a Na^+/K^+ dependent ATPase. Two such Na^+-dependent glucose transport proteins (sodium-glucose transporter, SGLT), identified as SGLT1 and SGLT2, have been located in the luminal membrane. Transport of glucose from the PCT cell into the bloodstream is mediated by the Na^+-independent glucose transporter-2 (GLUT2), similar to that found in the liver.

Tubular glucose reabsorption is usually very efficient and the urine is glucose-free. However, if plasma, and therefore tubular, glucose concentration rises above approximately 10 mmol/l, as occasionally occurs in diabetic individuals, the reabsorptive capacity is exceeded and glycosuria occurs. The condition known as renal glycosuria is, however, a non-pathological condition, in which the threshold for reabsorption is abnormally low and so even relatively modest rises in blood glucose concentration cause glucose loss in the urine. Because glucose is osmotically active in the nephron, excessive glycosuria from whatever cause can result in diuresis (large volume of urine).

Loop of Henle Although the cells of the loop of Henle and the distal regions of the nephron are thinner they are no less important in reabsorption and secretion. The descending limb of the loop is permeable to water but not to solutes, thus when the tubular fluid reaches the bend in the loop it has a higher osmolality than when it left the PCT. The ascending limb is thicker than the descending limb, reflecting its role in further reabsorption of solutes, for example sodium chloride, potassium, calcium and magnesium.

The distal tubule and collecting duct The volume of fluid entering the distal tubule is only about 90% of the original filtered load and is approximately isotonic with respect to plasma. It is the distal nephron that controls the fine tuning of the chemical composition of the blood by reabsorbing, for example, more or less sodium (under the influence of aldosterone), water (controlled by ADH, in the collecting duct), calcium (via parathyroid hormone, PTH). The DCT is also very important in achieving subtle pH regulation. A number of substances, for example some drugs and urate (uric acid) are secreted directly into the lumen of the distal tubule.

Sodium reabsorption Much less than 10% of the filtered load of NaCl reaches the distal nephron. Regulation of Na uptake, occurring mainly in the principal cells of the cortical collecting tubule, is controlled by the steroid hormone aldosterone (see Section 4.4). The net effect of aldosterone is the reclamation of NaCl and potassium excretion in to the luminal fluid.

Like all steroids, aldosterone enters the target cell and combines with cytosolic mineralocorticoid receptor. Such receptors are not entirely specific for aldosterone and will also bind cortisol, the principal glucocorticoid hormone. The receptors are 'protected' from cortisol activation by 11β hydroxysteroid dehydrogenase which converts cortisol into cortisone, a steroid which does not engage the aldosterone receptor and so does not influence sodium and potassium transport. In pathological situations of cortisol excess (Cushing's disease), the mineralocorticoid effect of cortisol becomes apparent as the excess hormone in effect swamps the ability of 11β hydroxysteroid dehydrogenase to inactivate it and patients exhibit hypertension due to sodium overload and hypokalaemia (low plasma potassium concentration). Conversely, in Addison's disease (inadequate cortisol) patients have potentially fatal hyperkalaemia because K^+ cannot be excreted properly.

Physiologists had postulated for a long time about the existence of a sodium excreting hormone to prevent Na overload and consequent deleterious effects of high blood pressure on the heart and vascular system. At least two such natriuretic factors have been described; atrial or A-type and brain or B-type natriuretic factors. Structurally, the natriuretic factors are peptides with a cysteine–cysteine disulfide bridge creating a characteristic 'loop', this is illustrated by Figure 8.8.

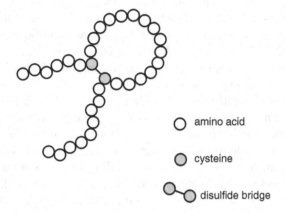

amino acid

cysteine

disulfide bridge

Figure 8.8 Conformation of B-type natriuretic peptide (BNP)

B-type natriuretic peptide (BNP) was first isolated from porcine brain in the late 1980s but is now known also to be secreted by the cardiac muscles in response to ventricular distension (Section 5.2). Such a situation arises when the blood volume, and therefore blood pressure, are increased (hypertension). BNP ameliorates the hypertension by (a) promoting the excretion of sodium (natriuresis) and water (diuresis) in the nephron and (b) by causing systemic vasodilation. The exact mechanism causing the release of BNP is uncertain but like other natriuretic factors, it operates on target tissues via a G-protein-linked rise in cytosolic cyclic AMP (cAMP) to antagonize the action of angiotensin in the sympathetic nerves innervating the arterial wall and in causing aldosterone release in from the adrenal cortex.

Water reabsorption and regulation of blood pressure   Approximately 7–7.5 l of fluid passes from the distal convoluted tubule into the collecting duct each day; substantially more than the 1.5 l of fluid which are excreted as urine. Reabsorption of water occurs

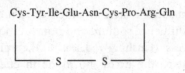

Cys-Tyr-Ile-Glu-Asn-Cys-Pro-Arg-Gln

Figure 8.9 Structure of ADH

via pores composed of aquaporins proteins within the collecting duct membrane. In the presence of ADH, the water conducting pores open allowing transfer from lumen to cell. The high osmolality of the tissue surrounding the collecting duct, created by the counter-current mechanism in the loop of Henle, ensures that the water moves by diffusion out of the basolateral side of the cells and in to the blood stream.

Anti-diuretic hormone is a small peptide shown as Figure 8.9, which is secreted by the pituitary gland located at the base of the brain. The cellular actions of ADH are mediated by activation of a G-protein linked receptor generating cAMP as second messenger. Absence of ADH or a functional defect in the action of ADH-stimulated water reabsorption in the collecting duct results in the condition diabetes insipidus, characterized by the passing of large volumes (= diabetes) of dilute (= insipidus) urine.

Oxytocin, a peptide which initiates uterine contractions during labour is identical in structure to ADH except at position 8 where a leucine residue replaces arginine. The close structural similarity but radically different biological functions, illustrate how specific some hormone receptors are in recognising only their own 'signal'.

Reabsorption of water is a fundamental function of the kidney because loss of fluid volume and reduction in blood pressure (hypotension) would have devastating consequences on all other tissues, possibly leading to severe metabolic disruption or even death. Blood pressure is monitored by the kidney and regulated by secretion of a proteolytic enzyme called renin, which initiates a cascade involving angiotensin and aldosterone to restore blood volume.

A fall in blood pressure causes local and sympathetic nerve driven secretion of renin, from a group of cells located in the macula densa. These cells are part of the juxtaglomerular apparatuss (JGA), a structure formed by the physically close association of cells of the proximal convoluted tubule and the afferent blood capillary. The natural substrate for renin is angiotensinogen. Structurally, angiotensinogen is a highly glycosylated protein of with 485 amino acid residues (molecular weight approximately 62 kDa) which belongs to the serpin family of proteinase inhibitors (serpin = *ser*ine *p*rotease *in*hibitor), although it has no inhibitory action itself. Secreted into the plasma by the liver, angiotensinogen is cleaved between two adjacent leucine residues near the N-terminal to produce angiotensin I (10 amino acids), which in turn has two further amino acids removed by angiotensin-converting enzyme (ACE) to produce angiotensin II (Figure 8.10).

ACE is bound to the luminal surface of vascular endothelium occurring in particularly high concentration in the pulmonary capillary bed. The product of the ACE reaction, angiotensin II, is a potent vasoconstrictor by its action on vascular smooth muscle cells allowing calcium influx and opposing the vasodilator effect of

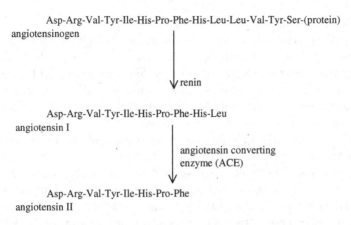

Asp-Arg-Val-Tyr-Ile-His-Pro-Phe-His-Leu-Leu-Val-Tyr-Ser-(protein)
angiotensinogen

↓ renin

Asp-Arg-Val-Tyr-Ile-His-Pro-Phe-His-Leu
angiotensin I

angiotensin converting
enzyme (ACE)
↓

Asp-Arg-Val-Tyr-Ile-His-Pro-Phe
angiotensin II

Figure 8.10 The enzymes renin and ACE cleave angiotensinogen

nitric oxide (NO) released from endothelial cells. Angiotensin II is also a stimulant for the secretion of aldosterone, from the glomerulosa of the adrenal cortex. As stated above, aldosterone promotes sodium reabsorption from the nephron; this in turn promotes water reabsorption so increasing blood volume. In short, the RAA axis acts to increase blood pressure but overactivity of the process leads to hypertension, which is just as damaging to the kidney and other tissues as is a low blood pressure. Compounds known as ACE inhibitors have been developed for use as antihypertensive drugs, for example captopril and enalapril.

Angiotensin II-induced stimulation of tubular cells by aldosterone causes an increase in plasma osmolality, that is the osmotic potential determined by the number and type of solutes present in the plasma. The rise in osmolality stimulates the release of ADH, (vasopressin) from the pituitary resulting in enhanced water reabsorption from the distal portions of the nephron and thus increases in blood volume and pressure. Interestingly, recent research has shown renin–angiotensinogen systems (RAS) occur in certain other tissues such as the pancreas and adipose tissue, presumably to regulate local blood dynamics. Such a localized RAS generates angiotensin, which operates in an autocrine or paracrine fashion and is independent of the classical so-called systemic RAS targeting the renal and vascular systems as described above. Angiotensin II also acts on vascular smooth muscle cells where it promotes an influx of calcium which initiates contraction thus opposing the vasodilatory effects of bradykinin-induced NO within endothelial cells.

Acidification of urine   The point was made earlier that the pH of the fluid in the PCT is approximately 6.8, only slightly below that of blood. In contrast, urinary pH may vary considerably but is usually approximately pH 6. Acidification of the urine occurs mainly in the distal portion of the nephron.

Depending on diet and systemic (whole body) metabolism, which produces so-called 'fixed acids' such as sulfate derived from sulfur-containing amino acids and phosphate, mainly from nucleic acid metabolism, there is a tendency for pH of body fluids to fall.

Additionally, the incomplete oxidation of fuel generates weak organic acid anions. Furthermore, there is loss of base in the faeces each day which, in effect, leaves the body with an excess of protons to be excreted. In total, about 70 milliequivalents (mEq) of acid require excretion each day. {Note: mEq is used to quantify the acid load because this takes account of the valency of the ion; 1 mmol of sulfate, SO_4^{2-}, for example, is 2 mEq of negative charge, requiring 2 mEq of protons for neutralization, but for monovalent ions, such as protons or bicarbonate, 1 mEq = 1 mmol}.

The addition of 70 mEq of protons to 1.5 l (typical 24 h urine volume) of water would give rise to a pH < 3; clearly below the physiological range and obviously therefore, the urine is buffered. As described above, filtered hydrogen phosphate, HPO_4^{2-}, and ammonia (generated in the PCT) contribute significantly to urine buffering. However, the so-called α intercalated cells (αIC) of the collecting duct possess an aldosterone-sensitive ATP-dependent proton translocase whose action can achieve acidification of the luminal fluid to a minimum of approximately pH 4.5. For each proton secreted, and buffered by HPO_4^{2-}, one bicarbonate ion is added to the blood via a basolateral HCO_3^-/Cl^- exchanger. This anion exchanger has been found to be homologous with a red cell membrane protein AE1, which mediates the chloride shift as described in Section 5.3.1. Proton secretion by the α IC cells ensures that the bicarbonate ions not reabsorbed in the PCT are reclaimed, thus urine is bicarbonate-free.

8.3 Metabolic pathways in the kidneys

Over 90% of the blood supplying the renal tissue is directed to the cortex, suggesting that metabolism in this area is aerobic whereas the medulla, with only at most 10% of the blood supply, is mainly anaerobic. The pathways that operate in these areas reflect the oxygen availability, that is cortical cells oxidize glucose, fatty acids and ketone bodies, whereas the renal medulla, being relatively oxygen deficient, relies on anaerobic glycolysis and so produces a significant amount of lactate. It is this compound that contributes to the body's ability to synthesize glucose via renal gluconeogenesis. Normally, it is the liver which contributes most glucose to the body via gluconeogenesis and in usual feeding/fasting cycles, kidney-derived lactate accounts for only about 10% of the gluconeogenic substrates. However, in times of starvation, this proportion rises significantly and the kidney becomes a net glucose provider.

Oxidative catabolism of glucose, fatty acids and produces acetyl-CoA which enters the Krebs TCA cycle. Generation of reduced coenzymes by isocitrate dehydrogenase, 2-oxoglutarate dehydrogenase, succinate dehydrogenase, and malate dehydrogenase supports mitochondrial oxidative phosphorylation and much of the ATP so generated is used to drive the active transport mechanisms involved with tubular function.

The kidney is an endocrine organ. The classification of calcitriol (vitamin D_3) as a vitamin is erroneous because the active molecule is synthesized *in vivo* by a process which usually provides an adequate amount to ensure correct calcium homeostasis. The final step in the biosynthesis of calcitriol (also known as 1,25 cholecalciferol or 1,25

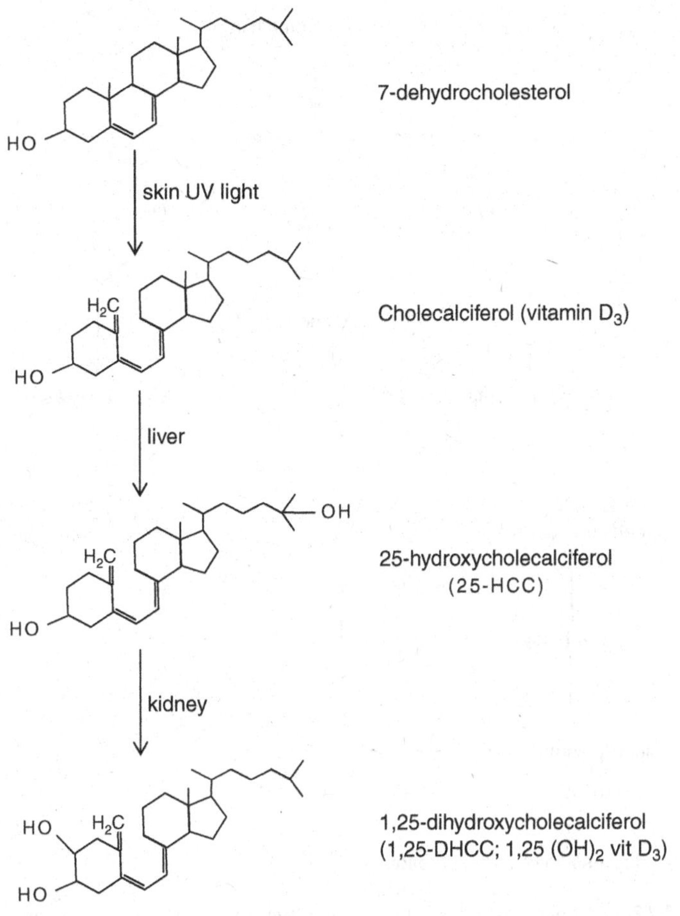

7-dehydrocholesterol

skin UV light

Cholecalciferol (vitamin D$_3$)

liver

25-hydroxycholecalciferol
(25-HCC)

kidney

1,25-dihydroxycholecalciferol
(1,25-DHCC; 1,25 (OH)$_2$ vit D$_3$)

Figure 8.11 1,25 Dihydroxy vitamin D$_3$ synthesis

(OH)$_2$ D$_3$ or dihydroxy vitamin D$_3$) occurs in mitochondria of the proximal tubules. The pathway of dihydroxy vitamin D$_3$ synthesis is shown in Figure 8.11.

The precursor, 7-dehydrocholesterol is converted by a non-enzymatic reaction to cholecalciferol (calciol). This reaction occurs in skin exposed to sunlight due to irradiation by UV-B light at a wavelength of about 300 nm. Cholecalciferol is transported via carrier proteins to the liver where hydroxylation at carbon-25 occurs in a reaction catalysed by a microsomal cytochrome P450 hydroxylase to form calcidiol. This compound travels to the kidney attached to specific binding proteins, where another cytochrome P450 enzyme, mitochondrial 1-α-hydroxylase, introduces a second hydroxyl group in to the molecule to form the active calcitriol.

The two hydroxylase enzymes can also utilize the plant-derived steroid, ergocalciferol, (vitamin D$_2$) as a substrate. The final product is biologically active and so food manufacturers often fortify their products with ergocalciferol to prevent the occurrence of vitamin D deficiency and consequent rickets in childhood or osteomalacia in adults.

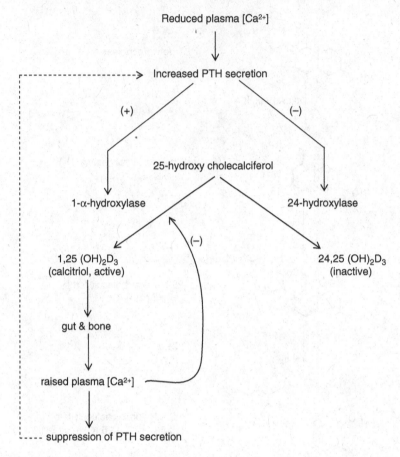

Figure 8.12   Feedback control of dihydroxy vitamin D_3 synthesis by ionized calcium

In addition to 1-α-hydroxylase, the kidney also possesses a 24-hydroxylase which uses calcidiol as substrate; the product of the reaction, 24,25 dihydroxy D_3, is biologically inactive. This represents an important control point in the pathway. The activity of the 1-α-hydroxylase is promoted by calcium ions and the action of PTH acting via a G-protein/cAMP cascade. However, calcitriol itself simultaneously induces the 24-hydroxylase and suppresses 1-α-hydroxylase creating an effective feedback loop (Figure 8.12).

Calcitriol's action primary function is in regulating plasma calcium concentration. In health, the plasma total calcium concentration is tightly controlled at 2.35–2.55 mmol/l. Only the ionized or 'free' fraction, amounting to about 50% of the total, is physiologically active in for example, maintenance of membrane electrical potential and bone formation. The hormone causes increased bone resorption via activation of osteoclasts (see Section 9.4) and increased intestinal absorption of calcium following the synthesis of a specific binding protein in mucosal cells. As described in Section 4.7, some hormones enter a target cell, then combine with their receptor and initiate DNA transcription by engaging with the promoter sites of particular genes. In order for the vitamin D receptor (VDR) to bind the vitamin D response element (VDRE) of appropriate genes, it must dimerize with retinoid X receptor (RXR) to create a fully functional transcription factor.

The other hormone of note synthesized by kidney is erythropoietin (EPO), a glycosylated peptide hormone (molecular weight approximately 50 000), which promotes red blood cell formation and is secreted in response to poor oxygen perfusion (hypoxia) of the kidney. This, along with the control of blood pressure via the RAA system illustrates the importance of the kidney in regulating aspects of the blood vascular system. Further details of EPO can be found in Chapter 5.

Chapter summary

A major biochemical contribution of the kidneys to the body's overall health is in maintaining a normal pH of the extracellular fluids; the combined actions of carbonic anhydrase, glutaminase and various translocase proteins regulate the generation and excretion of protons coupled with the reabsorption of bicarbonate ions. A significant amount of 'fixed acid' (e.g. organic acids such as lactate) is removed from the body each day by the kidneys. Carrier-mediated processes in the renal tubules also maintain ionic homeostasis, controlling as they do reabsorption or elimination of sodium, potassium, calcium and anions such as chloride and phosphate, usually in response to hormone stimulation. Renal tubules both synthesize and are responsive to dihydroxy vitamin D_3. Specific cells of the blood capillary draining the glomerulus monitor blood pressure and secrete renin, an enzyme which initiates release of aldosterone and vasoconstriction mediated by angiotensin II.

Case notes

1. Mrs Amin, a 55-year-old female had apparently enjoyed good health and had not seen a physician for a considerable period of time. Mr Amin requested a visit from her general practitioner (GP) because his wife had suffered diarrhoea and vomiting for over a week. The GP questioned her about her eating and drinking habits during her illness but she appeared confused and a little drowsy so was unable to answer coherently. The patient's husband stated that she had lost her appetite and had been unable to drink normally, although until the current illness started she had drunk 'a lot' of fluid each day. The doctor noticed that Mrs Amin looked dehydrated and arranged for her to be admitted to the local hospital where blood tests were performed. The laboratory results were as follows;

Chemistry		Reference range:
Plasma urea	29.2 mmol/l	3.5–7.5 mmol/l
Plasma sodium	126 mmol/l	136–147 mmol/l
Plasma potassium	6.1 mmol/l	3.6–4.7 mmol/l
Plasma creatinine	310 μmol/l	70–110 μmol/l

Plasma bicarbonate	17 mmol/l	23–28 mmol/l
Plasma glucose	14.2 mmol/l	3.5–5.5 (fasting)
Liver function tests	all normal	
Plasma total calcium	2.32 mmol/l	2.35–2.55 mmol/l

Haematology

White cell counts	normal;	
Red cell count	$3.1 \times 10^{12}/l$	$3.8–4.8 \times 10^{12}/l$
Haemoglobin	8.6 g/l	11–13 g/l

The estimated GFR, based on the creatinine concentration was very low.

A diagnosis of renal failure was made. Mrs Amin's results show poor glomerular function (high urea and creatinine concentrations) and tubular function (sodium, potassium and bicarbonate results). The slightly low plasma calcium points to a defect with vitamin D metabolism and the haematology data suggest erythropoietin deficiency. Given the patient's reluctance to eat, the blood glucose result may be taken to be a fasting value, which being very high along with the evidence from Mr Amin of increased thirst for some time raised the suspicion, later confirmed, of diabetes. This condition had been undiagnosed but possibly present for a long time.

Chronic renal failure is a common consequence of diabetes but this case is complicated by the loss of fluid and electrolytes (sodium and potassium) due to diarrhoea and vomiting. Normally, the kidneys would respond to such a challenge and maintain homeostasis but Mrs Amin's kidneys were unable to do so. Mrs Amin was put on haemodialysis and treated to control the diabetes.

2.   Alan Smith was a 58 year old who suffered chemical burns to 35% of his body following exposure to a phenol-containing industrial solvent. He was admitted to hospital for treatment of the burns and observation. Concerns arose, when despite intravenous fluids, he failed to pass urine and plasma concentrations of urea and creatinine began to rise.

Results of plasma analysis.

		Reference range
Sodium	125 mmol/l	136–147 mmol/l
Potassium	5.9 mmol/l	3.6–4.7 mmol/l
Bicarbonate	16 mmol/l	23–28 mmol/l
Calcium	2.24 mmol/l	2.35–2.55 mmol/l
Urea	8.1 mmol/l	3.5–7.5 mmol/l
Creatinine	190 μmol/l	70–110 μmol/l
Albumin	31 g/l	35–48 g/l
Blood pH	7.15	7.35–7.45

These results suggest acute renal failure (ARF) due to tubular necrosis caused by phenol. Plasma sodium is low due mainly to impaired reabsorption in the nephron, although the slightly low albumin suggests haemodilution possibly as a result of excessive i.v. fluids. Potassium is raised due to poor exchange with sodium in the distal tubule and the acidosis (low pH and low bicarbonate concentration) arises from defective acidification of the glomerular filtrate; acidosis is often associated with hyperkalaemia (raised plasma potassium). Elevated urea and creatinine concentrations are usually taken as indicators of glomerular damage but both compounds are also subject to tubular activity.

Mr Smith was placed on daily maintenance dialysis for several days until his urinary output showed signs of improvement. As is common in ARF, as tubular damage began to resolve, the patient went into a polyuric phase (increased urinary output) caused by an osmotic effect of the solute load passing through only a few functional nephrons. Following his discharge from hospital, Mr Smith's urine output remained slightly higher than normal for many months but he was able to lead a reasonably normal life.

3. W.B. is a 27 year old male who is HIV positive. His current treatment regime is a 'cocktail' of drugs including the antiretroviral agent tenofovir. Although effective against HIV, the use of such drugs as part of HAART (highly active antiretroviral therapy) is commonly associated with nephropathy. Some researchers argue that the virus itself rather than the drug may be the cause of the renal complications. As is often the case with drug-induced nephropathies, changes are seen in tubular, rather than glomerular, function. Patients present with Fanconi syndrome, a generalized reduction in proximal tubular function leading to increased urinary loss of amino acids, glucose, phosphate and bicarbonate. Laboratory monitoring of tubular function specifically often involves the analysis of urinary amino acids but this is both expensive and labour intensive.

A study was initiated to find a more convenient chemical marker, or panel of markers, that could be used to assess the rate of deterioration of renal function in subjects taking antiretrovirals. Although he was asymptomatic at the time, W.B. agreed to participate in the study. Each of the 86 participants in the study was asked to collect and deliver to the laboratory, a urine sample once a month for a year. On each monthly visit to the clinic, a blood sample was taken.

The concentrations of phosphate and glucose were measured on all samples and the urine samples were subject to detailed protein analysis; two small proteins, retinol binding globulin and β-2-microglobulin (β2M), received particular attention. Unfortunately, none of the markers alone or in combination proved to be entirely appropriate, but a rising concentration of β2M was able to identify nearly three-quarters of subjects who subsequently showed some degree of renal damage.

9

Biochemistry of connective tissue: bone and adipose

Overview of the chapter

Histologically, connective tissue is diverse but fulfils similar supporting and lubricating roles throughout the body. This chapter includes a discussion of metabolism of typical proteoglycans and proteins, the most significant being collagen, of the extracellular matrix, bone architecture and the roles of osteoblasts and osteoclasts in bone turnover. The importance of adipose tissue as a fuel storage organ and as an endocrine tissue influencing feeding and fuel disposition is reviewed.

Key pathways

Biosynthesis and degradation of glycosaminoglycans; biosynthesis of collagen, mineralization and demineralization of bone. Fatty acid synthesis and triglyceride storage in adipocytes promoted by insulin and triglyceride hydrolysis and fatty acid release stimulated by glucagon and adrenaline (epinephrine).

9.1 Introduction

Connective tissue has vital functions in supporting, lubricating, strengthening, protecting, insulating and connecting the major organ systems. Included in the generic term 'connective tissue', we find skin, bone, cartilage, adipose tissue and blood cells. Members of this functionally diverse group are related by their common mesenchymal origin. Histologically, connective tissue is unusual because the various cell-types present are more loosely organized than in a 'solid' organ such as, for example, the liver. The notable exception to this loose arrangement of cells is the adipose tissue where adipocytes are organized in a more congruent fashion. Individual cells of connective

Essential Physiological Biochemistry: An organ-based approach   Stephen Reed
© 2009 John Wiley & Sons, Ltd

tissue are embedded in a fibrous, gelatinous material called the extracellular matrix (ECM). This matrix is composed of proteins and carbohydrate-rich proteoglycans, which are secreted by the particular types of connective tissue cells. The protein matrix of bone is particular in that it is calcified to form the mechanically strong structure which is so familiar.

9.2 Histology of connective tissue

Slender spindle-shaped fibroblasts are the commonest cell to be found in connective tissue. These cells, which manufacture matrix collagen and proteoglycans, have an elliptical nucleus and a cytoplasm which is contains extensive rough endoplasmic reticulum and Golgi apparatus, which are typical features of secretory cells. Mast cells are found widely within connective tissues and like fibroblasts these are secretory cells and characterized by a very granular cytoplasm. Almost as numerous as the fibroblasts and cytologically very similar to them, are macrophages, also called histiocytes or littoral cells.

The various cell types are distributed differently within various types of connective tissue; fibroblasts often found lying close to collagen fibres, whilst macrophages and mast cells usually appear in clusters around blood vessels. Additionally, whereas fibroblasts are widespread, macrophages are absent from some types of connective tissue, tendons for example. Specialized cells also occur within particular locations for example osteoblasts and osteoclasts (bone), chondrocytes (cartilage), adipocytes (fat tissue) and haemopoietic cells (bone marrow).

Classification of connective tissues creates two major groups: specialized and non-specialized (Figure 9.1). The non-specialized connective tissues found in an adult human show marked architectural diversity, due largely to the proportion of cellular and non-cellular material present. Loose (or areolar) connective tissue, found widely, for example surrounding blood vessels and nerves and in the mucosa of various organs, contains more cells than collagen fibres. In contrast, dense connective tissues have abundant fibres of collagen arranged as straight thick bundles in either an ordered fashion to provide strength in tendons and ligaments, or an irregular arrangement where collagen fibres are found as wavy cords, as found in the submucosa of the gut and in the dermis of the skin.

9.3 The ECM of connective tissue

Chemically, the extracellular matrix (also called ground substance) consists of a mixture of fibrous proteins (mostly collagens, but also elastin, fibronectin and laminin in some cases), glycoproteins and proteoglycans. Mesh-work arrangements of the protein and heteropolysaccharide fibres within the ECM create a highly hydrated gel-like consistency, which accounts for the physical properties – that is lubricating, mechanical strength, 'shock absorbing' yet nutrient-permeable material.

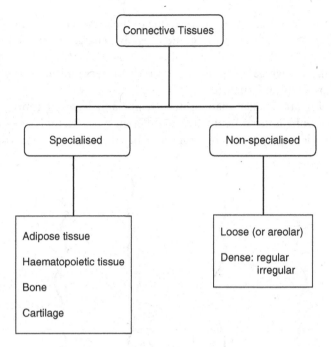

Figure 9.1 Classification of connective tissue

9.3.1 Protein-carbohydrate complexes of the ECM

Macromolecular complexes of proteins and carbohydrate present in the ECM serve not only as 'adhesive' keeping cells in their correct positions but also facilitate control of cell activity by signalling through membrane proteins such as the integrin family of receptors. *Glycoproteins* are mainly protein with covalently attached carbohydrate whereas *proteoglycans* are predominantly complex carbohydrates secured on a protein framework.

By far the largest, (molecular mass of up to 1 million), of the molecules found in the ECM are the proteoglycans (Figure 9.2). These molecules consist of three basic components:

- core proteins;

- linking proteins;

- glycosaminoglycans attached directly to the core proteins.

Glycosaminoglycans (GAGs, but still sometimes called mucopolysaccharides, MPS) are complex heteropolysaccharides constructed from repeat disaccharide units, usually of an *N*-acetyl amino sugar (either glucose or galactose) and hexuronic acid, most commonly glucuronic acid or iduronic acid, which are oxidized hexose sugar derivatives typically where a carboxyl group replaces a hydroxyl group.

The monosaccharides present are charged due to the presence of sulfate in addition to carboxylate groups, conferring a large net negative charge on the whole molecule. Numerous GAGs attach to the same hyaluronic acid (strictly speaking, hyaluronate) 'stem' to form an aggregate (Figure 9.2). Table 9.1 lists typical examples of GAGs; Figure 9.3 shows typical structures.

Glycoproteins are smaller than proteoglycans, and have far more variability in protein and carbohydrate content between different types. In both types of molecule, the oligosaccharide moieties are covalently attached to the protein via serine, threonine

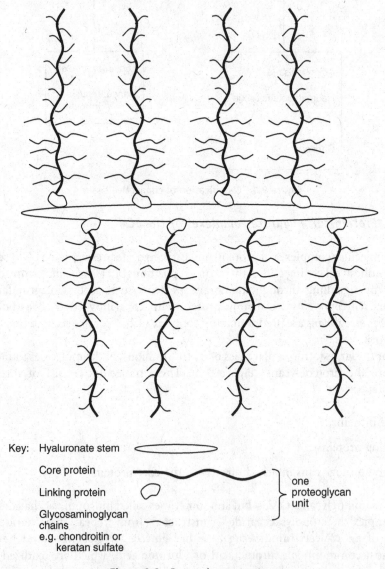

Key: Hyaluronate stem

      Core protein

      Linking protein

      Glycosaminoglycan
      chains
      e.g. chondroitin or
        keratan sulfate

      } one
        proteoglycan
        unit

Figure 9.2 Proteoglycan structure

Table 9.1 Example glycosaminoglycans (GAGs)

GAG	Component monosaccharide units	Repeat disaccharide structure n may range from 20 000 to 50 000	Typical location
Hyaluronate (hyaluronic acid)	Glucuronate (glucuronic acid) Abbreviated to GlcA; N-acetylglucosamine abbreviated to GlcNAc	$\{GlcA(\beta1\text{–}3)\ GlcNAc\ (\beta1\text{–}4)\ldots\}_n$	Ground substance of connective tissue; Vitreous humour of the eye; Umbilical cord; synovial fluid of joints
Keratan sulfate	Galactose Abbreviated to Gal; N-acetylglucosamine-6-sulfate Abbreviated to GlcNAc6S	$\{Gal(\beta1\text{–}4)\ GlcNAc6S\ (\beta1\text{–}3)\ldots\}_n$	Costal cartilage; Aorta; Cornea
Chondroitin-4-sulfate	Glucuronate (glucuronic acid) Abbreviated to GlcA; N-acetylgalactosamine-4-sulfate Abbreviated to GalNAc4S	$\{GlcA(\beta1\text{–}3)\ GalNAc4S\ (\beta1\text{–}4)\ldots\}_n$	Aorta; Cornea; Cartilage and bone
Dermatan sulfate	Iduronate (iduronic acid) Abbreviated to IdUA; N-acetylgalactosamine-6-sulfate abbreviated to GalNAc6S	$\{IdUA(\beta1\text{–}3)GalNAc6S\ (\beta1\text{–}4)\ldots\}_n$	Heart; Skin; Sclera

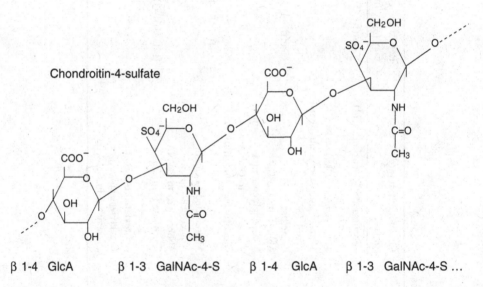

....β 1-4 GlcA   β 1-3 GlcNAc   β1-4 GlcA   β 1-3 GlcNAc ....

β 1-4 GlcA   β 1-3 GalNAc-4-S   β 1-4 GlcA   β 1-3 GalNAc-4-S ...

Figure 9.3 Glycosaminoglycan structure

or asparagine residues. Laminin and fibronectin are found widely in connective tissues but other glycoproteins, such as osteocalcin and elastin show a more restricted distribution.

9.3.2   GAG synthesis and degradation

Synthesis of GAGs requires the sequential attachment of an *N*-acetyl amino sugar or uronic acid to the protein core molecule (Figure 9.2), each step being catalysed by a specific glycosyl transferase. Only when the monosaccharides are in place does sulphation occur with phosphoadenosine phosphosulfate (PAPS, an analogue of ATP) acting as the donor. Monosaccharide raw materials for the polysaccharide chains, all of which must be 'activated' by the attachment of UDP, are derived from glucose (Figure 9.4).

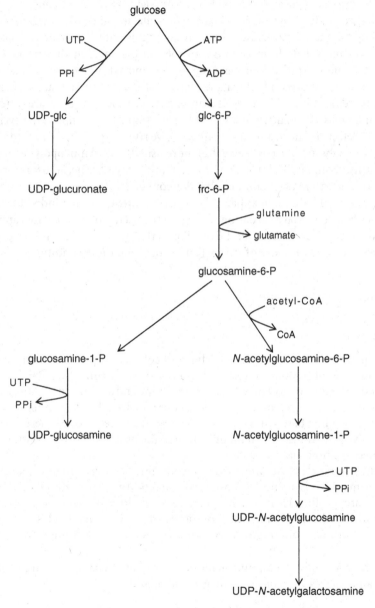

Figure 9.4 Derivation of acetyl monosaccharides and glucuronate for GAG synthesis

GAGs are degraded in lysosomes as part of normal cellular homeostasis and their constituents are recycled. Lysosomal exohydrolases progressively remove monosaccharides from the ends of the molecule; specific sulphatases remove the sulfate groups before cleavage of the glycosidic bond between adjacent monosaccharides. Figure 9.5 exemplifies the degradation of GAGs by showing the catabolism of heparan sulfate.

Heparan sulfate catabolism progresses systematically through the pathway cascade catalysed by three glycosidases (iduronidase, N-acetylglucosaminidase and glucuronidase) three sulfatases (iduronate sulfatase, heparan sulfatase and N-acetylglucosamine-6-sulfatase) and an acetyltransferase. The pathway begins with iduronate sulfatase (IDS), which cleaves the sulfate from carbon-2 of the iduronic acid at the end of the GAG molecule. The now exposed iduronate residue is removed by the exohydrolase idurionidase, followed by the second sulphatase. The next step is curious in that the glucosamine can only be removed, by N-acetylglucosaminidase if it has first been acetylated at the expense of acetyl-CoA. Glucuronidase and the last sulphatase complete the cycle.

Enzyme defects in the degradative pathway give rise to a group of conditions known as glycosaminoglycanoses, (see Table 9.2) more usually called mucopolysaccharidoses. Genetic transmission is, in all except X-linked Hunter's syndrome, autosomal recessive. Partially degraded GAGs accumulate in the connective tissue, skeleton and central nervous system and although apparently normal at birth, affected individuals exhibit varying degrees of mental retardation, slow growth, often grossly abnormal appearance and abnormalities of the heart and liver. Formerly, the facial deformities evident in most cases prompted the use of gargoylism to describe these conditions.

9.3.3 Proteins of the ECM

1. **Collagen**

   The many (possibly more than 30) types of collagens found in human connective tissues have substantially the same chemical structure consisting mainly of glycine with smaller amounts of proline and some lysine and alanine. In addition, there are two unusual amino acids, hydroxyproline and hydroxylysine, neither of which has a corresponding base-triplet or codon within the genetic code. There is therefore, extensive post-translational modification of the protein by hydroxylation and also by glycosylation reactions.

   Although the exact amino acid sequence differs between the various collagens, the primary structure usually conforms to a repeating tripeptide Gly-X-Y where X and Y are, proline, lysine, or hydroxyproline, hydroxylysine respectively. A single unit of collagen is a triple helix composed of three α *chains*. This conformation differs from the common α *helix* found in proteins in two important ways:

   a. there are only three amino acids per turn of an α chain, compared with 3.6 in an α helix of a typical globular protein, and

   b. each α chain is a left-handed helix, not right-handed.

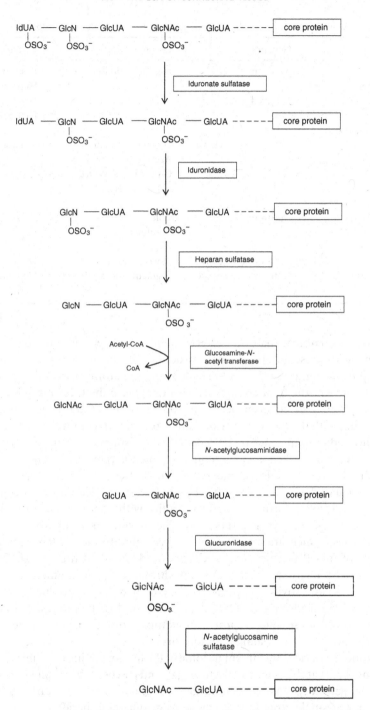

Figure 9.5 Degradation of heparan sulfate

Table 9.2 Biochemical features of glycosaminoglycanoses (mucopolysaccharidoses)

Type	Name	Genetics	Enzyme defect	Urinary excretion
I	Hurler	Autosomal recessive	α-L-iduronidase	Dermatan sulfate
Ia	Scheie			Heparan sulfate
II	Hunter	X-linked	iduronate sulphatase	Dermatan sulfate
				Heparan sulfate
III	Sanfilippo			
	III a	Autosomal recessive	Heparan sulphatase;	Heparan sulfate
	III b		Acetylglucosaminidase	
	III c		Glucosamine N-acetyl-transferase	
IV	Morquio			
	IVa	Autosomal	Galactosamine-6-sulphatase	Keratan sulfate
	IVb	Recessive	β-galactosidase	
VI	Maroteaux-Lamy	Autosomal recessive	Arylsulphatase	Dermatan sulfate
VII	(none)	Autosomal recessive	β-glucuronidase	Heparan sulfate
				Dermatan sulfate

The more compact nature of each α chain explains why we find so few different amino acids present in the primary structure. Glycine has the smallest side chain group of the amino acids, just a single hydrogen atom, as anything more bulky would prevent the tight coiling required to achieve the three residues per turn.

Biosynthesis of collagen begins in the ribosome of the rough endoplasmic reticulum with the manufacture of a single α chain. In common with many newly synthesized proteins, each α chain has at its N-terminal, a sequence of hydrophobic amino acids called a signal sequence which is removed by signal peptidases within the rough endoplasmic reticulum of the osteoblast or chondrocytes. Individual α chains are then subject to limited hydroxylation of selected proline and lysine residues forming 4-hydroxyproline with minor amounts of 3-hydroxyproline and hydroxylysine respectively. The enzymes prolyl hydroxylase and lysyl hydroxylase which are responsible for these modifications are examples of the family of mixed function oxidases, MFOs. Members of this group of enzymes contain a haem group, the iron atom of which must be maintained in the reduced state (Fe^{2+}) for activity. For prolyl and lysyl hydroxylase, the reduction is achieved with ascorbic acid (vitamin C) and deficiency of this vitamin results in impaired ability to synthesise collagen properly and this accounts for some of the symptoms such as poor wound healing seen in scurvy.

Monosaccharide attachment to some of the hydroxylysine residues within the α chains occurs at this stage as galactose (gal) and glucose (glc) residues are added by specific galactosyl transferase and glucosyl transferases. The carbohydrates are attached as single monosaccharides or as gal-glc disaccharides.

A molecule of procollagen is formed by the association of three α chains, twisted around each other to form a *right*-handed triple helix of approximately 300 nm in length, excluding non-helical domains, also called extension peptides, at the C- and N-terminals. Note here the opposite three-dimensional geometry of the 'super-coiled' triple chain helix compared with the single chain helices; an orientation which strengthens the whole assembly. The carboxyl terminal extension peptides are critical in the assembly of procollagen because inter-α-chain disulphide bonds act to anchor the chains allowing the winding of the three individual α peptides into the triple helix.

After the formation of triple helix, procollagen is transported into the Golgi apparatus and packaged into secretory vesicles, from where it is released from the cell by exocytosis. Enzymatic removal by procollagen peptidases, of both terminal non-helical extension domains forms tropocollagen molecules which spontaneously assemble into collagen fibres. The terminal globular domains known as P1CP and P1NP (procollagen type 1 C-terminal peptide and procollagen type 1 N-terminal peptide respectively) which are removed at this stage pass in to the blood stream. Measurements of the plasma concentrations of P1CP and especially P1NP have attracted clinical interest as potential markers of the rapidity of bone turnover.

Individual tropocollagen molecules aggregate forming collagen fibrils, stabilized by covalent cross-links forged by copper-containing lysyl oxidase between lysine and hydroxylysine residues. Additional cross-links at the terminals of each collagen molecule are made by pyridinoline (PYD) and deoxypyridinoline (DPD) molecules. When the fibrils are associated with appropriate proteoglycans, a complete collagen fibre is generated prior to the assembly of a collagen bundle (see Figure 9.6).

When we consider the complex cross-linked super-coiled assembly of collagen, it is no surprise that the molecule provides such great tensile strength, equivalent to steel wire of comparable dimensions. The importance of collagen's structure is made clear by the abnormalities which are seen when gene mutations arise. For example, Ehlers–Danlos syndrome of connective tissue and osteogenesis imperfecta of bone are both due to defects in the synthesis of collagen resulting from mutations in the *COL1* gene on chromosome 17. Mutations have also been found in genes directing the synthesis of types II and III collagen (*COL2* and *COL3* genes respectively) and alterations in the gene encoding the lysyl oxidase, the enzyme responsible for converting lysine into hydroxylysine, also occur.

In the case of osteogenesis imperfecta, also known also as 'brittle bone disease', approximately 80 individual mutations of collagen-directing genes have been discovered. The impact of the mutations varies greatly form mild to lethal, the more severe types being associated with protein defects at or near to the carboxyl end of the procollagen molecule, probably affecting its super-coiling. Commonly, in the abnormal forms, glycine is replaced by an amino acid such as serine with a bulkier side chain. Ehlers–Danlos syndrome is generic term for a group of at least 10 individual heritable conditions affecting collagen synthesis. The most common presentation of the disease is type IV due to mutation in *COL3A1*. As would be predicted, poor cross-linking of tropocollagen molecules or changes in the

(a)   single α-chain; gly-X-Y repeat sequence

N-terminal   C-terminal

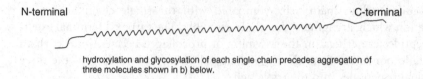

hydroxylation and glycosylation of each single chain precedes aggregation of
three molecules shown in b) below.

(b)   triple helix of procollagen

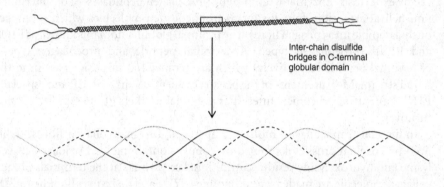

Inter-chain disulfide
bridges in C-terminal
globular domain

three left-handed helices inter-coiled in a right-handed helix

(c)   non-helical terminal regions removed by proteases after secretion from the
         chondrocyte or osteoblast. Tropocollagen formed

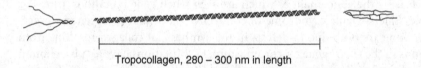

Tropocollagen, 280 – 300 nm in length

(d)   assembly of tropocollagen molecules into bundles, stabilised by cross-links (⌇).
         Note the staggered arrangement of individual tropocollagen molecules

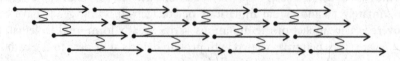

Figure 9.6 Collagen structure

cleavage site for the peptidases which remove the N-terminal portion of procollagen also result in weak and dysfunctional connective tissues.

Genetically determined abnormalities in GAG or collagen synthesis lead to joints
becoming easily dislocated, hyperextendable and hyperelastic ('double-jointed'),

bones which fracture easily and blood vessels become weakened and rupture; skin is brittle and easily damaged. Although many such conditions are compatible with life morbidity among those affected is high. Progressive deterioration in the quality of collagen is also responsible for many physical injuries suffered in old age.

2. **Elastin**

The fibrous protein elastin found extensively in connective tissues is unlike collagen in that it occurs in a less well ordered fashion, furthermore, there are quite marked differences seen between the chemical compositions of collagen and elastin. Whereas collagen comprises a very limited number of different amino acids, elastin contains a wider variety, the most abundant being glycine (approximately 30% dry weight), alanine (23%) valine (15%) and proline (12%).

9.4 Bone

The strength and apparent rigidity of bone belie its dynamic metabolic activity. In addition to its obvious supporting role, bone also acts as a reservoir for calcium, phosphate and other ions. In health, more than 15 mmol of calcium are exchanged between the bone and the plasma each day; this is equivalent to completely exchanging the entire plasma calcium content twice. Bone matrix can act as an ion-exchanger, for example by taking up protons in exchange for calcium or sodium during periods of chronic acidosis. Remodelling of bone occurs throughout the lifespan of an individual, but is clearly enhanced following traumatic injury or during certain disease states. Typically, a bone remodelling cycle consists of a phase of resorption lasting about a week, while reformation takes up to 3 months. Over the course of a year, approximately 20% of bone surface area is replaced whereas the bulk of the tissue which is compact bone turns over much more slowly at between 2 and 3% a year. The structure of bone is shown in Figure 9.7.

Bone tissue consists of two components: (i) the organic osteoid which is 90–95% type 1 collagen, the remainder being mostly proteoglycan and several strongly anionic (negatively charged) non-collagen proteins such as osteopontin, osteonectin and osteocalcin; (ii) the inorganic mineral, consisting mostly of calcium and phosphate, but also includes sodium, magnesium, fluoride and carbonate. The exact chemical composition of bone mineral is not entirely clear but it resembles needle-shaped crystals of hydroxyapatite, $Ca_3(PO_4)_2Ca(OH)_2$. In simple quantitative terms, bone contains 99% of total body calcium and over 85% of total body phosphate. The ionic nature of the apatite attracts a surface layer of water and counter ions such as sodium (approximately 30% of the total within the body) and citrate in what is known as an hydration shell. The water content contributes significantly to the flexibility of bone and one of the reasons for fractures occurring in elderly people is due to the reduction of the mineral and water content.

Articulated joints between bones, for example at the knee, are covered in a capsule enclosing a space, which contains synovial fluid. The lining of the capsule is composed by the synovial membrane; it is this synovium that becomes inflamed in rheumatoid arthritis (RA). Secretions produced by inflammatory cells (lymphocytes, macrophages

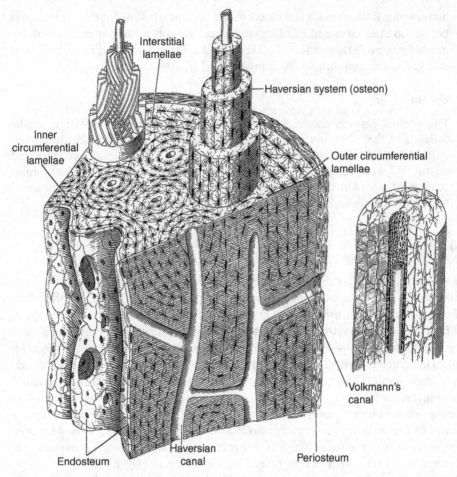

Figure 9.7   Bone structure. (Reproduced with permission from Basic Histology (1995) Junquiera LC, Carneiro J, and Kelley RO, McGraw-Hill)

and plasma cells) may lead to secondary damage to the cartilage within the joint. In the long term, sufferers often have characteristic joint deformities especially of the hands, wrists, feet and ankles. There may be non-articular features with involvement of eyes, spleen, lungs and nervous system. The cause of RA is unknown, but a particular protein of the major histocompatibility complex, HLA-DR4 occurs frequently in affected individuals. RA may develop at almost any time of life but is far more common in women than men.

9.4.1   Bone cells

Embedded within the protein–mineral structure are the three types of bone cell: osteoblasts ('bone builders' which form new bone matrix), osteoclasts (which degrade existing bone) and osteocytes whose function is not known for sure but assumed to be related to maintenance of normal bone turnover and metabolism. The coordinated action of all three cell types is required for normal processes of bone growth, remodelling and repair. Bone cells have different origins: osteoclasts are from the monocyte–macrophage lineage whilst osteoblasts arise from a pluripotent mesenchymal progenitor, which may differentiate along an osteogenic or haematopoietic line. Recall that blood, which may also be classified as a connective tissue, is formed in the marrow in bone cavities (see Section 5.3).

Cytologically, osteoblasts are similar to fibroblasts because they too contain an extensive protein synthesing machinery (rough ER and Golgi), indicating that these cells secrete the collagen-based osteoid matrix. At rest, osteoblasts are flat, spindle shaped cells but at sites of bone repair and regeneration, they enlarge and form a single columnar-like layer. The structure of bone collagen is identical with collagen elsewhere in the body yet bone is the only site to undergo extensive calcification. This suggests the presence of a natural inhibitor whose effect within the bone matrix must be overcome by osteoblasts to allow mineralization to occur normally; pyrophosphate (PPi) is believed to act as just such as inhibitor, but osteoblasts secrete alkaline phosphatase to hydrolyse PPi and so relieve the inhibition during periods of bone growth. Measurement of the activity bone alkaline phosphatase in plasma acts a marker of active bone formation.

The mechanical strength of bone approaches that of cast iron and granite. The tissue can withstand tremendous forces (several hundred kilograms in some cases in young adults), but unlike metals, bone has elasticity and flexibility allowing it to withstand twisting and bending so reducing the risk of fracture. Within reasonable limits, once a pressure has been removed from bone, it will resume its correct structure, whereas an iron bar will not spontaneously 'un-bend'. The physical properties of bone are due to the layered arrangement of collagen into approximately circular structures called osteons or Haversian systems. Layers of collagen bundles in an osteon are arranged as helices but successive layers are aligned in alternating orientations, clockwise/anticlockwise/clockwise/anticlockwise and so on (Figure 9.7).

As they secrete osteoid matrix around themselves, osteoblasts eventually become entrapped within lacunae ('spaces') and mature into osteocytes. Although topographically isolated from each other, osteocytes maintain contact with neighbours by cell processes which run through tiny interconnecting fluid-filled canaliculi (channels) within the matrix.

Osteoclasts sit in a shallow cavity on the surface of bone, as illustrated in Figure 9.8. These cells are large, multinucleated and contain numerous mitochondria. The differentiation of osteoclasts from macrophages is regulated by osteoblasts in two inter-related ways:(i) intercellular signalling and (ii) direct cell–cell contact. Signalling by osteoblast-derived macrophage colony-stimulating factor (M-CSF) induces the expression of a surface protein called RANK. Cell-cell contact occurs when this protein engages its ligand (RANKL) expressed on the surface of the osteoblast. This engagement allows terminal differentiation of the macrophage into a fully functional osteoclast. However, the RANK/RANKL binding can be prevented by osteoprotegerin,

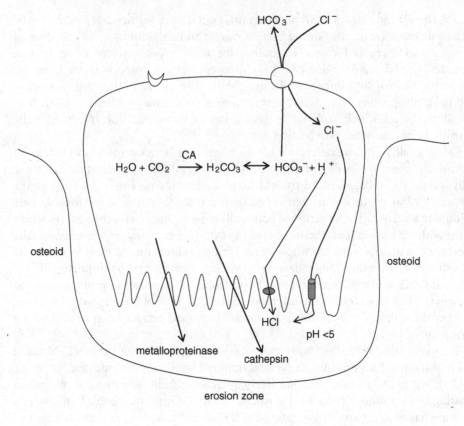

$$H_2O + CO_2 \xrightarrow{CA} H_2CO_3 \longleftrightarrow HCO_3^- + H^+$$

osteoid

osteoid

metalloproteinase

cathepsin

HCl

pH <5

erosion zone

KEY

hormone receptor

HCO_3^-/Cl^- exchanger

Cl^- channel

Proton pump

metalloproteinase
cathepsin
} Degradative
enzymes

Redrawn from figure 4-24 Histology and Cell Biology. An Introduction to Pathology by
Abraham L Kierszenbaum pub Mosby 2002

Figure 9.8 Osteoclast

a glycoprotein secreted by osteoblasts. Thus, the osteoblast both initiates and regulates
the production of mature osteoclasts. Parathyroid hormone (PTH) shifts the balance in
favour of osteoclast production by stimulating the production of M-CSF, RANKL and
osteoprotegerin, the net effect being enhanced osteoclastic activity and an increase in
bone demineralization (resorption).

Part of the metabolic machinery of an osteoclast resembles the red cell and the renal tubule cells because all of these cell types contain the enzyme carbonic anhydrase (carbonate dehydratase) which generates acid, that is protons, and have ion pumps in their plasma membranes. The mechanism of bone resorption requires the action of cathepsin and metalloproteinase-9 working in an acidic environment (Figure 9.8).

The proton which is generated is secreted from the osteoclasts by an ATP-dependent pump and chloride follows via a specific ion channel. Cytosolic concentration of chloride is maintained by an anion exchanger that mediates influx of chloride and efflux of bicarbonate (see Section 8.2.2 for comparison).

Several hormones other than PTH mentioned above are implicated in normal bone turnover, that is setting the balance between bone synthesis and demineralization. Thyroid hormones increase osteoclast activity whereas calcitonin which reduces plasma calcium concentration, suppresses bone resorption. Physiologically, calcitonin is not significant in humans except possibly during growth spurts in childhood and during pregnancy to ensure adequate skeletal formation. Sex hormones, in particular oestrogens, have a role in the regulation of bone remodelling. Although both men and women may suffer from bone thinning (osteoporosis) due to demineralization, the condition is most commonly associated with menopausal changes. Relative oestrogen lack causes a net loss of bone mineral due to increased osteoclastic activity resulting in bone fracture.

9.4.2 Calcium and bone mineralization

Calcium is the major mineral component of bone and normal repair and remodelling of bone is reliant on an adequate supply of this mineral. Calcium uptake in the gut, loss through the kidneys and turnover within the body are controlled by hormones, notably PTH and 1,25 dihydroxy cholecalciferol (1,25 DHCC or 1,25 dihydroxy vitamin D_3 or calcitriol). Refer to Figure 8.12 for a summary of the involvement of PTH and vitamin D_3 in controlling plasma calcium concentration. These two major hormones have complementary actions to raise plasma calcium concentration by promoting uptake in the gut, reabsorption in the nephron and bone resorption. Other hormones such as thyroxine, sex steroids and glucocorticoids (e.g. cortisol) influence the distribution of calcium.

Parathyroid hormone is a peptide of 84 amino acids. The main physiological action of PTH is to raise the plasma concentration of calcium by, (i) increasing renal reabsorption (of calcium *and* magnesium), (ii) increasing the synthesis of 1,25 dihydroxy vitamin D_3 and (iii) promoting calcium loss from bone mineral by stimulating osteoclasts and suppressing the activity of osteoblast cells. PTH secretion is controlled by a negative feedback effect on the parathyroid glands exerted by ionized calcium (Ca^{2+}). Approximately half of the plasma total calcium concentration is ionized, the remainder being mostly protein bound.

Vitamin D_3 is a fat-soluble steroid-like molecule which elevates plasma calcium concentration. A typical diet contains some preformed vitamin D_2 (from plant sources)

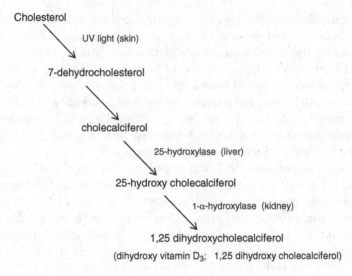

Cholesterol

UV light (skin)

7-dehydrocholesterol

cholecalciferol

25-hydroxylase (liver)

25-hydroxy cholecalciferol

1-α-hydroxylase (kidney)

1,25 dihydroxycholecalciferol

(dihydroxy vitamin D_3; 1,25 dihydroxy cholecalciferol)

Figure 9.9 Outline of vitamin D_3 synthesis

and vitamin D_3 but most of our DHCC requirements are met by endogenous synthesis which begins with the conversion of 7-dehydrocholecalciferol mediated by ultra violet radiation in the skin into vitamin D_3. Two hydroxylation reactions then occur; firstly in liver mitochondria (25 hydroxylase) and in the kidney (1αhydroxylase). An outline of vitamin D_3 synthesis is shown in Figure 9.9. Both of the hydroxylation steps are catalysed by reduced nicotinamide adenine dinucleotide phosphate and oxygen-dependent enzymes which are members of the cytochrome P450 family of mixed function oxygenases. Transport of 25-hydroxy vitamin D through the blood stream is mediated by specific binding proteins and the plasma concentration is maintained in the range of 3–30 µg/l. Typically of steroids and steroid-like hormones, 1,25 dihydroxy vitamin D_3 acts on its target cells by binding with nuclear receptors.

There seems to be no metabolic control exerted on hepatic 25-hydroxylase and so all of the available cholecalciferol is converted. Hydroxylation in the kidney however is an important control point being regulated by PTH, and indirectly therefore by calcium and phosphate concentrations. Stimulation of 1α-hydroxylase by PTH is via a cyclic AMP (cAMP)-dependent mechanism and longer-term regulation of the activity of this enzyme is via induction mediated by other hormones such as oestrogens, cortisol and growth hormone. Typically, the plasma concentration of 1,25 dihydroxy vitamin D is in the range 20–60 ng/l, that is approximately 1000-times lower than that of its precursor.

When 1,25 DHCC provision is adequate or when plasma calcium concentration is above approximately 2.20 mmol/l, 1α hydroxylase activity is suppressed and 25-hydroxy vitamin D_3 is converted by 24-hydroxylase into 24,25 dihydroxy vitamin D_3 a metabolite whose true role is uncertain but one which seems to have little if any physiological activity. Renal 24-hydroxylase does have a role to play in the *de*activation of 1,25 dihydroxy vitamin D; the major metabolite of the vitamin being 1,24,25 trihydroxy vitamin D.

9.5 Cartilage

In addition to bone, the skeletal system is composed of cartilage; indeed it is hyaline cartilage which is laid down first in the embryo most of which is ossified (made into bone) beginning at about 2 months after conception. Some cartilage remains into adulthood and can also be found in number of skeletal and non-skeletal sites around the body. The structure of cartilage is similar to non-calcified bone, that is fibrous proteins, elastin and type I collagen in hyaline cartilage and elastic cartilage or type II collagen in fibrocartilage, embedded within a highly hydrated gel-like proteoglycan matrix, rich in chondroitin sulfate and keratan sulfate. Unlike bone however, cartilage does not have a blood supply.

The physical properties of various types of cartilage include flexibility, non-compressibilty and slipperyness. These characteristics reflect the roles played by the tissue as tendons and ligaments, lining joints, intervertebral discs, and in several sites within the respiratory system.

Osteoarthritis (OA), one of the most common and most debilitating diseases affecting the skeletal system is due to erosion of the cartilage within joints. The condition is slowly progressive, possibly affecting all joints eventually although it is the weight-bearing joints which bear the brunt of the damage. The pathogenesis of OA is believed to be due to release of proteolytic enzymes from chondrocytes causing partial destruction of collagen and proteoglycans within the cartilage. The resultant inflammation is associated with pain and swelling of the joint, especially at night-time and typically localized to the hands, knees, feet and hips. Swelling causes stiffness of joints after periods of inactivity especially sleeping.

9.6 Adipose tissue

A little over 20% of the 70 kg body weight of the typical lean adult male is adipose tissue. Not only does the embryological origin of adipose tissue allow it to be classified as a connective tissue but it too acts as a 'packing' material around internal organs providing mechanical support, thermal insulation, and in neonates especially, heat generation. Unlike other types of connective tissue however, it is not the secretions (e.g. ground substance of loose connective tissue or osteoid of bone) which are of prime importance, but the fact that adipose tissue stores fuel in the form of triglycerides. The 13 kg of adipose tissue in 'Mr Average' provides sufficient energy reserves to sustain basal activity for a period of about 3 months. As discussed below, it is now increasingly apparent that adipose is also an endocrine tissue.

Adipocytes, the fat storage cells, are well supplied with blood vessels. The first recognizable cell type of the lineage is the preadipocyte. This cell may differentiate along one of two adipogenic pathways for the formation of either:

1. white fat, the predominant form in adults, and

2. brown fat, which occurs only to any significant extent in the newborn and in young children.

White fat tissue may actually appear slightly yellow due to the accumulation of various pigments. Lipid contained in white fat is found as one large globule (unilocular) not bounded by a membrane. The organelles of white adipocytes are all marginalized by the presence of the single globule. White adipose tissue is central in regulating the lipid removal from, and addition to, the bloodstream.

The appearance of brown fat is due to the extraordinarily high number of cytochrome-containing mitochondria found within the cells. The mitochondria are positioned around and between the numerous individual fat globules (multilocular), which are characteristic of this tissue. The larger surface area of fat globules and the numerous mitochondria indicate that these cells are suited to rapid turnover of stored triglyceride. Moreover, brown fat contains thermogenin (also called UCP), a protein which uncouples electron transport from oxidative phosphorylation by dissipating the proton gradient across the inner mitochondrial membrane. The effect is to produce more heat than ATP.

9.6.1   Fat storage in adipose tissue

In the fed state, insulin promotes the uptake and utilization of glucose by adipocytes. The glucose enters the cell via specific glucose transporter proteins (e.g. GLUT-4) and is metabolized through glycolysis to produce pyruvate and subsequently acetyl-CoA, the starting substrate of fatty acid synthesis (lipogenesis). Unlike most cells, adipocytes contain glycerol-3-phosphate dehydrogenase which is able to divert some of the triose phosphate dihydroxyacetone phosphate (DHAP) formed in glycolysis to glycerol-3-phosphate. Glycerol-3-phosphate is essential as it provides the glycerol backbone for the synthesis of triglyceride (Figure 9.10).

As a result of our relatively fat-rich diet, the significance of the *de novo* synthesis of triglyceride from glucose described above is probably quite small. Most of the triglyceride laid down in adipose tissue is derived from preformed fatty acid delivered to the adipocyte by chylomicrons and very low-density lipoproteins (VLDL). Insulin has a role to play here also because it (a) inhibits the hydrolysis and mobilization of stored triglycerides, and (b) promotes the expression of endothelial lipoprotein lipase (LPL). Dimeric LPL is anchored by heparan sulfate (a GAG, see Section 9.3 above) to the luminal surface of endothelial cells of the vessels perfusing the adipose tissue. In this location, the enzyme is best placed to arrest, and begin the hydrolysis of, circulating triglyceride-rich apoprotein C_{II}-containing lipoproteins.

The substrate for the LPL is the triglyceride contained within the oily core of VLDL particles and chylomicrons. The fatty acids and monoglycerides liberated from triglyceride hydrolysis are taken into the adipocytes and reformed into neutral fat for calorie storage (Figure 9.11). The ultimate result of the delipidation of VLDL and chylomicrons to the formation of low-density lipoproteins (LDL) as described in Section 5.5.

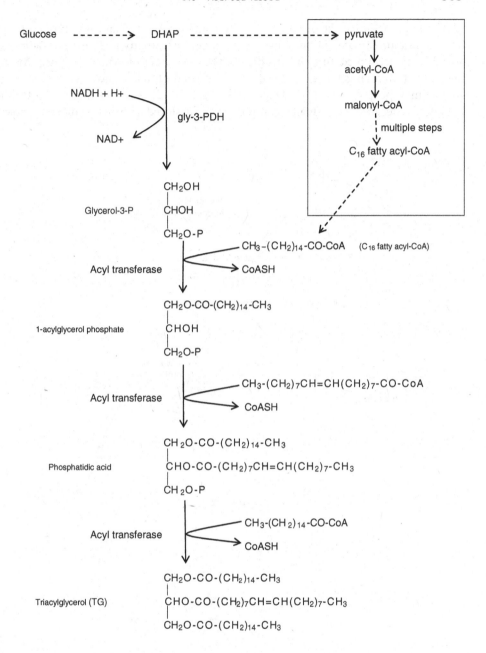

KEY:

Gly-3-PDH	glycerol-3-phosphate dehydrogenase
DHAP	dihydroxyacetone phosphate
CoASH	Coenzyme A

The box shows fatty acid synthesis which may occur within the adipocyte but most of the fatty acids required for TG synthesis are pre-formed in liver and delivered by lipoproteins.

Figure 9.10 Triglyceride synthesis in adipocytes

The resynthesized triglycerides invariably contain saturated and unsaturated fatty acids in an approximate 2 : 1 ratio. Typically, nearly all of the fatty acid content of adipose tissue can be accounted for by just six different types of molecule with palmitate (C16:0) and oleate (C18:1) together contributing over 75% of the total. Turnover studies suggest that for most people, much of the fat is metabolically relatively inert acting as depot with a long half-life, and only a smaller component of the stored fat being readily accessible.

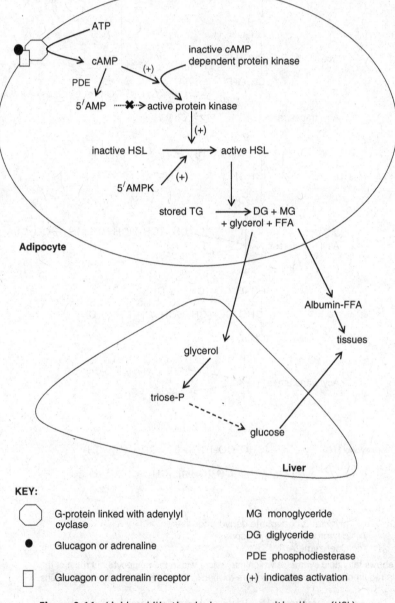

Figure 9.11 Lipid mobilization by hormone sensitive lipase (HSL)

9.6.2 Release of free fatty acids from storage

Adipose tissue provides fuel for many tissues during periods of fasting or starvation or when there is a 'metabolic stress' placed upon the body for example during exercise. Changes in the physiological status of the individual are communicated to the adipose tissue by both neural (noradrenaline; norepinephrine) and hormonal messages; glucagon (the body's signal of fasting) and adrenalin and cortisol (stress).

The key enzymes in fatty acid liberation from adipose are (i) hormone sensitive lipase (HSL) which uses tri- and di-glycerides a substrates and a (ii) monoglyceride lipase. Like many other enzymes involved with fuel storage and release, HSL is controlled by covalent modification; the phosphorylated form being active and the dephosphorylated form inactive. Glucagon or adrenaline (epinephrine) stimulation of adipocytes via cell surface G-protein linked receptors causes the production of cAMP in the cytosol. As occurs with hepatic glycogen metabolism (see Section 6.3.3), cAMP initiates a series of protein phosphorylations by activating protein kinase A (PKA). Inactive HSL becomes activated and triglyceride hydrolysis begins. Fatty acids released from adipocytes are transported through the plasma by attachment to albumin for delivery to tissues and the glycerol backbone of the triglycerides can be used by the liver for gluconeogenesis (Figure 9.11).

Thus, and importantly, fuel mobilization from hepatic glycogen stores and triglyceride in adipose tissue are co-ordinated events. Glucagon stimulation initiates a phosphorylation cascade in both tissue types simultaneously. Cyclic AMP (cAMP) activates a protein kinase, which in turn activates the enzyme responsible for hydrolysis of the bonds holding the stored fuel, fatty acid or glucose, in triglyceride and glycogen respectively. Conversely, insulin, the fed-state hormone, (1) promotes glycogen synthesis (see section 6.3) by activation of protein phosphatase-1 (PP-1), which dephosphorylates and activates glycogen synthase and (2) inhibits triglyceride degradation by activating a phosphodiesterase (PDE), which converts $3'5'$cAMP to $5'$AMP thus relieving the activation of PKA. Insulin therefore co-ordinates the storage of fuel as carbohydrate and fat.

9.6.3 Adipose as an endocrine tissue

Adipocytes have an important secretory function. Numerous factors (collectively termed adipokines or adipocytokines), mostly peptides but also eicosanoids are produced by preadipocytes and mature adipocytes (Table 9.3). Some of these factors act in an autocrine or paracrine fashion to regulate adipogenesis, that is differentiation and maturation of adipocytes themselves, whilst others, notably, leptin, adiponectin and some cytokines act in truly endocrine way, having effects on the brain, endothelial cells, liver and skeletal muscle. Disturbance in secretion from adipocytes is associated with eating disorders and metabolic syndrome.

9.6.3.1 Leptin
Leptin is a 16 kDa protein first described in experimental mice in the mid-1990s, which regulates energy balance. The protein is encoded by the *OB* gene and leptin deficient

Table 9.3 Examples of adipokines

Cell Of Origin	Name	Action	Target cells or tissues	Effect
Preadipocyte	Prostacyclin (PGI$_2$)	Autocrine/paracrine	Preadipocytes	Promotes adipogenesis
	Leukaemia inhibitory factor (LIF)	Autocrine/paracrine	Preadipocytes	Promotes adipogenesis
	Fibroblast growth factor-10 (FGF-10)	Autocrine/paracrine	Preadipocytes	Promotes adipogenesis
Mature adipocyte	Leptin	Endocrine	CNS, blood vessels, ? peripheral tissues	Appetite regulation; angiogenic
	Adiponectin	Endocrine	Liver skeletal muscle	Insulin-sensitization
	Plasminogen activator inhibitor-1 (PAI-1)	Systemic	Haemostatic system	Inhibits fibrinolysis
	Visfatin	Systemic	Peripheral tissues	Insulin-sensitization?
	Lipoprotein lipase (LPL)	Local enzymatic	Endothelial cells within adipose tissue	Lipoprotein–triglyceride hydrolysis

mice (*ob/ob* mutants) have voracious appetites and become obese. Plasma leptin concentration is determined by the size of body fat stores: in health, as fat is deposited during the post prandial phase, plasma leptin concentration rises and the drive to eat falls, thus leptin is an anorexigenic agent or 'eating inhibitor'. The stimulus to leptin secretion is intake of carbohydrate, but not fat, suggesting a role for insulin in triggering the adipokine's release. Leptin, and possibly insulin, act on particular cells within the hypothalamus to alter the synthesis of neuropeptides which control eating behaviour; details of this are given in Section 9.7.2.

Leptin signalling is via monomeric receptors in the brain. A short-form of the leptin receptor (Lep-R) is required to transport the hormone across the blood–brain barrier and a long-form Lep-R is located in the hypothalamus. The long-form is functionally linked with a particular type of receptor-associated tyrosine kinase called Janus kinase (JAK, see Section 4.7) whose function is to phosphorylate a STAT (*s*ignal *t*ransducer and *a*ctivator of *t*ranscription) protein a similar mechanism to that often associated with signalling by inflammatory cytokines.

Obesity is associated with higher circulating concentrations of leptin that would be found in non-obese individuals. Although this effect can be explained by expansion of adipose tissue mass, there is reason to believe that obesity may also be due to leptin insensitivity in the target tissues.

9.6.3.2 Adiponectin

Adiponectin was discovered in the mid-1990s at about the same time leptin. Unlike leptin, there is an inverse correlation between high body mass index (BMI, a commonly used measure of obesity) and the plasma concentration of adiponectin. Although secreted as a 30 kDa monomer, most of the circulating adiponectin is in the form of multimeric complexes, for example trimers and hexamers. Each peptide chain consists of a globular head at the C-terminal, a collagen-like spindle domain of 65 amino acids and a short N-terminal sequence. Formation of the trimeric conformation occurs when individual peptides entwine their collagen domains; two trimers may then attach together via the short N-terminal sequence to form a hexamer. It is likely that the different molecular forms of the hormone target different tissues, such as the central nervous system, liver, adipose tissue and muscle. In contrast to leptin, adiponectin signals through a cAMP-linked mechanism rather than a tyrosine kinase-dependent process.

The overall action of adiponectin is the promotion of eating behaviour and therefore it is not surprising that its metabolic effects are to inhibit gluconeogenesis, a pathway associated with fasting and physical exercise, and promote the uptake and oxidation of glucose and fatty acid in skeletal muscle. These effects are brought about by increasing the insulin sensitivity of the target tissue but the precise mechanism of this is uncertain.

9.6.4 Obesity and appetite control

In the early twenty first century, obesity has reached almost epidemic proportions in Western societies. An estimated 20% of the population is considered to be clinically obese as measured by the BMI ($>30 \, kg/m^2$, compared with normal $<25 \, kg/m^2$) and a further 30% are overweight (BMI 25–30 kg/m^2). It should be noted that the use of BMI to assess obesity is somewhat contentious and other measures such as skin-fold thickness, various imaging techniques and skin electrical impedance are also used. Furthermore, it is apparent that the distribution, and not just amount, of adipose tissue in the body is a significant factor in health assessment and waist-to-hip ratio and waist circumference are also useful measures.

A traditional view that obesity is due simply to over indulgence and a lack of self control is no longer valid as numerous studies, often comparing twins, or following patterns within families and studies of adopted children, have shown the importance of gene-environment interactions. Although the central role of the hypothalamus in eating control was first suggested in the 1940s, the complexity of the process only came to light with the identification of leptin and adiponectin.

Adipocyte-derived leptin targets particular neurones in a region of the hypothalamus called the arcuate nucleus via a JAK/STAT pathway. Within this region, neurones designated NPY are suppressed by leptin (and insulin) whilst another cluster cells, the proopiomelanocortin (POMC) related neurons, are stimulated. The NPY cells secrete a peptide called agouti-related peptide (AGRP) whilst the major product of the POMC-related cells is POMC, a large peptide which undergoes post-translational processing to generate a number of active hormones, including α and β forms of melanocyte-stimulating hormone (α and β MSH).

The NPY and POMC cells act as the 'switching centre' for appetite control because both cell types are in contact with secretory cells. A region of the brain called the lateral hypothalamic area (LHA) consists of cells that secrete orexigenic (eating promoting) factors whilst cells of the paraventricular nucleus (PVN) release anorexigenic factors, which both suppress the urge to eat and promote energy utilization. Receptors, designated MC3R and MC4R are highly expressed in cells of the PVN and LHA. MSH is the activating ligand for MC3R and MC4R but AGRP is an antagonist ('competitive inhibitor') for MC4R. Research suggests that it is the balance between MSH and AGRP that determines the urge to eat. Leptin-derived production of MSH (i.e. stimulation of the POMC cells) simultaneously inhibits LHA output and increase PVN activity, so generating a *'Don't Eat'* message. However, this effect can be overcome if secretion of AGRP from NPY cells rises and blocks the action of MSH.

Chapter summary

Connective tissue is composed of apparently very different cells; metabolically it is 'dynamic' undergoing continual turnover and so to maintain health means that a balance must be achieved between biosynthesis and degradation. Cells in connective tissue are usually found embedded within a matrix composed of proteins with variable amounts of proteoglycan and genetically determined enzymatic defects in the production of the matrix may result in often serious pathologies. Metabolism in both bone and adipose tissues is subject to hormonal control, whilst secretions from adipose tissue also help to regulate eating habits and fuel metabolism in other tissues, notably muscle.

Case notes

1. **Two cases of hypercalcaemia**

   a. Mr Harrison was a 53-year-old who presented with severe and recurrent 'sharp' back pain. A blood test revealed the following results;

plasma total calcium	2.59 mmol/l (reference 2.25–2.55 mmol/l)
parathyroid hormone	5.8 pmol/l (reference 1–6 pmol/l)

   The calcium value is only slightly raised but the PTH concentration, although within the stated reference range, is inappropriately high for the stated calcium concentration. Normally, PTH and calcium concentrations are inversely related. A diagnosis of primary hyperparathyroidism was subsequently confirmed. The role of PTH is to raise plasma calcium concentration, partly, by enhancing osteoclast activity.

   The increased load of calcium presented to the kidneys had begun to precipitate from the glomerular filtrate and renal stones (or calculi) had formed within the renal tissue; it was these that were causing the back pain. Often, the calcium-containing crystals are small enough to be passed with the urine; alternatively, the patient may notice fine gravel in the urine, but occasionally the stone must be either removed surgically or disintegrated by the use of ultrasonic waves, allowing the fragments to be passed.

   Had it been possible to collect a sample of the calculi from Mr Harrison, the probability is that the chemical composition would have been almost entirely calcium phosphate, a poorly soluble salt.

   b. Miss Johnson was 64 years old when she was admitted to hospital following a minor accident at home which resulted in a broken leg. Investigations included an X-ray, which showed several patches of demineralization of the bone; a plasma calcium concentration of 3.20 mmol/l and a plasma total protein concentration of 95 g/l (reference range 55–70 g/l).

   A diagnosis of multiple myelomatosis (myeloma) was made. This is a malignant disease of white blood cell production occurring within the bone marrow. The condition is characterized by the over production of immunoglobulin, usually IgG, less frequently IgA, and often free immunoglobulin light chains by lymphocytes. The bone becomes weakened by loss of mineral (calcium apatite) due to the production of an osteoclast activating factor (OAF) by the tumour cell line and this leads to pathological fractures. Loss of calcium salts from bone causes hypercalcaemia. The prognosis (natural course of the condition) is usually poor with many fatalities being associated with renal damage which is a consequence of the hyperproteinaemia.

2.   **Osteoporosis and osteomalacia**
     Both of these conditions result in weakened bones but the origins of the problems are different.

a.   Mrs Grant, a 62-year-old lady presented with a fracture to the distal end of the radius (Colles' fracture). Clinically, she had thyroid goitre, a slightly irregular heart rhythm and appeared malnourished although she claimed to have a good appetite. Routine investigations revealed the following;

'renal profile'

sodium and potassium	within reference ranges
urea	8.1 mmol/l (reference 3.0–7.5 mmol/l)
creatinine	55 μmol/l (ref. 60–120 μmol/l)

'bone profile'

calcium and phosphate	both within reference appropriate range
alkaline phosphatase	295 U/l (ref. <250 U/l)
plasma proteins	all within appropriate reference range

thyroid hormones

total thyroxine	300 nmol/l (ref. 50–150 nmol/l)
free thyroxine	75 pmol/l (ref. 10–25 pmol/l)
thyroid stimulating hormone	<0.1 U/l (ref. 0.5–4.5 U/l)
plain X-ray	No abnormality detected

A diagnosis of hyperthyroidism was made but the bone symptoms prompted further investigation whilst Mrs Grant was an inpatient.

bone densitometry   reduced bone mass

These results indicated osteoporosis. In this condition, which often appears secondary to another pathology such as an endocrinopathy, chronic renal failure or following long term immobilization, bone architecture is normal but its mass is reduced relative to its volume, that is there is normal mineralization but the amount of osteoid matrix is reduced. Treatment is with bone resorption inhibitors such as the bisphosphonate group of drugs, for example alendronate.

Increased concentrations in plasma of markers such as P1NP or cross-linked C-terminal telopeptides (CTx), or urinary excretion of DPD, indicate increased bone turnover but are generally not useful for initial diagnosis of osteoporosis. Changes in plasma concentrations or urinary excretion of bone markers may be useful for monitoring patients response to therapy.

b.   Mrs Al-Ameri a 55-year-old lady who moved to northern Europe from Saudi Arabia. For some time she had experienced generalized bone pain and some muscle weakness and walked with a rolling gait. There had been no medical history of fractures or evidence of gastrointestinal disturbance.

On examination, there was obvious tenderness in the arms and some slight deformity of the spine and chest. Mrs Al-Ameri's general practitioner ordered blood tests; the results were as follows;

haemoglobin	10.8 g/dl	(female ref. 12–14 g/dl)
white cells	NAD	
blood film	mild red cell hypochromia	
renal profile	all values within reference ranges	
liver profile	all values within reference ranges	
bone profile		
calcium	2.20 mmol/l	(ref. 2.25–2.55 mmol/l)
phosphate	0.4 mmol/l	(ref. 0.8–1.5 mmol/l)
total plasma proteins	60 g/l	(ref. 55–70 g/l)
plasma albumin	36 g/l	(ref 35–45 g/l)

The GP made a provisional diagnosis of osteomalacia and prescribed vitamin D supplements. Vitamin D measurements are not performed routinely, but the assumption is that a low result would have been obtained on the blood sample. Most of the vitamin D necessary to maintain normal calcium homeostasis is derived from endogenous synthesis by reactions in the skin (which require UV radiation from sunlight), liver and kidney. The cultural habits of Mrs Al-Ameri required her to dress in a burqah and niqab whenever she left the home, meaning that very little of her skin was exposed to daylight.

Vitamin D deficiency may also occur through inadequate dietary intake, gut (poor absorption), renal disease (1-hydroxylase deficiency or failure to reclaim calcium from the glomerular filtrate), or liver disease (25-hydroxylase deficiency). The slightly low haemoglobin concentration and pale stained (hypochromic) red cells suggested a coincident mild iron deficiency.

Bone scans were not performed, but typically in osteomalacia there is widespread under mineralization of osteoid, so unlike osteoporosis, osteomalacia is due to *poor quality* bone causing a loss of rigidity and strength, rather than a *quantitative* reduction of bone. A condition in children, which is biochemically similar to osteomalacia, is rickets. Sufferers have misshapen bones of the head, spine and chest and typically bowed legs.

Appendix 1: Answers to problems

Chapter 1

1. Distinguish between: free energy, entropy and enthalpy

    Free energy (G) is used to perform useful biological work such as active secretion, muscle contraction, powering biosyntheses.

    Entropy (S) is energy that cannot be harnessed for work, it is wasted energy associated with chaos and disorder.

    Enthalpy (H) is total energy in a system

2. Define the terms endergonic and exergonic

    Endergonic: a reaction in which the free energy of the products is greater than the free energy of the reactants. Energy must be added to the reaction for it to operate.

    Exergonic: a reaction in which the free energy of the products is lower than the free energy of the reactants. These reactions are likely to be spontaneous, favourable . . . once given an appropriate 'push' by the enzyme.

3. What information is given by the sign (+ or −) of the free energy value?

    $+\Delta G$ is an endergonic reaction

    $-\Delta G$ is an exergonic reaction

4. Why does metabolism NOT grind to a resounding halt when an endergonic reaction occurs within a pathway?

    Because, energy from an exergonic reaction can be used to drive forward an endergonic reaction providing they share a common intermediate; this is reaction coupling. Alternatively, ATP or another high energy compound can be used to overcome the endergonic 'hump'. In effect this is also coupling of two reactions; an endergonic reaction with hydrolysis of ATP.

Essential Physiological Biochemistry: An organ-based approach   Stephen Reed
© 2009 John Wiley & Sons, Ltd

5. WITHOUT performing any calculation, state with reasons if the following reactions are likely to be strongly exergonic, weakly exergonic, strongly endergonic or weakly endergonic:

    (i) $R \rightarrow P$ $K'_{eq} = 0.005$

    (ii) $R \rightarrow P$ $K'_{eq} = 127$

    (iii) $R \rightarrow P$ $K'_{eq} = 2.5 \times 10^{-4}$

    (iv) $R \rightarrow P$ $K'_{eq} = 0.79$

    (v) $R \rightarrow P$ $K'_{eq} = 1.27$

    If K'_{eq} is greater than 1, the reaction is likely to be spontaneous and exergonic (reactions (ii) and (v)); if K'_{eq} is less than 1, it is likely to be endergonic. Reactions with a K'_{eq} close to zero (reaction (iv)) will have little energy change associated with them; reactions far from equilibrium ((ii) and (iii)) will be strongly endergonic or exergonic.

6. Like glc-6-P, pyruvate and acetyl-CoA are at metabolic cross-roads. Consult a metabolic map and identify these important compounds and note the ways in which they may be formed and metabolized.

    Pyruvate is derived from phosphoenolpyruvate may be inter-converted into lactate, alanine, oxaloacetate. Formation of acetyl-CoA from pyruvate is essentially irreversible

    Acetyl-CoA is at the product of fatty acid catabolism and may be derived from amino acids and carbohydrates (via pyruvate). Acetyl-CoA is the precursor of fatty acids, cholesterol and ketone bodies.

7. (a) G-6-P $\rightarrow$ F-6-P

    This is an isomerization

    (b) F-6-P + ATP $\rightarrow$ F-1, 6 bis phosphate + ADP

    This is phosphate transfer

    (c) pyruvate + CO_2 $\rightarrow$ oxaloacetate

    An addition reaction

    (d) F-1, 6-bis phosphate + H_2O $\rightarrow$ F-6-P + Pi

    This is hydrolysis

Chapter 2

1. '$\Delta G'$ not $\Delta G'^{\circ}$ is the appropriate criterion on which to judge the spontaneity of a reaction'. Explain.

$\Delta G'^\circ$ indicates the free energy change under defined physiological conditions but such conditions may not truly reflect those inside a cell. $\Delta G'$ is the free energy change under 'real' cellular conditions so the value is a better reflection of the spontaneity or probability of the reaction occurring.

2. Calculate K'_{eq} and $\Delta G'^\circ$ for each of the following reactions where [r] and [p] represent equilibrium concentrations of reactant and product; r1, r2, p1 and p2 indicate different reactants and products respectively in the reaction

   (a) [r] = 3.4 mmol/l
        [p] = 2.5 mmol/l
        at 37°C
        $K'_{eq} = 0.74$ and $\Delta G'^\circ = +0.7$ kJ/mol

   (b) [r] = 0.18 mmol/l
        [p] = 5.4 mmol/l
        at 37°C
        $K'_{eq} = 30.0$ and $\Delta G'^\circ = -8.8$ kJ/mol

   (c) [r] = 0.95 μ mol/l
        [p] = 1.05 μ mol/l
        at 30°C
        $K'_{eq} = 1.11$ and $\Delta G'^\circ = -0.25$ kJ/mol

   (d) [r₁] = 1.25 μ mol/l   [r₂] = 0.85 μ mol/l
        [p] = 4.55 μ mol/l
        at 25°C
        $K'_{eq} = 4.28$ and $\Delta G'^\circ = -3.60$ kJ/mol

   (e) [r] = 65.8 μ mol/l
        [p₁] = 21.5 μ mol/l   [p₂] = 3.5 μ mol/l
        at 37°C
        $K'_{eq} = 1.14$ and $\Delta G'^\circ = -0.35$ kJ/mol

   (f) [r₁] = 4.75 mmol/l [r₂] = 2.5 mmol/l
        l[p₁] = 8.6 mmol/l [p₂] = 1.1 mmol/l
        at 37°C
        $K'_{eq} = 0.80$ and $\Delta G'^\circ = +0.59$ kJ/mol

3. Using your values for $\Delta G'^\circ$ for (a) to (f) above, calculate $\Delta G'$ for each reaction given the following data

   (a) $K'_{eq} = 3.65$; temperature $= 37\,^\circ C$

   $\Delta G' = -0.2\,kJ/mol$

   (b) $K'_{eq} = 0.3$; temperature $= 37\,^\circ C$

   $\Delta G' = -11.8\,kJ/mol$

   (c) $K'_{eq} = 28.5$; temperature $= 37\,^\circ C$

   $\Delta G' = +8.19\,kJ/mol$

   (d) $K'_{eq} = 15.7$; temperature $= 37\,^\circ C$

   $\Delta G' = +3.3\,kJ/mol$

   (e) $K'_{eq} = 0.015$; temperature $= 37\,^\circ C$

   $\Delta G' = -10.9\,kJ/mol$

   (f) $K'_{eq} = 32.5$; temperature $= 37\,^\circ C$

   $\Delta G' = +9.4\,kJ/mol$

4(a). Given that the K'_{eq} for a reaction is 1.45, which of the following statements is/are true?

   - the forward reaction is faster than the reverse reaction;
   - the concentration of p is higher than the concentration of r; *TRUE*
   - the forward reaction is slower than the reverse reaction;
   - the concentration of r is lower than the concentration of p;

4(b). What biochemical change might cause K'_{eq} to shift to the left (net formation of r), that is the reverse reaction to accelerate?

   If the product concentration rises the reaction may be forced backwards, if K'_{eq} is close to 1

5. The standard free energy change for the hydrolysis of ATP is estimated to be $-30.5\,kJ/mol$.

$$ATP + H_2O \rightarrow ADP + Pi \quad \text{(Pi stands for inorganic phosphate)}$$

   Estimate the relative concentrations (i.e. ratio) of ATP to ADP and Pi at $37\,^\circ C$. Assume $[ADP] = [Pi]$

$$-30.5 = -2.303 \times RT \log K_{eq}$$

$$-30.5/-2.303\,RT = \log K_{eq}$$

$$\therefore -30.5/-5.80 = 5.25$$

$$\text{antilog } 5.25 = 177828$$

6. Fill in the missing values.

	v_0 (µmol/min)	K_m (mmol/l)	V_{max} (µmol/min)	[S] (mM)	v_i (µmol/min)	[I] (mM)	K_i (mmol/l)	Type of inhibitor
a	15.1	0.825	25	1.25				
b	0.5	2.5	1.0	2.5				
c	1.82	1.5	4	1.25				
d	6.0	10	18.0	5.0				
e	1.20	2.0	2.0	3.0	1.06	0.2	0.6	Competitive
f	10	15	25	10	6.25	3.5	3.5	Non-competitive
g	0.86	2.0	2.0	1.5	0.57	0.04	0.6	Non-competitive

7. A 75 kg rower uses 52.5 kJ per minute in a competitive race. Assuming that the race lasts 18 min, (i) calculate the total energy consumption during the race.

$$18 \times 52.5 = 945\,\text{kJ}$$

Given that the standard free energy change for the hydrolysis of ATP is -30.5 kJ/mol, (ii) calculate the total amount (in grams) of ATP required to furnish the energy to sustain the competitor during the race.

(HINT: The molecular weight of ATP $= 507$; to convert grams to moles and back use: Actual weight of substance $=$ molecular wt. of the substance $\times$ number of moles)

$945/30.5 = {\sim}31$ moles of ATP required if 1 mole of ATP $= 507$ g (by definition)

31 moles of ATP $= 507 \times 31 = 15\,709$ g $= 15$ kg (compare this with body weight of a 'typical' adult male ${\sim}70$ kg)

Given that the maximum energy liberated from the complete oxidation of glucose to CO_2 and H_2O is 2866 kJ/mol and that the molecular weight of glucose is 180 (iii) calculate the mass of glucose (in grams) which would have been oxidized during the race assuming that the actual energy yield from glucose is 40% of the theoretical maximum (i.e. 2866 kJ/mol).

$40\% \times 2866 = 1146.4$ kJ/mol as the actual energy yield

$\therefore = 1146.4$ kJ/180 grams of glucose

$= 6.34$ kJ of energy liberated per gram of glucose oxidized

945 kJ are consumed during the race, so

$945/1146.4 \times 180 = 148.4$ g

This assumes that no fat is oxidized and standard conditions.

Chapter 3

1. Control points often occur at or near to the start of a pathway or a branch points in a pathway. Why?

    Controlling an enzyme near the beginning will necessarily slow down the whole pathway, possibly allowing initial substrates to be used in other ways, that is diverted through alternative pathways where they be used to better effect.

    Control at branch points acts like diversion signs on the road side or points on railways, directing substrates through preferred routes.

2. Verify the figure for the number of moles of ATP generated per mole of glucose. You will need to consider ATP generated from substrate level phosphorylation in glycolysis *and* from oxidative phosphorylation. Assume each NADH generates three ATP and each $FADH_2$ generates two ATP. (You may need to refer to the diagrams showing glycolysis and the TCA cycle).

    For each glucose molecule which begins its journey along glycolysis, there is a net gain of two ATP (phosphoglycerate kinase and pyruvate kinase), but there is also generation of NADH at the glyceraldehyde-3-phosphate dehydrogenase step. This can be reoxidized by a shuttle system which transfers hydrogen atoms across the mitochondrial membrane and so generates three ATP. Because one molecule of glucose equates to two molecules of glyceraldehyde-3-phosphate and therefore two NADH, our running total at this stage is eight ATP.

    Pyruvate formed by glycolysis is converted into acetyl-CoA; pyruvate dehydrogenase also generates one NADH per pyruvate so the net gain is $2 \times 3\,ATP = 6$; so far, $8 + 6 = 14\,ATP$.

    Each acetyl-CoA which enters the TCA cycle generates three NADH ($= 9\,ATP$) and one $FADH_2$ ($= 2\,ATP$), that is an overall gain of $2 \times (9 + 2) = 22\,ATP$. Lastly, each turn of the TCA generates one GTP which may also be converted into ATP so another two ATP per molecule of glucose. Potentially, 24 ATP produced per turn of the TCA cycle.

    Overall, $14 + 24 = 38\,ATP$ per molecule of glucose.

3. What is the biological 'logic' in ATP, ADP, AMP, F-2,6 bisphosphate and citrate acting s regulators of such a key enzyme (PFK-1) in glycolysis?

    High cellular concentrations of ATP and citrate partially inhibit PFK; ATP represents readily available energy and a high citrate concentration indicates that the TCA cycle is working efficiently. Citrate is normally located within mitochondria, but it may also be synthesized in the cytosol which is of course where glycolysis is occurring. Fructose-2,6-bisphosphate is formed when fructose-1,6 bisphosphate concentration begins to build up, so by activating PFK, there will be greater 'flow' through this point and fructose-1,6 bisphosphate concentration will begin to fall.

Cytosolic concentrations of AMP and ADP will rise when ATP is being utilized, signalling the need for an increase in glycolysis, and therefore TCA and oxidative phosphorylation, to replenish ATP.

4. Why is it desirable to have an enzyme in heart muscle which works efficiently to remove lactate, that is an isoenzyme of lactate dehydrogenase with a low K_m for lactate?

Lactate accumulation in muscle can cause cramp; not a good thing to happen to the heart! Cardiac LD has a low K_m for lactate ensuring that the concentration does not rise too high. Also, by converting lactate into pyruvate, more fuel can be made available for the heart muscle in times of greater need, for example exercise.

5. What would be the effect of (a) a competitive inhibitor and (b) a non-competitive inhibitor on K_m and V_{max} of an enzyme catalysed reaction? Explain your answer.

Competitive inhibitors increase K_m by preventing substrate access to the active site for binding.

Non-competitive inhibitors form inactive ESI complexes so less product is released. Substrate binding is not affected so K_m is unaltered but V_{max} is reduced.

6. Study the graph below which shows (a) the response of a typical allosteric enzyme in the absence of an inhibitor and (b) the same enzyme in the presence of an inhibitor.

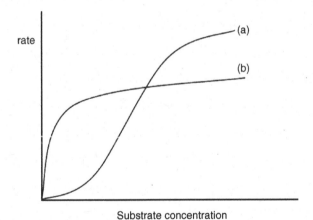

Substrate concentration

What conclusions can be drawn about the effect of the inhibitor on the enzyme?

The effect of the inhibitor has resulted in a loss of the cooperative effect normally seen in allosteric enzymes so the graph looks like a simple Michaelis–Menten plot.

In effect, the inhibitor is 'locking' the enzyme in its T conformation so substrate binding is impaired, rather like the effect of a competitive inhibitor.

7. Covalent modification of enzymes (molecular weight of several hundreds or thousands) by the incorporation of inorganic phosphate in the form of PO_3^{2-} (formula weight $= 85$), seems to represent a small chemical change in the enzyme yet is an important control mechanism of enzyme activity. Explain how phosphorylation can exert its controlling effect on the activity of the enzyme.

   Phosphate is charged $(2-)$ so when it is incorporated into an enzyme, alterations in the electrostatic attractions between parts of the enzyme molecule will occur causing a change in the three-dimensional conformation of the protein. The effect may be to 'expose' the active site to allow substrate binding (if phosphorylation activates the enzyme) or may 'hide' the active site, so switching off the enzyme.

8. In the fed state when the insulin : glucagon ratio is high, would you expect glycogen synthase, PK and PDH to be phosphorylated or dephosphorylated? Give reasons to support your answer.

   Glucagon stimulation of liver cells in particular leads to phosphorylation of regulatory enzymes whereas insulin has the opposite effect. So, after a meal, we would expect glycolysis and glycogen synthesis to operate very efficiently so the control enzymes will be dephosphorylated.

Appendix 2: Table of important Metabolic Pathways

Some important metabolic pathways

Pathway	Principal tissue/organ	Sub-cellular location	Section
Glycolysis	All cells	Cytosol	3.3.1
TCA	All cells except RBC	Mitochondria	3.3.2
Oxidative phosphorylation	All cells except RBC	Mitochondria	2.4.1.2
Gluconeogenesis	Liver and kidney	Cytosol	6.5.2
Glycogen metabolism	Liver and muscle	Cytosol	6.3.2.4, 6.5.1, 7.3
TG synthesis	Adipose and liver	Cytosol	6.3.2.1, 9.6.1
Cholesterol synthesis	All cells, liver is especially important	Mainly cytosol but acetyl-CoA is exported from mitochondria	6.3.2.3
TG hydrolysis	Stored TG: adipose, liver and muscle. Dietary TG: gut lumen	Cytosol	9.6.2
Transamination	Muscle and liver	Cytosol and mitochondria	6.3.1.1
Haem synthesis	All cells, liver and red blood cell precursors especially	Partly in cytosol and partly within mitochondria	5.3.1.3, 6.3.3
Haem catabolism	Liver	Cytosol	6.4.2.2
β-oxidation of fatty acids	All tissues except RBC, especially important in muscle	Mitochondria	7.5
Fatty acid synthesis	Adipose, liver and muscle	Cytosol	6.3.2.1
Pentose phosphate pathway	Liver, adipose, RBC	Cytosol	5.3.1.6
Ketogenesis	Liver	Cytosol, but acetyl-CoA is exported from mitochondria	6.3.2.3
Urea synthesis	Liver	Part cytosolic and part mitochondrial	6.2.1

RBC, red blood cells; TG, triglyceride.

Essential Physiological Biochemistry: An organ-based approach   Stephen Reed
© 2009 John Wiley & Sons, Ltd

Index

Bold font = Figure
Italic font = Table

Essential Physiological Biochemistry: An organ-based approach   Stephen Reed
© 2009 John Wiley & Sons, Ltd